Home Satellite TV Installation & Troubleshooting Manual

Third Edition

Frank Baylin

contributions by

Brent Gale and Ron Long

Baylin Publications

Boulder and London

ACKNOWLEDGEMENTS

Since first published in 1985, this book was thoroughly revised for its second edition, and now, due to the rapid developments in satellite communications technology, we have created an essentially new third edition. We could not have compiled such an up-to-the-minute guide without the help of many people.

Sincere thanks to Brent Gale of Echosphere Corporation whose participation was instrumental in preparing the first two editions. Ron Long's contributions were essential to this completely revised edition. Again thanks to H. Taylor Howard, Michael L. Gustafson and Blair Gilbert who all proofed the first edition, to Richard Zlotky, president of Earthbound, Inc. who provided the useful information about wind loading and to Mark Widner who supplied the installer's pre- and post-installation checklists.

Thanks to Jennifer Stewart-Laing who produced all the maps, to Rod Schubert who was instrumental in aiding us to desktop publish this third edition, to Amy Lockart who prepared some of the line drawings and to Bruce Frehner and Gay Lang who assisted in preparing the computer generated artwork.

For speaking engagements or technical consulting services contact:

Frank Baylin
1905 Mariposa
Boulder, CO 80302
Tel: (303) 449-4551
FAX: (303) 939-8720

Brent Gale
2748 Winding Trail Drive
Boulder, Colorado 80302
(303) 449-0122

Ron Long
Longfellows Antenna Service
942 North Gracia Street
Camarillo, California 93010
(805) 484-2432

Artists: Amy Lockart, Rod Schubert, Gay Lang and Bruce Frehner
Maps: Jennifer Stewart-Laing
Cover Illustration: Peter Stallard

First Edition - September 1985   Second Edition - August 1987   Third Edition - January 1991

ISBN: 0-917893-12-3
LCCCN: 85-71662

DEDICATION

This book is dedicated to a hope, a hope that satellite communications technology can serve to educate people of the world quickly enough to reverse the pervasive damage to our environment that is occurring this very moment.

Brief Table of Contents

Appendices

Table of Contents

I. SATELLITE COMMUNICATION THEORY

A. INTRODUCTION

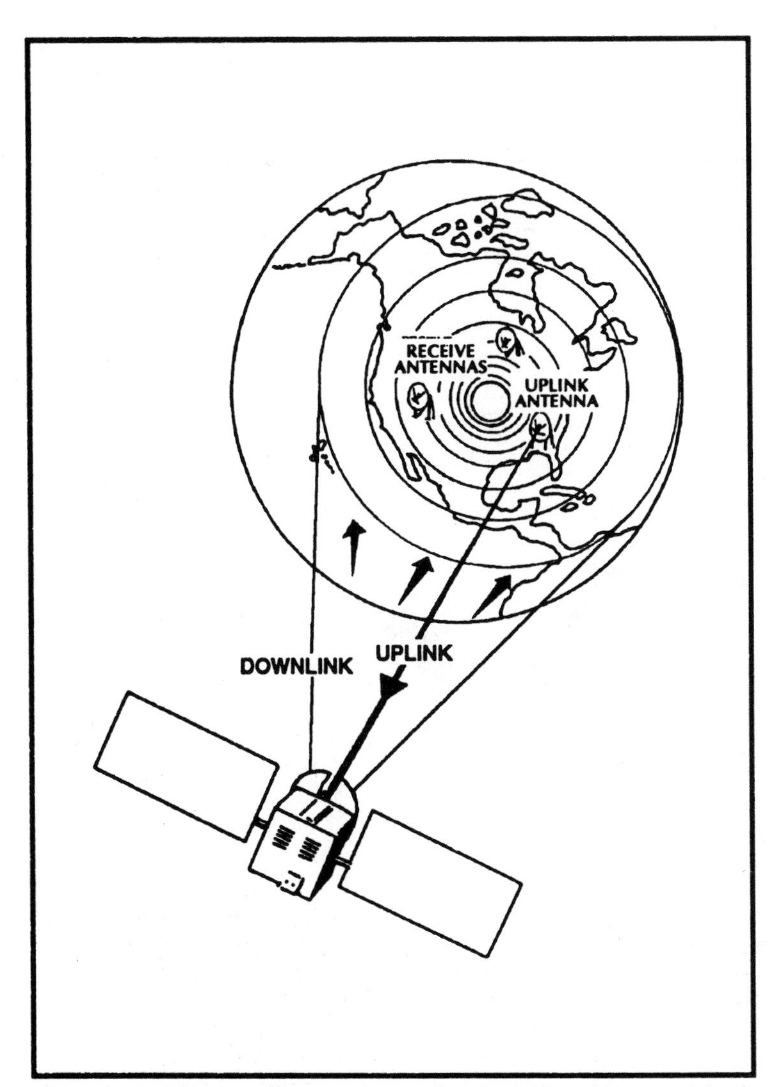

Figure 1-1. The Satellite Communication Circuit. *Signals are relayed to a satellite by one uplink antenna and then rebroadcast to an unlimited number of receiving stations that fall within the beam of the downlink antenna.*

Modern satellite communication has been made possible by combining the skills and knowledge of space technology with those of the microelectronics industry. This has allowed increasingly heavier, more sophisticated satellites to be launched into orbit for ever lower costs. The evolution of satellites is a perfect example of how the boundary between the computer and communication industries is quickly dissolving.

Satellite communications is one of our most rapidly growing and evolving technologies bringing with it a multitude of business opportunities in the decades to come.

The concepts underlying satellite broadcasting, first stated in Arthur C. Clarke's ground-breaking article in the October 1945 edition of *Wireless World,* are rather simple. Signals are beamed into space by an "uplink" antenna, received by an orbiting satellite, electronically processed, broadcast back to earth by a "downlink" antenna and received by an earth station located anywhere in the satellite's "footprint" (see Figure 1-1).

Most communication satellites are parked in the "Clarke Belt" or the "geosynchronous" arc at 22,247 miles directly above the equator. This circle around the earth is unique because in this orbit the velocity of satellites matches that of the surface of the earth below (see Figure 1-2). Each satellite thus appears to be in a fixed orbital posiition in the sky. This allows a stationary antenna to be permanently aimed towards any chosen geosynchronous satellite.

Pioneer communication satellites were placed into lower, more complex elliptical orbits because sufficiently powerful launching vehicles were not available to lift them into the distant geosynchronous arc. One of the earliest vehicles, Telstar, had to be tracked by very costly and bulky equipment mounted on rails to allow movement. Some special military communication satellites are still launched into hard-to-track elliptical orbits, generally for security reasons. The Soviet Molniya television broadcast satellites are also positioned in elliptical paths to more easily serve the far northern reaches of the Soviet Union.

The satellite communication system must generally meet all requirements that apply to radio communications. In the United States, for example, the Federal Communications Commission (FCC), mandates these standards. These include power output, frequency and bandwidth allocations, type of transmission (digital or analog) as well as polarization and modulation methods. A brief introduction to microwaves and communication fundamentals is presented next to explain some of these concepts.

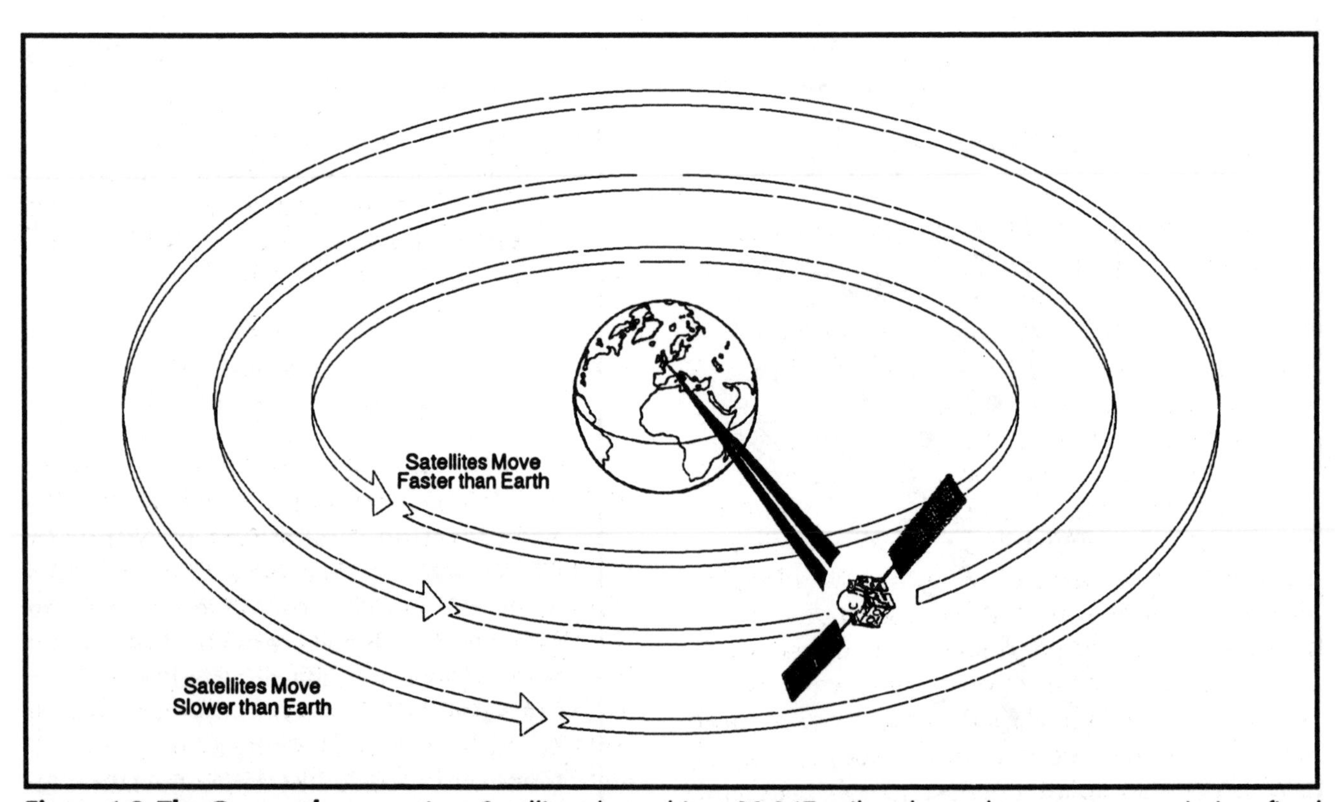

Figure 1-2. The Geosynchronous Arc. *Satellites that orbit at 22,247 miles above the equator remain in a fixed position relative to an observer on earth below. Satellites above or below this orbit rotate more slowly or more rapidly, respectively, than the speed of the earth below and therefore are not geostationary.*

B. COMMUNICATION FUNDAMENTALS

The same principles underlie all forms of man-made communication. First, information such as voice, data or music is changed into electronic form. The process of transforming this information into an electrical signal is known as coding. The signal is then added or modulated onto a carrier waveform having an assigned frequency. This modulated signal is relayed within an assigned bandwidth at a predetermined power by a transmitter. The carrier wave can be polarized into a variety of forms to allow the assigned frequencies to be used to their fullest extent. When the signal is received, it is usually amplified, extracted from the carrier by a process called demodulation, and filtered to remove noise. This reconstructed form of the original signal is then processed into audio, video or data as required.

Figure 1-3. Wave Motion in Pond. *Electromagnetic waves can be visualized by considering the wave motion in a pond caused when a pebble disturbs the water.*

Electromagnetic Waves

The media behind the entertainment and information relayed by satellite broadcasting are invisible, extremely low power microwaves. Microwaves and radio waves, the agents by which radio, conventional television and other man-made devices work, are called electromagnetic waves. If we could see such a wave, it would look like the ripples that travel outward in concentric circles when a pebble is tossed into a pond (see Figure 1-3). Radio and microwaves are similar in concept to the waves of vibrating air molecules that we know as sound. While sound travels at a plodding 760 miles per hour (1223 km per hour), radio waves and all electromagnetic waves travel at the speed of light, 186,000 miles per second (299,339 km per second). At this speed, a signal travels from an uplink antenna to a satellite and back again in about 0.4 second.

Power

One important property of electromagnetic waves is their power or strength, measured in units such as watts per square meter. Therefore, to illustrate, 10 watts per square meter means that the power passing through each square meter is ten watts. Satellite TV broadcasts are usually received by an antenna at powers less than one billionth of one billionth of a watt per square meter!

Frequency

A second important property of radio waves is their frequency, the number of vibrations that occur every second. Just as the frequency of sound vibrations determines whether a musical note is soprano or bass, the frequency of radio waves determines whether they are used for regular AM radio broadcasts or for satellite television relays. Microwaves have frequencies in excess of one billion cycles per second (known as one gigaHertz and abbreviated 1 GHz). By comparison, the electricity from a wall outlet has a frequency of 60 Hertz, or 60 cycles per second, so each second the voltage changes from positive to negative 60 times.

Many seemingly different phenomena encountered in nature including light, X rays, infrared heat rays, microwaves used for both cooking and communicating and gamma rays from the cosmos are electromagnetic waves (see Figure 1-4). Surprisingly, the only difference among them is their frequency. Since all electromagnetic waves travel at the same speed, the speed of light, as the frequency increases the wavelength must decrease. For example, the wavelength of visible light is comparable to the dimensions of atoms and molecules, and its frequency is many billions of billions of cycles per second. In contrast, microwaves have lower frequencies of one to 50 gigaHertz (GHz) and wavelengths ranging from one foot to fractions of an inch. Radio waves can have wavelengths which are miles long and frequencies up to millions of cycle per second (megaHertz or MHz).

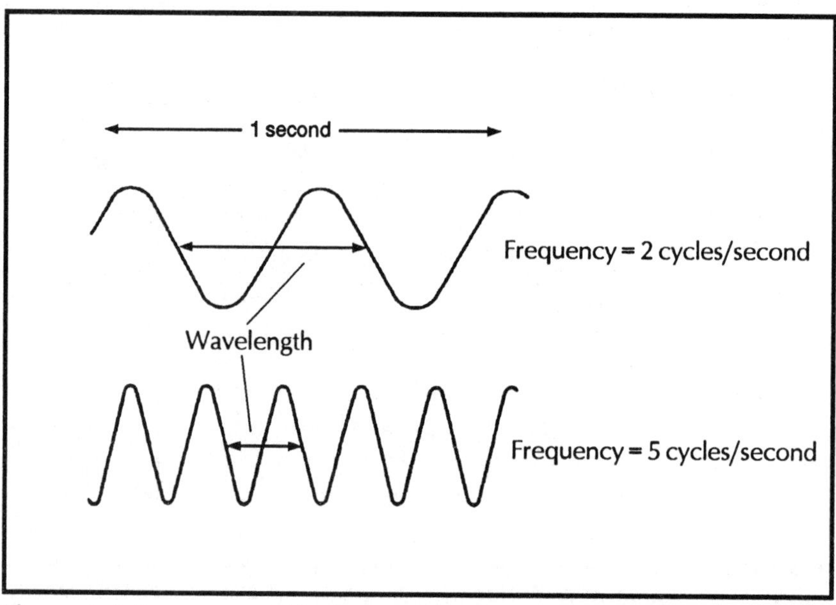

Figure 1-4. Frequency. *All electromagnetic waves are defined by how many vibrations occur each second, the frequency. Higher frequency waves have shorter wavelengths and vice versa. For example, a 4 GHz microwave signal has a 2.95 inch (7.5 cm) wavelength.*

Polarization

A third important property of all electromagnetic waves is polarization. To understand this idea we can imagine a car driving along a highway. It can travel to the same destination by following a curving road along flat ground or by following a straight road over rolling hills. Horizontally polarized waves vibrate in a horizontal plane like the car traveling along a curving road. Vertically polarized waves vibrate in a vertical plane (see Figures 1-5a and 1- 5b).

Microwaves can also be circularly polarized. Such formats, though rarely used in domestic U.S. broadcasts, are often used in international satellite communication. A left-hand circularly polarized wave has its plane of vibration following a left hand circular motion as it travels. Right-handed circularly polarized waves would follow a right circular moving plane of motion (see Figure 1-6).

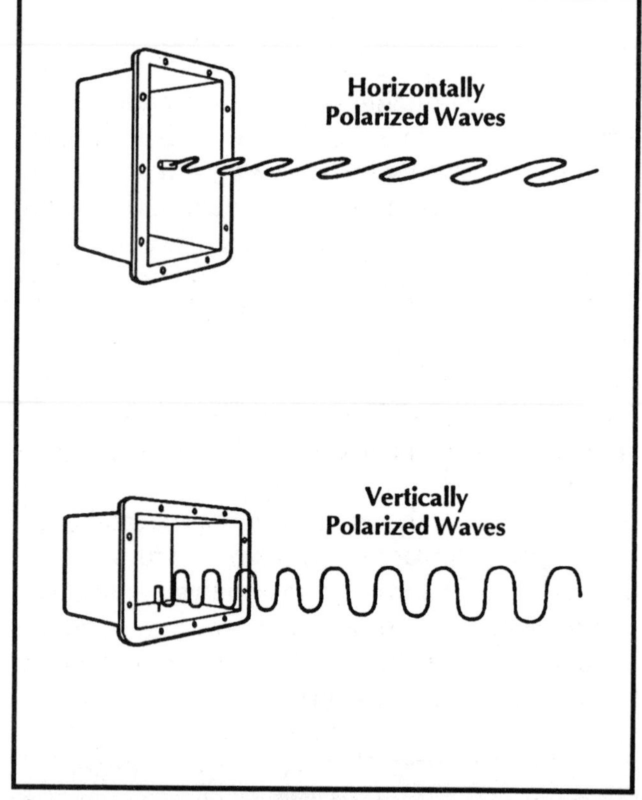

Figure 1-5a. Detecting Polarized Waves. *Polarized waves are detected by an antenna oriented parallel to the direction of polarization. These drawings, a view into a feedhorn, show the true antenna in a satellite system, the probe within its throat. Horizontally and vertically polarized waves are detected by this probe.*

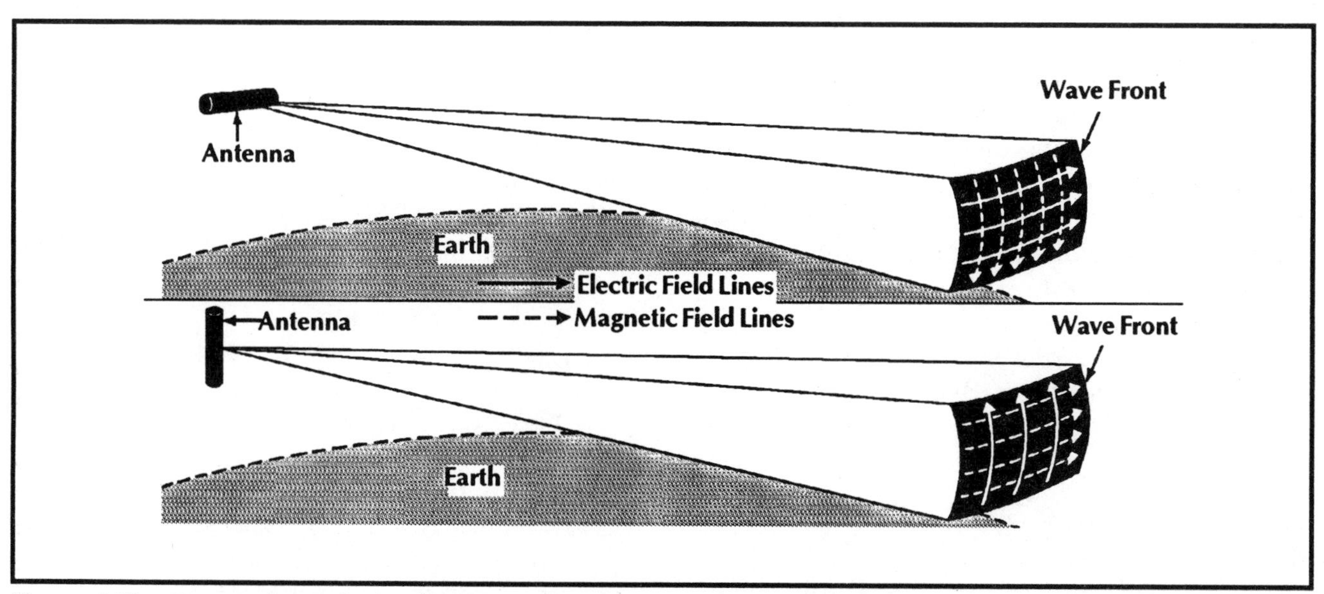

Figure 1-5b. Generating Polarized Waves. *A vertically or horizontally oriented antenna generates electromagnetic waves that are vertically and horizontally polarized, respectively. The direction of polarization is defined by the electric field (solid lines). The magnetic field lines (dotted) are always at right angles to the electric field.*

Coding the Message – Analog and Digital Signals

Any message, whether it is the image and voice of an entertainer or details of stock-market transactions, must first be changed into a form that can be relayed by radio waves. Analog coding methods mimic the pattern of a message by changes in electrical voltages. For example, a voice can be changed into an analog signal by a microphone that creates a voltage pattern determined by the loudness and frequency of the sound. The louder the sound, the higher the voltage. The higher the sound frequency, the more rapid changes occur in this voltage.

In contrast, the digital coding method uses only the numbers 0 and 1 to convey all information about voltage levels and frequencies. For example, a photograph can be described by a long series of 1s and 0s that are coded so that some impart information about the location of the dots composing the picture and others determine the brightness and color of the dots. Computers, for example, use digitally coded messages exclusively.

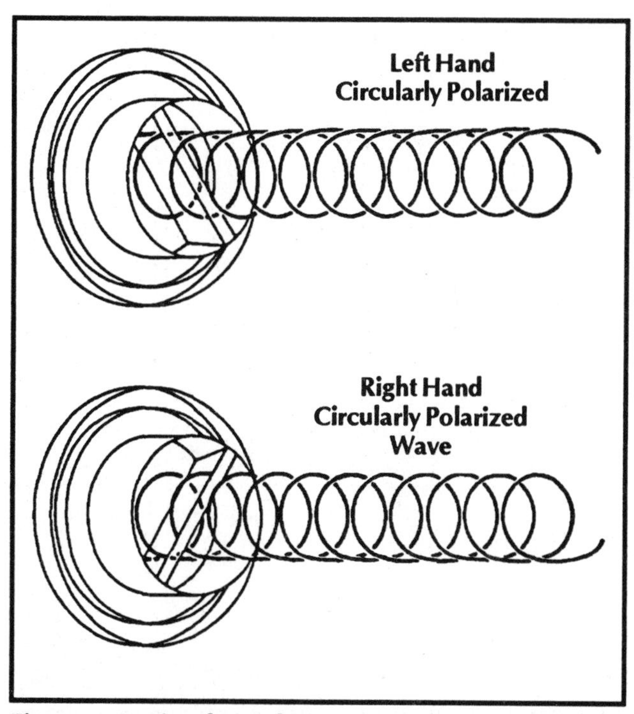

Figure 1-6. Circular Polarization. *Satellite signals are occasionally transmitted in a circularly polarized format. Instead of orienting the microwave energy in either a vertical or horizontal plane, circularly polarized signals are transmitted in a spiral pattern like a coiled spring. The direction of rotation of the electric field vibrations can follow either a clockwise or counterclockwise motion. The two senses, right hand circular polarization (RHCP) and left hand circular polarization (LHCP) are typically used by Intelsat spacecraft.*

Uplinks can relay either digital or analog forms of the same message. Converters that can translate between these two languages are available. For example, while most TV broadcasts are expressed in analog form, the trend is to use digital transmissions as satellites are packed with more sophisticated electronics allowing more information to be transmitted.

Modulation – Adding the Message to a Carrier Wave

Analog or digital signals are impressed upon radio or microwaves by a process called "modulation." Once the message is modulated onto the carrier wave of an assigned frequency it can be relayed from a sending to a receiving location whether it is via a satellite link, off-air or along a cable route. Radios, televisions and other communication equipment demodulate or extract the original message from the carrier wave.

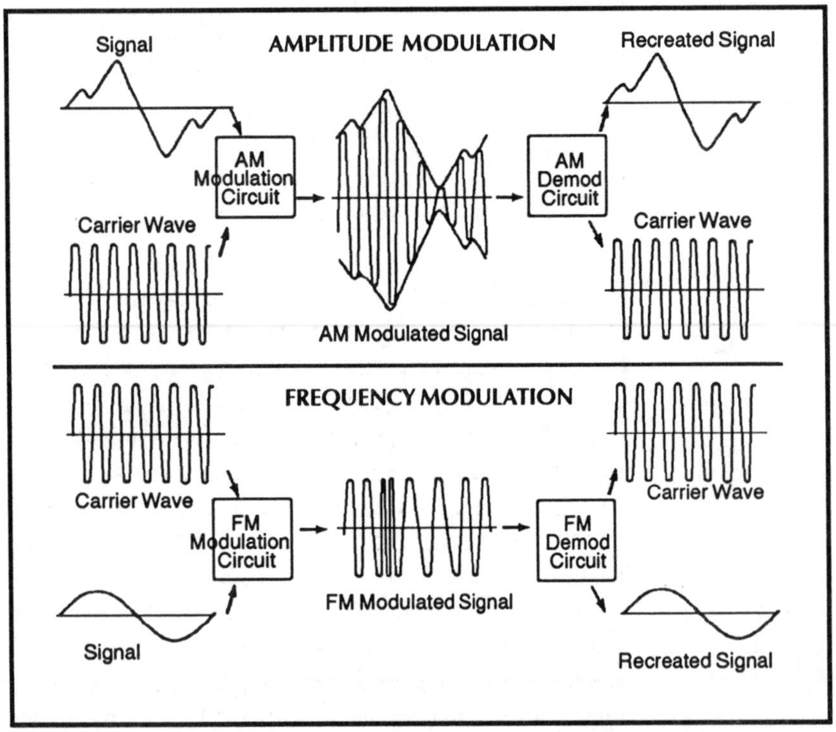

Figure 1-7. Modulation. *Two formats used to impress an audio, video or data signal onto a carrier wave are amplitude and frequency modulation. The modulated signal, the carrier wave, can then be relayed by cables, over-the-air, by satellite or any other transmission media to its final destination.*

The simplest method to modulate a carrier wave is to switch it on and off. For example, Morse code can be relayed as a series of dots and dashes by turning the carrier wave on and off. The most familiar methods of modulation are amplitude modulation (AM) and frequency modulation (FM) as encountered in AM and FM radio broadcasts (see Figure 1-7). Amplitude modulation varies the power of a carrier wave in accordance with the voltage level of the message being carried while frequency modulation varies the frequency of the carrier in response to the message.

Each modulation method has advantages and disadvantages. AM messages must have relatively high power to be capable of traveling long distances without being weakened by atmospheric disturbances and other forms of noise too severely for clear reception. Such relays are also more prone to picking up static than are FM messages. FM signals need relatively low power for successful, long distance transmission, but must use a substantially wider range of frequencies than AM messages to carry the same amount of information. Satellite messages are frequency modulated for these reasons. As signals travel the long distances between uplinks, satellites and earth receiving stations, their power becomes so low that AM transmissions would be unusable. Also, since extremely high frequencies are used in satellite communication the very wide bandwidth required by FM broadcasts is available.

Today, other methods of signal modulation have been designed to allow any given carrier wave to transmit a maximum amount of information over as narrow a range of frequencies and with as low a power as possible. Such efforts to conserve the available resources are quite sophisticated and are often used in conjunction with the two basic types of modulation.

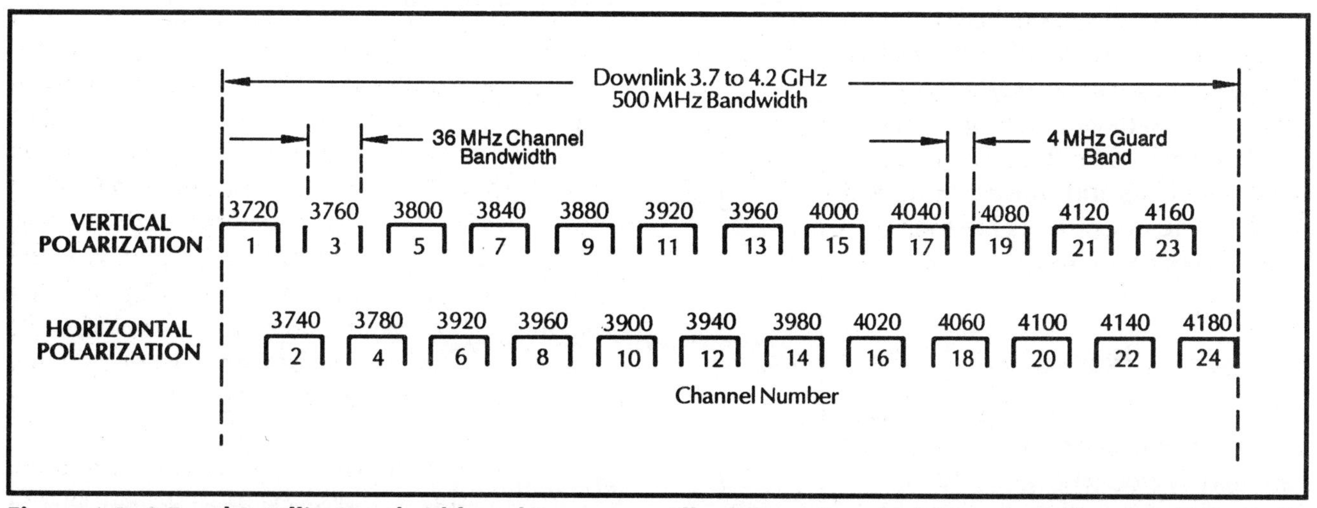

Figure 1-8. C-Band Satellite Bandwidth and Frequency Allocations. *A typical C-band satellite downlinks in the 3.7 to 4.2 GHz band having a bandwidth of 500 MHz. Twelve channels with 36 MHz bandwidths are relayed with vertical polarization and twelve 36 MHz channels are transmitted with horizontal polarization.*

Bandwidth – How Much Information Can be Carried?

Just as a large-diameter pipe can carry more water than a small one, a signal spanning a wide band of frequencies can carry more information than can one using a narrow band. This range of frequencies is termed the "bandwidth." For example, a TV message relayed in the frequency range from 54 to 58.2 MHz has a bandwidth of 4.2 MHz (megaHertz or million cycles per second). This is spread out to as much as 36 MHz in a satellite FM broadcast (see Figures 1-8 and 1-15).

Each communication medium requires a characteristic bandwidth. A device such as a television needs a substantially wider bandwidth than do radio or telephone because much more information is necessary to recreate a picture than music or a voice. To illustrate, channel one on U.S. broadcast satellites is located between 3,702 and 3,738 MHz and has a 36 MHz bandwidth (3,738 minus 3,702 MHz equals 36 MHz). Voice channels, however, normally require a bandwidth of only 3,000 to 4,000 cycles per second for quality sound reproduction.

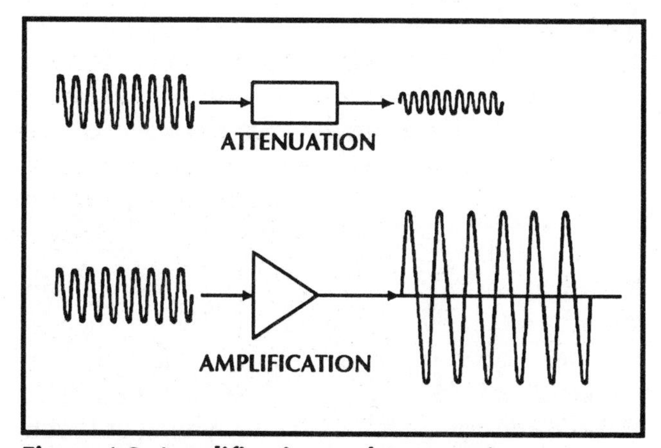

Figure 1-9. Amplification and Attenuation.
Amplification refers to an increase in a signal's peak-to-peak level. Attenuation refers to a decrease in level.

Amplification and Attenuation

Signals used for communication must often be amplified during their voyage from sender to receiver to preserve the information because their power is usually weakened or attenuated (see Figure 1-9). Just as a photograph can be enlarged but not changed, correct amplification retains the original message. All televisions, radios, stereos and other communication devices amplify a signal before demodulation occurs. For example, a signal beamed into space from an uplink antenna is weakened on its voyage to a satellite because it spreads out and is absorbed by water vapor, clouds and other atmospheric materials. The purpose of a satellite receiving antenna is to collect and concentrate this weak signal just as a magnifying glass focuses light to a point.

Occasionally, a signal will be intentionally attenuated. For example, a cable TV headend may deliver an excessively powerful signal to a feeder line which could over-drive connected TVs and cause distortion. Pads or line attenuators are then inserted to reduce the signal power.

Noise

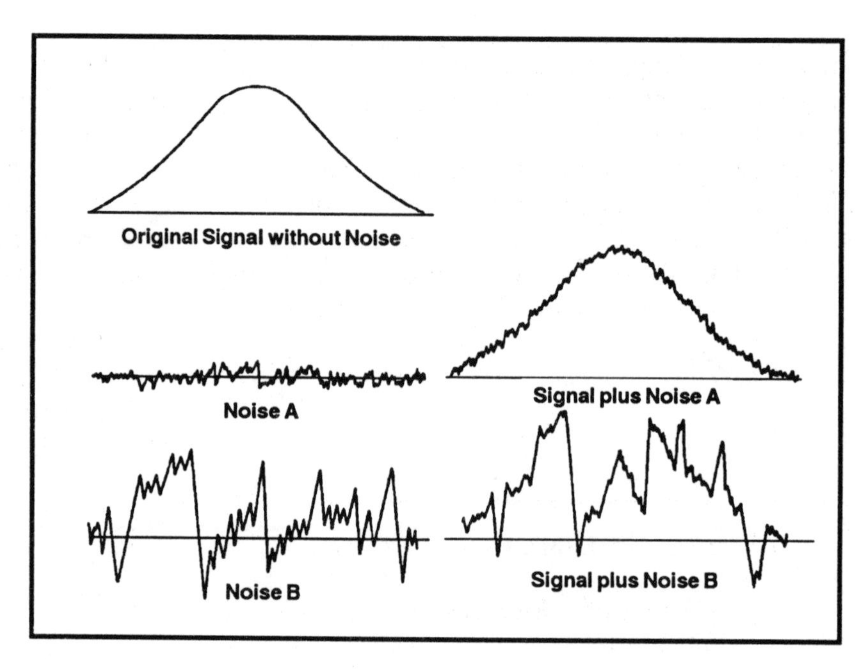

Figure 1-10. Noise. *Excessive amounts of noise (noise B) can completely garble an audio or video signal. Moderate amounts of noise (noise A) are manageable.*

In a perfect communication system signals would be relayed with no interference or noise. However, noise is present in all matter at temperatures above absolute zero, °K, the temperature at which all molecular motion ceases. (There is no temperature colder than absolute zero! The symbol °K is an abbreviation for degrees Kelvin above absolute zero. These units have the same magnitude as degrees Celsius). Noise is caused by the endless motion of the molecules that compose all matter. These small, vibrating charged particles generate electromagnetic waves that can mask the organized signal sent by man-made devices. Noise from the environment becomes stronger as the temperature increases (see Figure 1-10).

Satellite dishes detect noise from the warm ground. The temperature on an average day, 62°F, is 290°K. Outer space also generates noise at about 4°K, as a result of the "big bang." So an antenna pointed straight towards earth like one on a satellite, will pick up an additional 290°K; one on earth pointed straight into space will see approximately 10°K of noise, about 4°K from space and the rest from the ground. Receiving antennas also pick up more of this environmental noise as the signal bandwidth increases. Furthermore, noise is generated by internal heat in amplifiers, receivers and other electronic equipment.

Man-made signals not intended for the receiving station are also considered a type of noise called terrestrial interference, known as TI. TI results when earth-based communications other than those from satellite broadcasters are inadvertently received by an earth station.

Noise is always present in satellite communication systems. The quality of a communication link is determined by the ratio of signal to noise power (S/N). For example, if a signal of 10 watts is received along with 5 watts of noise, a S/N of 2, the picture quality could be poorer than if a signal of 4 watts was received with only 1 watt of noise, a S/N of 4. Typically, televisions must receive a signal having power greater than 63,000 times the accompanying noise in order for a "high-quality" picture to result

The Decibel Scale

The decibel scale was devised by Alexander Graham Bell to describe the enormous changes in power at various stages of the communication chain by relatively small numbers. This was necessary because components like amplifiers and antennas often increase power levels many hundreds of thousands of times: constantly writing out such numbers could be very cumbersome. A useful fact to remember is that small changes in decibels mean much larger changes in power levels. Thus, for example, a change of 3 dB means a doubling of power; a change of 30 dB means an increase of power by a factor of 1,000. Note that dBw means decibels relative to one watt; dBm means decibels relative to one milliwatt. So, by comparison with Table 1-1, 3 dBw means 2 watts and 0 dBm means 1 milliwatt (see Appendix A for more details on the decibel scale).

TABLE 1-1. THE DECIBEL NOTATION

Number of Decibels	Relative Increase in Power
0	1
1	1.26
3	2
10	10
20	100
30	1,000
50	100,000
100	10,000,000,000

Frequency Allocations

In the United States, the Federal Radio Commission, and later the Federal Communications Commission (the FCC) in concert with the International Telecommunication Union have kept order in the airways by successfully assigning portions of the radio wave spectrum to different communication media and users. Similar allocations of frequency space have occurred in nations throughout the world (see Figure 1-11).

In fact, the history of communication is unfolded by an exploration of these frequency assignments. As technology progressed, man has become capable of using higher and higher frequencies. Wire transmissions were first relayed with relatively low frequencies, since the radio waves used by the pioneers were limited in frequency by the rather primitive available electronics. When radio waves above 1.5 megaHertz frequency were first produced by man, the FCC delegated them to the "hams" because, at that time, they could not see a use for this region of the spectrum. As technology progressed, coaxial cable transmission, microwave relays and then satellite communication were allocated successively higher frequencies.

Even with all the new man-made forms of communication, a relatively small portion of the electromagnetic spectrum is presently being used. However, in those ranges where frequency space is being used, these resources are in heavy use and competition for allocated space can at times be fierce. As a result, innovative methods have been developed to "re-use" or to have more than one user simultaneously share the same portion of rare spectrum.

TABLE 1-2. FCC ASSIGNMENT OF SOME RADIO FREQUENCIES

Frequency (MHz)	FCC Assignment
3-54	Mobile Radio
54-72	VHF TV Channels 2-4
72-76	Radio Services
76-88	VHF TV Channels 5 & 6
88-108	FM Radio
108-120	Aeronautical
120-136	Aeronautical
136-144	Government
144-148	Amateur Radio
148-151	Radio Navigation
151-174	Land, Mobile, Maritime
174-216	VHF TV Channels 7-13
216-329	Government
329-890	UHF TV Channels 14-83

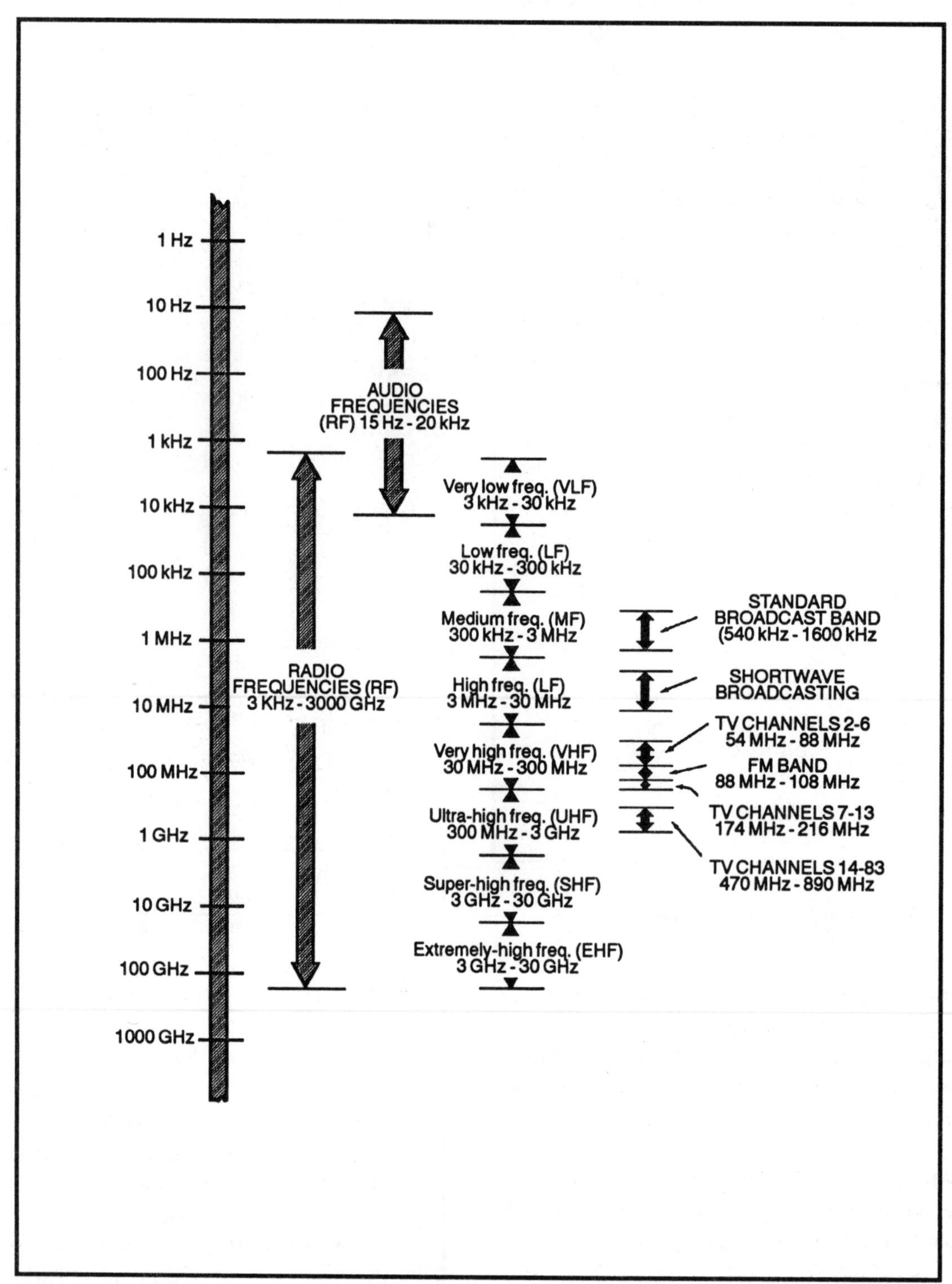

Figure 1-11. Allocation of Audio and Radio Frequencies. *The allocation of frequencies for conventional radio wave communications is illustrated here.*

C. MICROWAVES

Microwaves have been used in satellite communication for five specific reasons. First, higher frequency electromagnetic waves have the potential for relaying larger quantities of information because, as the frequency increases, any given bandwidth becomes a smaller fraction of the FM modulated carrier wave frequency. To illustrate, a 1 MHz wide band of signals imposed on a 10 MHz carrier of the spectrum modulates the carrier by 10 percent while the same bandwidth modulates a 10 GHz carrier by 0.1 percent. Since more bandwidth is available, wider bands with higher information capacities can be used at the higher microwave frequencies. Therefore, microwaves can relay more information per satellite than lower frequency signals and thus can pay off the expensive investment in satellite launching, operation and maintenance more quickly.

A second reason for using microwaves stems from the requirement for uplink antennas to aim a highly directional beam towards an extremely small target in space. Physics dictates that electromagnetic waves can be better focused by an antenna that is substantially larger than the wavelength of radiation it is managing. For example, sending a directional beam of an AM radio signal having a 100 meter wavelength would require an extremely large, cumbersome and expensive antenna. Since 6 GHz microwaves have wavelengths of approximately 2 inches (5 cm), a 15-foot uplink dish can aim most of its radiation into a very narrow beam, and relatively low power can be used.

Third, microwave transmissions to satellites or between earth-based, line-of-sight relay stations are not as susceptible to noise from atmospheric disturbances as are lower frequency transmissions. To illustrate, several times each year, for periods as long as two or three days, short-wave radio is useless for long-distance communication because sun-spot activity disturbs the required reflection of these relatively low frequency radio waves by the upper atmosphere.

Fourth, the most important property of microwaves that determines their use in satellite communication is their ability to pass through the upper atmosphere into outer space. Below frequencies of approximately 30 MHz, a radio wave will be reflected back from the ionosphere layer in the atmosphere towards earth. Since microwave frequencies are far above the 30 MHz range, they easily pass through the ionosphere shield.

Fifth, the microwave region of the electromagnetic spectrum was a virgin territory during the late 1950s and 60s when frequency spectrum was being allocated by the FCC and the International Telecommunications Union. Lower frequency space was already occupied by many different communication media and users.

As geosynchronous orbital space has become increasingly populated, progressively higher microwave frequencies have been allocated to satellite communications. Until the early 1980s most satellite broadcasters used C-band frequencies. Today, portions of the Ku-band are employed and higher frequency bands are being eyed by numerous potential users. However, a technical difficulty lies in this path. Microwaves are depolarized and more strongly absorbed by water vapor in the atmosphere at higher frequencies. So higher power transmissions must be used to counteract this effect.

TABLE 1-3. MICROWAVE FREQUENCY BANDS

Band Name	Bandwidth (GHz)
L-band	0.39 to 1.55
S-band	1.55 to 5.20
C-band	3.70 to 6.20
X-band	5.20 to 10.9
K-band	10.9 to 36.0

If satellite broadcasters had been allocated space in the S-band in the pioneer days of satellite TV (see Table 1-3), perhaps lower cost, more readily available techniques could have been used. The equivalent of the FCC in India recognized this loss caused by absorption in the atmosphere. They now use 2 GHz, S-band satellite relays.

D. SYSTEM OPERATION

The satellite communication circuit consists of an uplink, a communication satellite and an unlimited number of earth-based receiving stations (see Figure 1-12). The overwhelming strength of satellite broadcasting is based on its ability to reach a large number of customers irrespective of their geographic location.

The uplink is a complex system using many hundreds of watts of power to send a beam of microwaves to a pin-point target in space. Uplink antennas operate like car headlights which have a small light at the center and a parabolic reflector. High power microwaves aimed towards the antenna surface are reflected into a beam directed into space. Because of the power levels (typically about 1000 watts) involved, the potential for interference with other communicators, and the fact that just one antenna relayed to any chosen satellite can be detected by millions of receiving stations, uplinks are carefully regulated by the FCC. A "pirate" uplink could certainly create havoc (as the infamous Captain Midnight actually did in the United States).

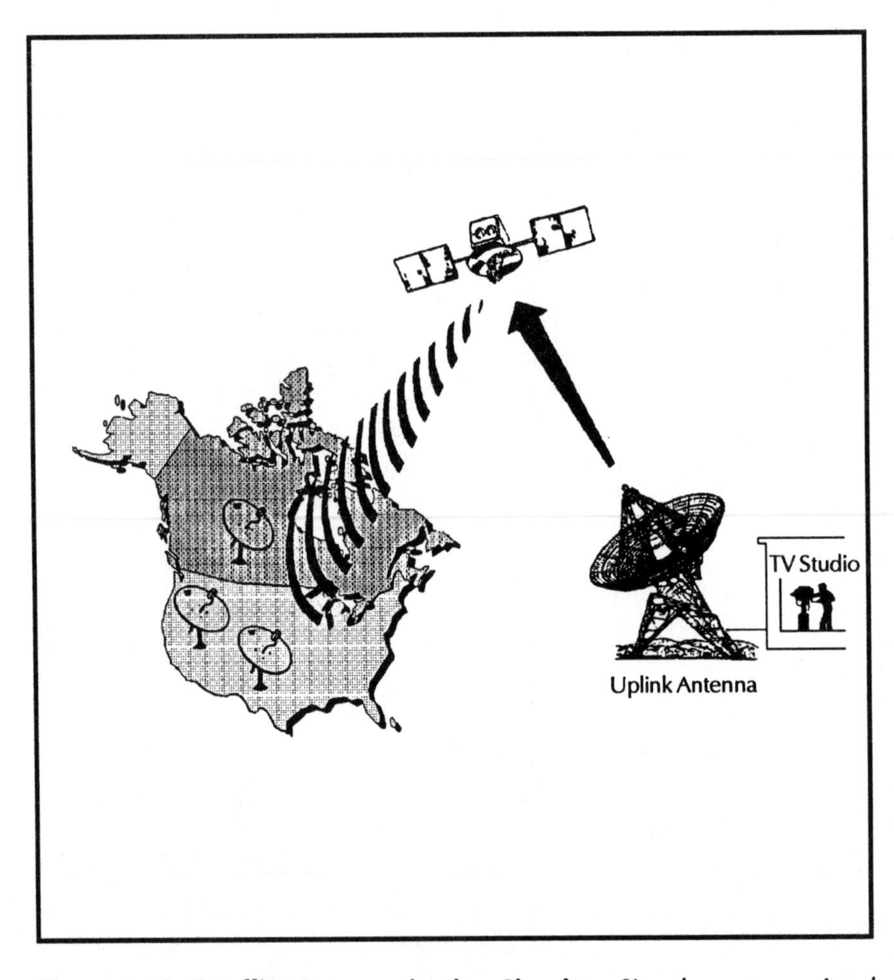

Figure 1-12. Satellite Communication Circuit. *Signals are transmitted by an uplink to a satellite, then re-transmitted at a lower frequency to an unlimited number of receiving stations.*

Uplinks are used by many segments of the business community including TV and radio stations, telephone companies, and data networks. In many cases, an on-site uplink is fed by direct cable, microwave or fiber optic links. Many television stations relay signals by conventional, off-air methods to distant uplinks and thus to communication satellites for rebroadcast. Groups needing uplink facilities on an occasional basis for teleconferencing, one-time sporting events or other special happenings have the option of "taking Mohammed to the mountain," i.e. of bringing the equipment directly to the action. This is often a viable economical alternative to off-air broadcasting to a fixed uplink site. This leasing of portable uplinks complete with all equipment and operating crews is a sensible alternative in view of the average $600,000 price tag for a fully operable on-site uplink.

Geosynchronous broadcast satellites receive the uplinked signal, change the frequency of the message and then broadcast it to any chosen geographic area below. Downlink antennas can target up to 40 percent of the earth's surface with so-called global beams, can broadcast zone beams to selected countries or continents, or can pinpoint smaller areas with spot beams. To illustrate (see Figure 1- 13), many American broadcast satellites have one antenna which blankets the continental United States and a second smaller one which directs a more localized beam to the Hawaiian islands. Another good example of a satellite that relays two separate beams is Satcom F-5, also known as Aurora.

The earth station utilizes a large antenna to collect and concentrate as much of the very weak downlinked signal as possible to its focus. The feedhorn, located precisely at the focus, channels radiation reflected and concentrated by the antenna into the first active component, the low noise amplifier. This component is abbreviated LNA or LNB for low noise amplifier/block converter. The signal is then sent down a coaxial cable line to an indoors satellite receiver and processed into a form understandable by either a television, stereo or data processor. An earth receiving station is in essence an uplink operating in reverse.

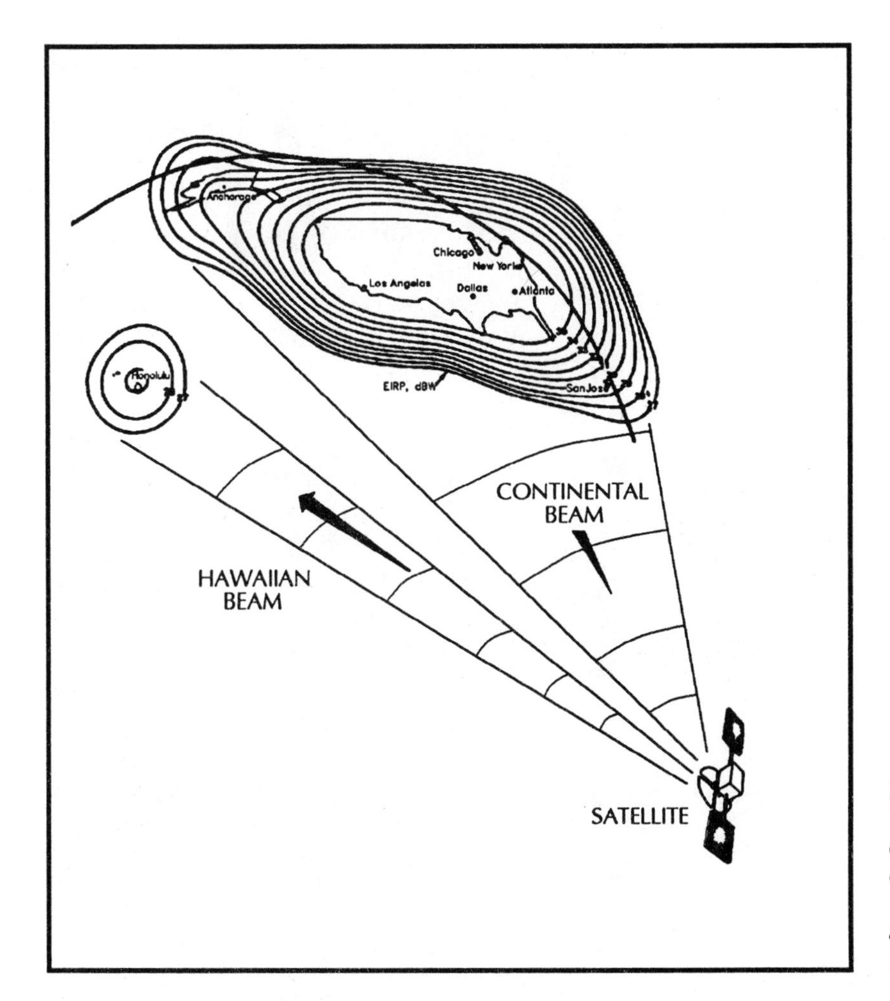

Figure 1-13. Galaxy I Footprint Map. *This footprint map shows the contour of equal effective isotropic radiated power, expressed in dBw. In this case, only the 12 horizontally polarized transponders are transmitted via a spot beam to the Hawaiian islands.*

E. SATELLITES

Satellites are the key component in the telecommunication revolution. Our communication system is now greatly improved because any locations within a satellite's "view" can be linked without the use of expensive cables or line-of-sight relay towers. Furthermore, a communication satellite operating as a relay in space can serve vast areas of our globe at once.

Satellite Operation – Audio and Video Channels

Uplinked signals have powers of fractions of a millionth of a watt when received by geosynchronous satellites. In the heart of this communication vehicle, typically about the size of a small truck, these signals are amplified many thousands of times, shifted to a lower frequency range and then re-transmitted back to earth.

This frequency conversion reduces terrestrial interference at the receiving station below. If this was not the case, some of the uplinked signal could be inadvertently detected by receiving stations near uplink sites along with the desired signal. American C-band circuits use an uplink having a 500 megaHertz bandwidth spanning the range from 5.925 to 6.425 gigaHertz and the same bandwidth shifted to the lower range of 3.7 to 4.2 GHz for the downlink. (Ku-band relays typically uplink signals from 13.7 to 14.2 Hz and downlink from 11.7 to 12.2 GHz, both ranges also spanning a 500 MHz bandwidth).

All the power to carry out these electronic functions is provided by the sun's energy and captured by arrays of solar or photovoltaic cells. The illustration of Satcom I (see Figure 1-14) shows these arrays as large wing-like structures. The smaller central features are the receiving and downlink antennas.

Television Channel Formats

The number of television channels, telephone conversations or the amount of data transmitted is related to the electronic design of a satellite. The early Western Union vehicles such as Westar I and II relayed twelve television programs simultaneously; the RCA series of Satcom and most modern C-band broadcast satellites handle at least twenty four channels.

How is the number of channels determined? The 500 MHz microwave band can be subdivided into twelve 40 MHz segments plus a remainder of 20 MHz. Since a 36 MHz bandwidth is sufficient to

Figure 1-14. Satcom Satellites. *The large wing-like structures on this satellite are solar cells which provide on-board power. The downlink and uplink antennas receive and re-transmit power, respectively. (Courtesy of RCA Americom)*

broadcast a high quality television picture, Western Union designed their early satellites to carry 12 channels having 36 MHz bandwidths with 4 MHz protection regions, guard bands, between each channel to eliminate the possibility of cross-talk. Each channel was handled separately on-board the satellite by a device called a transponder.

Engineers designing the Satcom I vehicle were somewhat more creative (see Figures 1-15 and 1-8). They doubled the number of channels which could be relayed by this 500 MHz total bandwidth by a technique called frequency re-use. All even channels were transmitted earthward with horizontal polarization; all odd channels were sent with vertical polarization; and the frequency centers of these cross polarized channels were offset from each other for further security against cross-talk. Since each earth station was equipped to detect only vertical or horizontal polarization at one time there could be some overlap between the frequencies used for the odd or even channels without interference between channels. The center frequencies of each channel are listed in Table 1-4. Of course, when an earth station is required to simultaneously receive all 24 channels it must be capable of detecting both horizontal and vertical channels at the same time.

Note that the format used for U.S. broadcast satellites is by no means the only one accepted by other nations. For example, European satellites relay their television broadcasts over 700 megahertz bandwidths. Twelve C-band channels on Morelos have 36 MHz bandwidths while six transponders have 72 MHz bandwidths.

Audio Channel Formats

Each transponder manages a 36 MHz wide band of frequencies. When an earth station receives and processes this information, the resulting signal is contained in a band of frequencies from near zero to about 10 MHz. The video signal

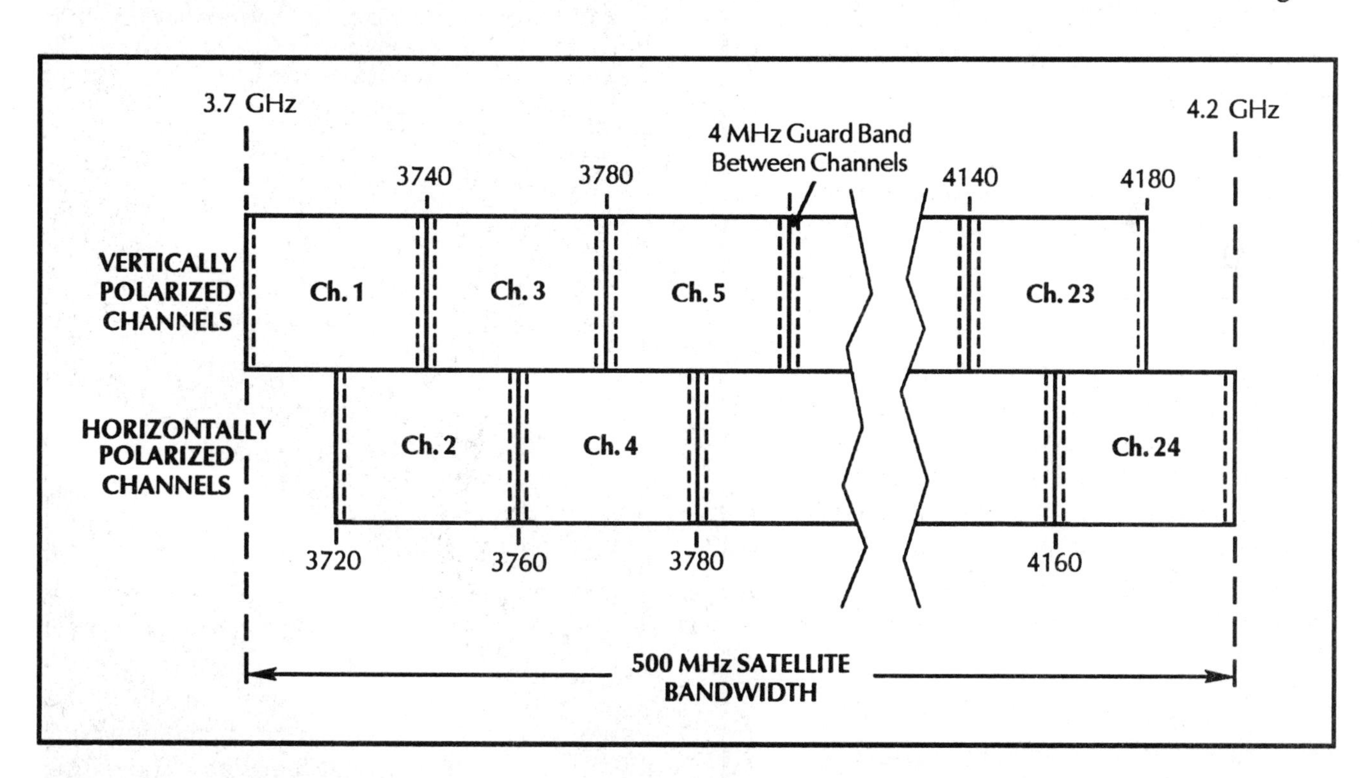

Figure 1-15. Video Channel Formats. *Most C-band broadcast satellites are designed to relay 24 channels each having a maximum bandwidth of 36 MHz. Twelve channels each are relayed by vertically and horizontally polarized waves. The Satcom, Telstar and Comstar series of satellites have odd vertically polarized and even horizontally polarized channels, as in this illustration. The Galaxy, Westar, Spacenet and Anik spacecraft have a reversed polarity scheme.*

is contained between zero and 4.6 MHz. All the remaining space can be used for audio channels, some of which carry stereo or mono sound that matches the television picture while others carry totally separate messages.

These audio signals are carried on "audio subcarriers." Most U.S. television broadcasts relay the audio information on a 6.8 MHz subcarrier or occasionally, on both 6.2 and 6.8 MHz subcarriers for transmitting stereo sound (see Figure 1-16).

Some satellite transponders operate in the single channel per carrier mode, known as SCPC, where only audio information is relayed. For example, National Public Radio dedicates an entire transponder (Westar IV, transponder III) exclusively to their audio broadcasts. (For further information see *Hidden Signals on Satellite TV* by Thomas Harrington).

TABLE 1-4. SATELLITE CHANNEL CENTER FREQUENCIES

Downlink Transponder Number	Frequency (MHz)
1	3720
2	3740
3	3760
4	3780
5	3800
6	3820
7	3840
8	3860
9	3880
10	3900
11	3920
12	3940
13	3960
14	3980
15	4000
16	4020
17	4040
18	4060
19	4080
20	4100
21	4120
22	4140
23	4160
24	4180

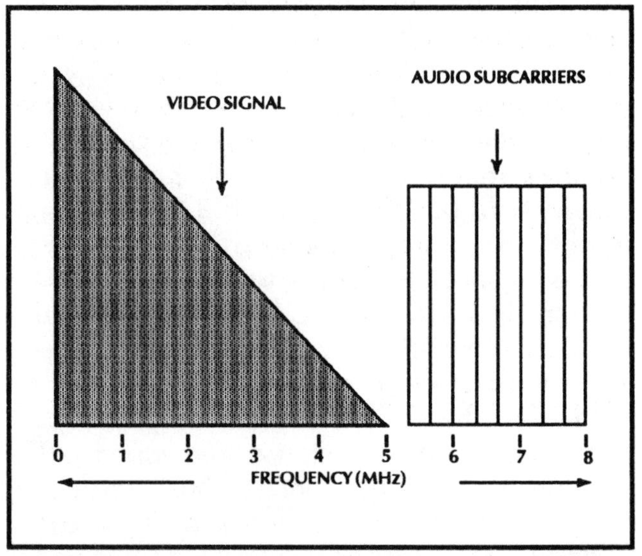

Figure 1-16. Audio Channel Format. *Audio information is typically relayed on subcarriers having center frequencies ranging from 5.0 to 8.5 MHz. The sound accompanying TV broadcasts often is located on a 6.8 MHz subcarrier.*

The Downlink Path and Satellite Footprints

Satellite downlink antennas broadcast all these processed microwave signals to any chosen geographic region below. The design of such antennas is in itself a science and an art. The satellite footprint is determined by the downlink antenna's geographic coverage and the microwave power generated by each on-board transponder. As would be expected, power levels are higher in regions targeted by the main axis of the downlink antenna and weaker in off-target areas. Thus larger dishes are required in Florida than in Nebraska for equivalent reception from a satellite that targets the whole continental United States. Nebraska is at the center of the footprint, directly along the downlink antenna "boresight." It also explains why extremely large antennas, often in excess of 30-feet in diameter, are necessary for adequate reception of American satellites in countries such as Argentina or Brazil, which are far removed from the downlink antenna's boresight.

Although published footprint maps are available for each satellite there is also some variation between transponders on any given satellite. Satellite TV dealers and enthusiasts know well which transponders on-board an older and somewhat weaker satellite such as Satcom III have the lowest power. In fact, if clear pictures can be received from the weakest transponders then a given earth station is performing well for its intended use.

Understanding Footprint Maps

Satellite footprint maps provide valuable information in sizing components of an earth receiving station. Power levels on these maps are termed effective isotropic radiated power, EIRP, a weighty sounding term. And EIRPs are measured in units called decibels above one watt (dBw). A footprint map is constructed by joining all those points on the map having equal EIRP by continuous lines. So a distinctive "footprint" is published for every orbiting satellite as a series of contour lines superimposed upon the map of the region served (see Figure 1-17).

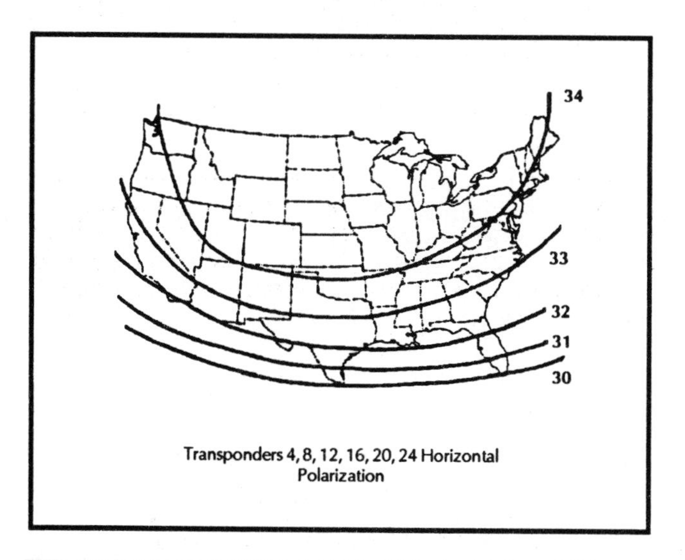

Transponders 4, 8, 12, 16, 20, 24 Horizontal Polarization

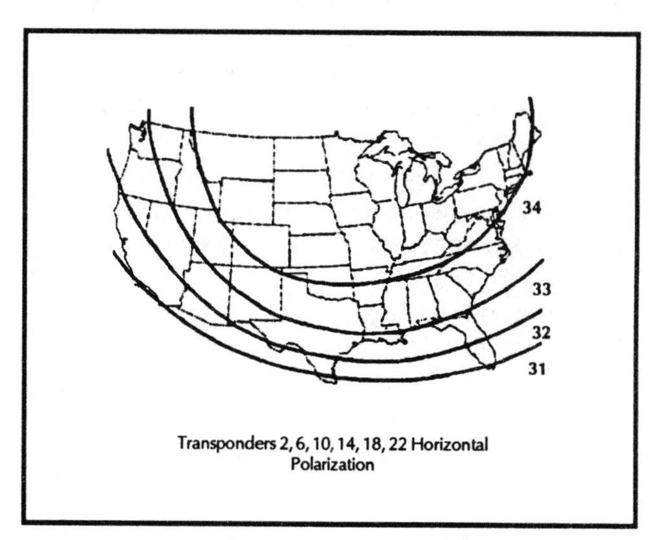

Transponders 2, 6, 10, 14, 18, 22 Horizontal Polarization

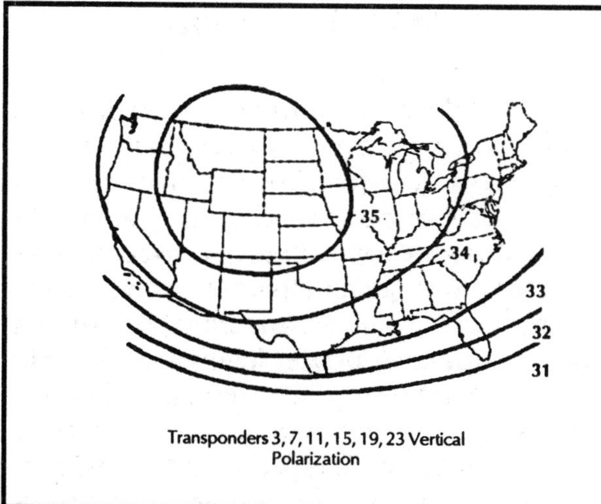

Transponders 3, 7, 11, 15, 19, 23 Vertical Polarization

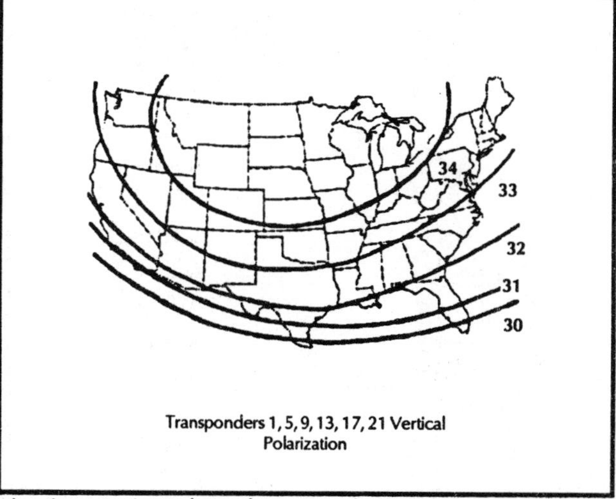

Transponders 1, 5, 9, 13, 17, 21 Vertical Polarization

Figure 1-17. Effective Isotropic Radiated Power. *The effective isotropic radiated power, the EIRP, is a measure of the relative strength of a satellite downlink signal, expressed in dBw. Each satellite has a different coverage pattern and, in many cases, different transponders on a given satellite have slightly different EIRPs. The area where the signal is strongest is referred to as the boresight.*

Footprint maps that show small changes in decibel levels demonstrate how widely signal powers actually change. Remember that the signal strength on any footprint map is always highest right along the central axis of the downlink antenna. For example, Satcom I relayed 33 dB to Anchorage, Alaska, roughly half that received in Denver, Colorado at 36 dBw because of the 3 dB difference in downlink antenna beam pattern.

Satellites broadcasting to a more limited geographic region have a footprint covering a smaller area. These "spot beams" have proportionately higher EIRPs than those of a satellite targeting a much larger area with either hemispherical, zone or global beams because the same power is more concentrated (see Figures 1-18 and 1-19). Global beams cover the maximum 42.4% of the earth's surface as viewed from a communication satellite. Hemispherical and zone beams target 20 and 10 percent of the globe, respectively. Therefore, each footprint map is determined by both the shape and orientation of the downlink antenna as well as by the power generated by each transponder.

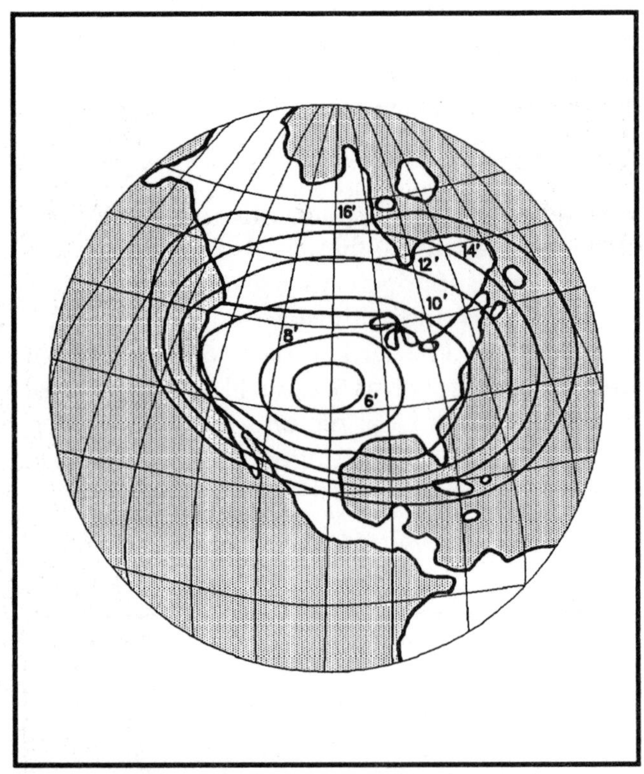

Figure 1-19. Footprint Map and Antenna Size. *Many footprint maps are drawn showing a minimum required antenna diameter instead of EIRP for ease of use.*

Why are these power levels named effective isotropic radiated power? Isotropic means equal in all directions. And effective isotropic radiated power means the power levels that would be received at any location if an antenna were radiating equally in all directions. So 33 dBw read from a footprint map means that such a perfect antenna would direct 33dBw or 2000 watts per square meter in all directions.

Figure 1-18. A Spot Beam. *This satellite aims a spot beam at the center of the continental United States.*

EIRP levels are measured as the power levels of signals leaving the downlink antenna. In the example above, 2000 watts per square meter are directed towards a location on earth. However, the signal leaving this satellite spreads out in a cone-like beam as it travels. This weakening or dilution in power as it moves further and further away from the satellite is called the "free space path loss" or "spreading loss." The longer this distance, called the "slant range," the greater the free space path losses.

Also contributing to signal losses incurred on the homeward voyage is absorption by molecules of the atmosphere. Water vapor is the main culprit in the attenuation of downlinked signals. In fact, during a severe rainstorm, the power received on

Figure 1-20. Intelsat VI. *The Intelsat VI satellite is equipped to handle both C and Ku-band transmissions. It stands nearly 39 feet high with its antennas unfolded and its aft solar panel extended. (Courtesy of Hughes Aircraft Company)*

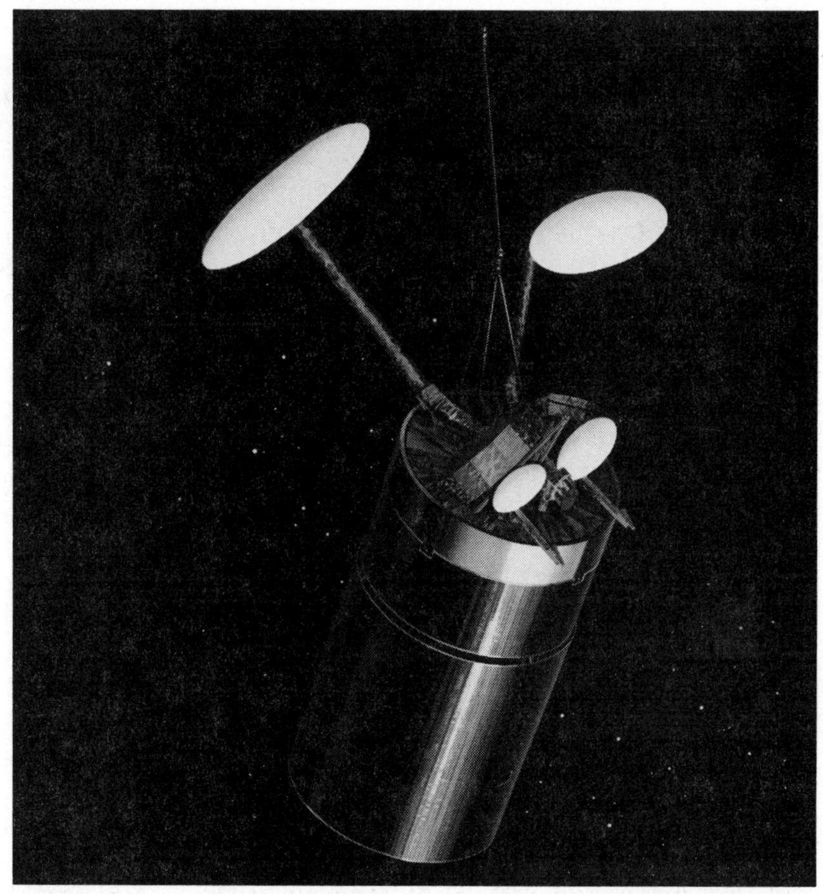

the surface of the earth can be reduced by as much as 3 dB or 50% at higher Ku-band frequencies.

Free space path losses and atmospheric absorption explain why a transponder operating with only 5 to 9 watts total power that sends a signal via a downlink antenna into a beam of 2000 watts per square meter is detected on earth at a strength of less than approximately one tenth of one billionth of a billionth of a watt per square meter! Note that detecting a 5 watt transponder is like seeing an average Christmas light bulb from a distance of 22,300 miles. Or it is like receiving a CB radio transmission, which is designed to have a 10 to 15 mile range, from 22,300 miles away. The C-band satellite broadcast operates at a frequency 131 times higher than CB radio.

Ku-Band Satellite Broadcasts

The move towards higher frequency satellite communications has occurred for a number of reasons. This technical frontier offers a large amount of available frequency spectrum which can potentially lower communication and broadcasting costs. In addition, C and Ku-band signals do not interfere with each other so one satellite can relay both types of messages. In fact, many modern satellites are now equipped to relay both C and Ku-band transmissions (see Figure 1-20).

Microwave antennas are also more efficient at higher frequencies and satellites are permitted to transmit higher power signals. As a result, relatively small antennas can be used for excellent reception of Ku-band transmissions.

The frequency and polarization formats used by Ku-band satellites are not standardized as for C-band relays. Channel bandwidths as well as the center frequencies vary between satellites. The interested reader can find more information in *Ku-Band Satellite TV - Theory, Installation and Repair,* also published by Baylin Publications.

Satellite Launching and Maintenance

The excellence of a satellite communication system depends upon launching long-lived vehicles into stable geosynchronous orbits. Since Sputnik's successful flight in 1957, man has been able to lift heavier and more sophisticated payloads into orbit. During the 1960's booster rockets were inadequate to place satellites directly into geosynchronous orbit. As a result, extra rockets in the body of a telecommunication vehicle were used for repositioning it from its initial low circular orbit into its final geosynchronous position. The engine and fuel of the satellite then accounted for nearly half its total weight.

Today, similar techniques for boosting satellites into space are used even though current communication vehicles having greater masses can be lifted into space by much more powerful crafts. Earlier Thor-Delta launch rockets have been replaced by more powerful boosters such as the Delta-2914, the Delta-3912 and the Atlas Centaur. The European Space Agency's Ariane rockets as well as American private vehicles now launch the lions share of geosynchronous satellites.

Communication satellites would remain stationary forever above a chosen location over the equator if there were no extra gravitational forces from the sun and moon, if solar winds did not sweep past our globe, and if the earth were perfectly spherical. This is not the case, so these unbalanced forces cause satellites to drift slowly away from their assigned locations. There are, however, two "zero-pull" locations at approximately 104.5 degrees West and 75.5 degrees East on the exact opposite side of our globe where these forces balance so a geostationary body will remain stationary. The Canadian satellite Anik D1, located at 104 degrees West is closest to the western hemisphere zero-pull location.

Since all satellites to the west of 104.5 degrees West will drift towards the east and those located to the east of this point will drift westward, ground controllers periodically adjust satellite positions to counteract these forces. All communication vehicles are equipped with small hydrazine gas charged "thrusters" that are fired whenever necessary.

Satellite antennas and solar cell arrays must also be periodically realigned with their targets. In order to relay a beam having a width of approximately 4 degrees (as is typically required to cover the continental United States) towards a chosen location on the earth, downlink antennas must be accurately pointed within 0.2 degrees. This is very important because most satellites, even when accurately positioned, move in a small figure-eight pattern, usually about 15 miles across. Even this very small movement can impair reception of signals on earth by large, narrow view antennas.

Also, in some satellites the solar cells must also be accurately pointed towards the sun so that they will provide as much power as possible. Satellites are aligned, for both these reasons, by spinning their whole body in order to create a stabilizing gyroscopic effect similar to the forces that keep a rapidly spinning top upright. The improved stability of present-day vehicles permits use of larger antennas, capable of focusing higher-power microwaves and housing larger solar cell arrays to increase satellite power.

Satellites do not live forever. Their life expectancy is determined by how long adequate power and stability can be maintained. Repositioning rockets usually run out of hydrazine gas before surviving ten years in space. Also, solar cells, constantly bombarded with micrometeorites and ultraviolet rays from the sun, slowly wear out. Either a 30% reduction in solar cell power or expiration of the hydrazine fuel supply is a signal to retire a satellite from active duty. For example, Satcom III, the time-tested workhorse of broadcast satellite TV, will soon be decommissioned due to slowly decaying transponder powers. When ground controllers put a satellite to rest it is boosted into a slightly higher, unstable, non-geosynchronous orbit where, in time, it will eventually re-enter the atmosphere and burn up. Recently, the space shuttle has recovered some damaged satellites. This maneuver has saved 50 to 80 million dollars in hardware and launch costs and may someday be used to recapture and re-use satellites.

Future Trends in Satellite Design and Operation

Satellite technology is still in its infancy. The next decade should bring some rather incredible advances in design and operation of the space-based communication system. Some of these improvements will come in response to the need for additional orbital space, the most obvious move being the use of higher frequency bands.

Satellites can be co-located in the geosynchronous arc if they operate in different frequency bands and can thus be effectively invisible to each other. Already some satellites, for example Spacenet I, have both C and K-band transmitting antennas installed. Clusters of non-interfering, co-located satellites could receive signals from just one uplink facility. In addition, all these vehicles could further conserve resources by sharing a single platform, positioning system and power supply. Such a rigid platform housing all the satellites would probably also improve the overall stability of the downlink antennas and could therefore enhance performance of each.

Satellite circuits could be used more effectively by employing communication directly between satellites with microwave or laser beams. This method works very well in outer space because there is no atmosphere to weaken signals bouncing between satellites. In fact, the FCC and ITU have already allocated frequency bands for inter- satellite links. Using such relays could make communication between very distant countries less expensive by eliminating an intermediate link. For example, a message uplinked in San Francisco could be received by a American satellite, relayed directly to one positioned over the Mediterranean and subsequently downlinked to Saudi Arabia. The alternative would be to use an intermediate Intelsat (International Telecommunications Consortium) satellite or a submarine cable, which would probably increase the cost of communication.

Satellites could switch between differently targeted antennas on command or have movable, narrowly aimed antennas called steerable spot beams to select from pre-defined footprints. This would allow their available allocated frequencies to be re-used in different geographic regions. Such vehicles could employ the available power more effectively by restricting their output to limited areas and thus increasing EIRPs. The Intelsat V commercial series of satellites was the first to accomplish this task.

One other technique that has recently been developed involves using more efficient coding methods so, for example, one transponder previously used to relay a single broadcast can now relay up to ten television channels.

II. SYSTEM COMPONENTS

A. ANTENNAS

Purpose

A satellite antenna intercepts the extremely weak microwave transmission from a targeted satellite and reflects the signal to its focal point, where the feedhorn is placed. This is the process that concentrates the signal so that the necessary power is available for subsequent electronic components.

The quality of a satellite antenna, often simply called a dish, is determined by how well it targets a satellite and concentrates the desired signal and by how well it ignores unwanted noise and interference. These seemingly simple design objectives have been the subject of a well developed scientific and engineering discipline for many years.

Dishes must be durable and able to withstand winds as well as other natural and man-made forces. In order to be able to compete in the marketplace, they also must be aesthetically pleasing and affordably priced. Ever since the October 1979 FCC decision in the United States which deregulated receive-only systems and spurred the rapid growth of the home satellite TV industry, many small and large businesses have tried to realize all these objectives.

Antenna Designs

Most microwave antennas used today in satellite earth receiving stations have designs based upon combinations of circular or parabolic surfaces, which are familiar geometrical shapes originally discovered by the ancient Greeks (see Figure 2-1). Any microwaves reflected off these surfaces will be concentrated to a point or to a series of points called "focal points."

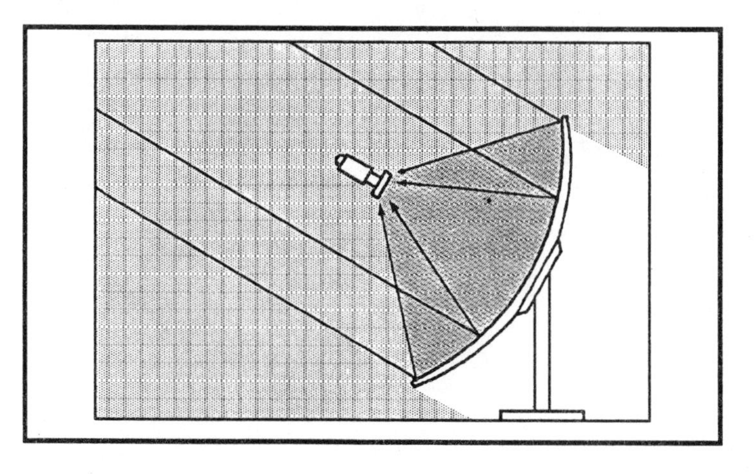

Figure 2-1. A Parabolic Antenna. *Signals are focused to a feedhorn and low noise amplifier at the focal point of the parabolic reflector.*

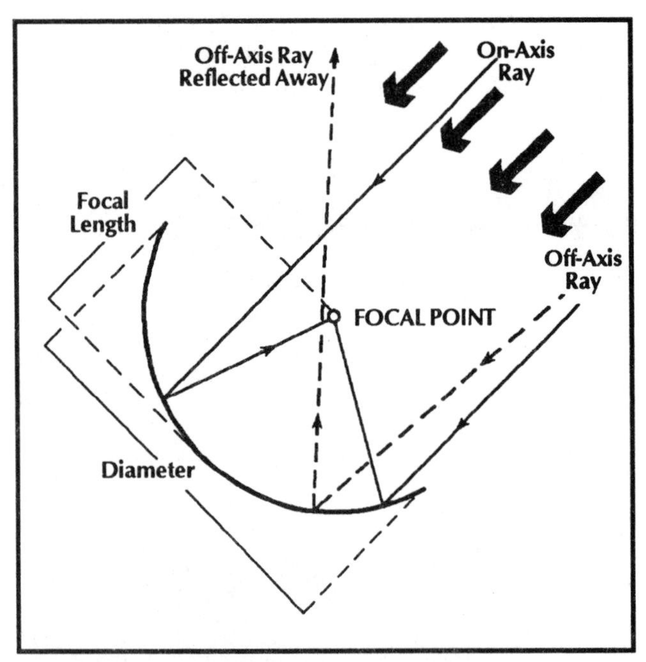

Figure 2-2. Parabolic Antenna Geometry. *In the prime-focus configuration all rays are directed via one reflection to a feedhorn at the focal point.*

Prime-Focus Parabola

By far the most common dish in the satellite TV industry, the prime-focus parabola, theoretically focuses all incoming signals that are parallel to its axis to a single point. Any signals arriving from a direction other than that of the targeted satellite will be reflected away from this focal point (see Figure 2-2).

In practice, this dish does not behave as predicted by theory for three reasons. First, the equipment mounted at the antenna focus is spread around the focal point so that it can intercept some microwaves from directions slightly off target. Second, irregularities in the surface shape cause errors so that some off-axis signals are detected and some targeted signals pass by without being observed. Third, dishes simply do not behave as perfectly as ray-tracing geometry would dictate because the radiation they intercept behaves according to principles of waves and some spreading of the signal occurs around their edges (see Figure 2-3).

Offset Fed Parabola

An offset fed parabolic antenna is constructed by using only a section of a larger prime-focus parabola. The feed assembly still remains at the focus of the original larger parabolic dish and appears to be offset from the portion of the reflective surface is use. In this configuration, the feed assembly does not block any incoming signals (see Figure 2-4).

While offset fed dishes have been rarely used in North American home satellite TV installations, this variety has become the most popular forr receiving Ku-band broadcasts in Europe (see Figure 2-5). These antennas range from 30 to 120 centimeters (1 to 4 feet) in diameter and are less cumbersome than their larger C-band counterparts. Such construction is easier because the structure which holds the offset feed is substantially shorter and can be made sufficiently rigid.

Offset dishes are aimed towards the arc of satellites at a steeper angle than the prime-focus variety. They therefore have the advantage of being better able to shed snow.

Figure 2-3. Prime Focus Parabolic Antenna. *This Paraclipse 3 meter parabolic antenna is constructed with from 1 mm thick (0.040 inch) perforated aluminum with a hole diameter of just under 2 mm. (Courtesy of Paraclipse Manufacturing, Inc.)*

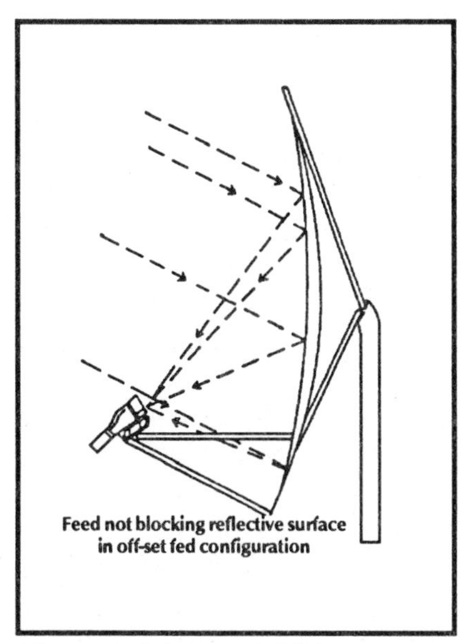

Figure 2-4. Beamwidth Patterns of an Offset Fed Antenna. *The beamwidth pattern of an offset fed antenna is determined in part by its shape. If the dimension parallel to the arc of satellites is longer than that at right angles, the beam pattern would be compressed into a vertical elongated ellipse. This occurs because beamwidth decreases as the diameter increases. In this case, the reflector has a relatively larger diameter along the arc than perpendicular to it.*

Figure 2-5. Maspro Antenna and LNB. *This stamped antenna is made of galvanized steel with baked-on paint to extend its life expectancy. (Courtesy of Maspro)*

Cassegrain Parabolic

The Cassegrain antenna also has a parabolic surface but redirects radiation via a second reflective surface, the hyperbolic subreflector, down a waveguide to an LNB mounted behind the dish. In extremely hot climates, earth stations using Cassegrains can perform more effectively than prime focus antennas because the LNB is installed behind the antenna and is therefore protected from direct exposure to solar energy (see Figures 2-6 and 2-7).

Cassegrain antennas are generally more expensive than the ordinary prime-focus variety and are usually found in commercial installations.

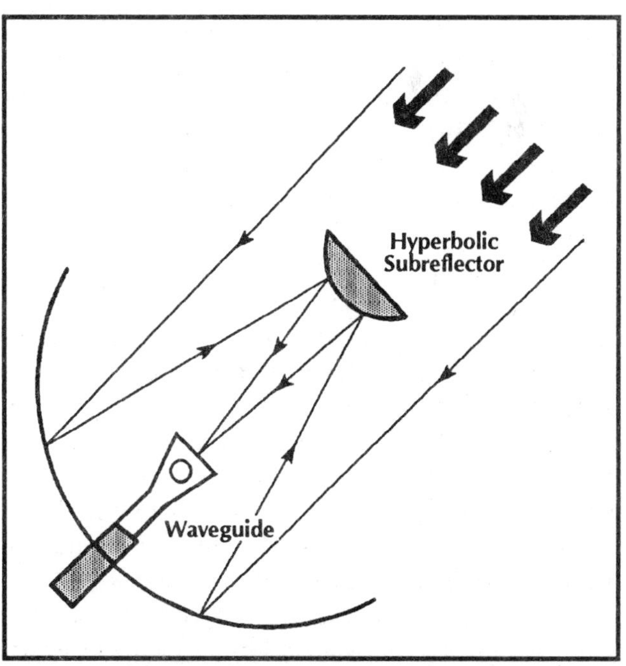

Figure 2-6. Cassegrain Antenna Geometry. *A Cassegrain parabolic antenna employs a second reflector, the hyperbolic subreflector, to direct microwaves to a feed located behind the reflective surface.*

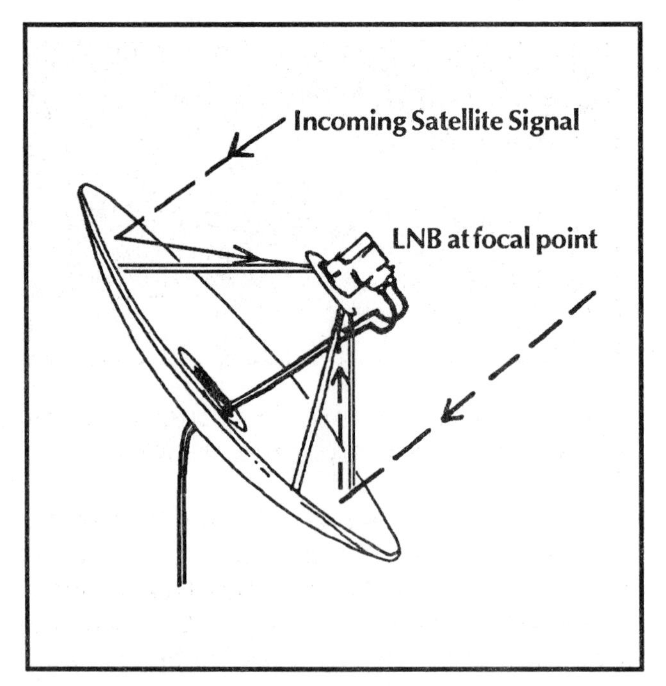

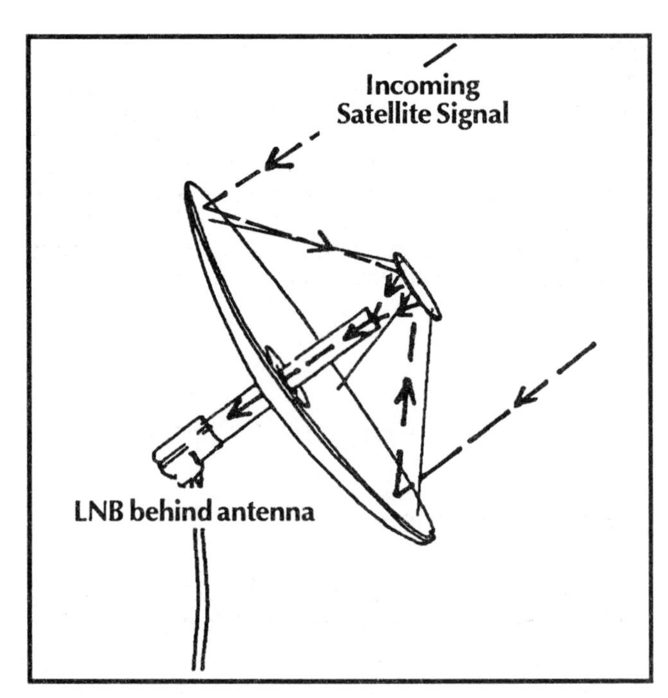

Figure 2-7. Prime Focus versus Cassegrain Feeds. *In a prime focus configuration, the low noise amplifier is located at the antenna focal point. In contrast, in a Cassegrain system, the amplifier is in a more protected rear position.*

Flat-Plate Antennas

While flat plate antennas have not yet been widely used in home satellite systems, some designs are commercially available. Some of these innovative antennas are based upon the optical technique known as Fresnel reflections. A series of concentric rings that are opaque to radiation overlaid onto a transparent sheet creates a lens which is capable of focusing microwaves. If the concentric rings are elongated into a set of ellipses, signals arriving from off-boresight can still be focused to a point behind the Fresnel pattern.

If a negative pattern is used whereby concentric rings are made transparent and the space between the rings is opaque, then a reflective sheet laid behind this assembly will reflect satellite signals to a feedhorn in front of the antenna. Both these configurations are outlined in Figures 2-8 and 2-9.

Both designs have some advantages. The same elliptical pattern can be used within a radius of approximately 300 miles. Slightly different elliptical patterns can be created for regions further removed. Given the appropriate pattern, both the transmission and reflection version can be installed flush to the mounting surface.

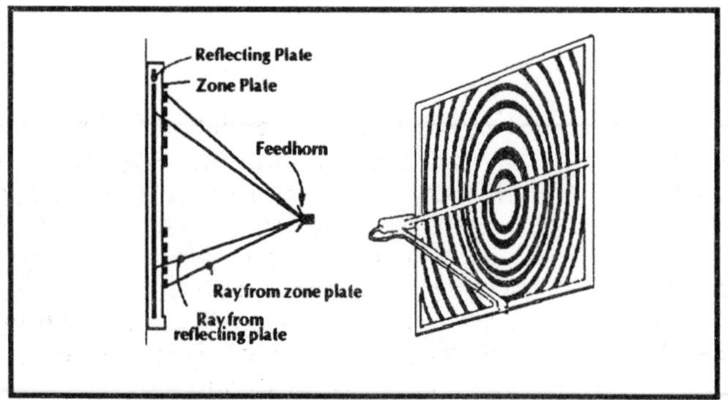

Figure 2-8. Operation of the Mawzones Flat Plate Antenna. *The Mawzones antenna is designed so that impinging signals are reflected to the feed by the Fresnel rings. (Courtesy of WV Publications)*

Figure 2-9. The Matsushita Flat Plate Antenna. *The flat plate antenna, a recent introduction, is lightweight, does not require a feedhorn/LNB and its support structure, and can be easily packaged and installed. While this brand is non-steerable, flat antennas which electronically target a satellite will probably be marketed within ten years. The two square antennas in this figure are 34.5 and 72 cm in width and weight approximately 2.3 and 9 kg, respectively. (Courtesy of Matsushita Electric Works, Ltd.)*

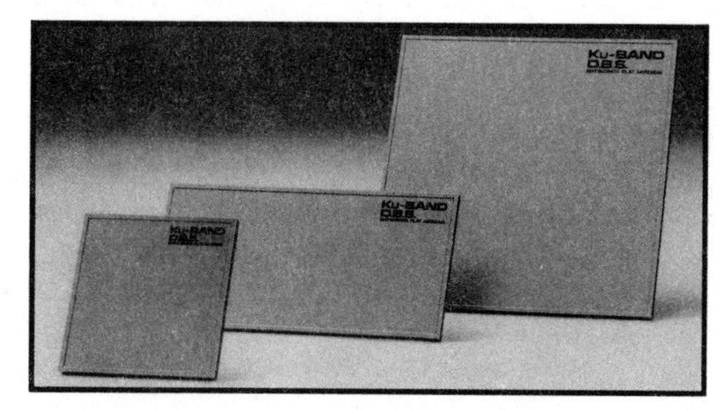

Special Purpose Antennas

Multi-Focus Antenna

Multi-focus dishes were developed to allow more than one satellite to be simultaneously detected by a fixed antenna. In contrast, most single-focus dish are repositioned by actuators to receive signals from one satellite at a time.

Commonly encountered multi-focus antennas incorporate variations of spherical and parabolic surfaces (see Figure 2-10). In all cases, these dishes are aimed at the arc of satellites, and the reflected signals from each space vehicle are focused to a series of feedhorns. To illustrate, the 5 meter Simulsat antenna (see Figure 2-11), which is cut from a rectangular section of a sphere with feedhorns mounted in a long box at the focal line, simultaneously detects up to 35 satellites within a 70 degree arc (at a 44 dB gain). Another example is the Torus antenna, which is a dual-curvature reflector having a circular crossection along the satellite arc and a parabolic contour at right angles.

A standard prime-focus parabolic dish can also be used in a multi-focus configuration where one or two extra feedhorns and LNBs are offset on either side of the central feed. In this case, the reception of signals is not as "clean" because reflections from different parts of the antenna travel slightly different distances to reach these additional feeds. This is the case with all multi-focus dishes. The resultant performance decrease can be partially overcome by using a slightly larger reflector. So an antenna of 12 to 16 feet diameter can then be capable of simultaneously seeing more than one satellite with minimum degradation of performance.

Multi-focus antennas are almost always found in commercial installations such as cable TV company headends or apartment complexes. They are a curiosity for home satellite TV enthusiasts.

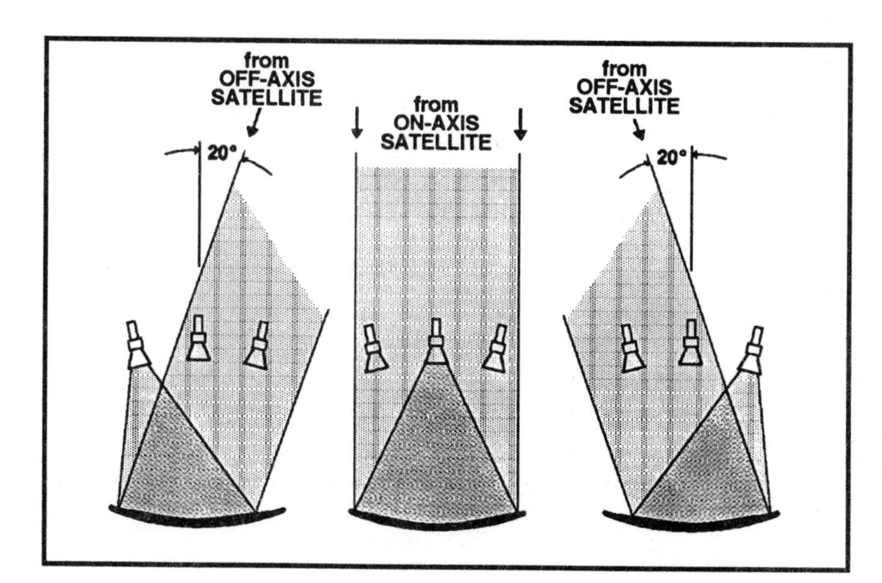

Figure 2-10. Spherical Antenna Geometry. *A spherical reflective surface can be used to focus signals from neighboring satellites to a series of focal points. The feedhorn is located further from the surface of the dish than in a prime focus configuration. In addition, some deterioration of signal, known as spherical aberration, occurs.*

Figure 2-11. The Simulsat Multi-focus Antenna. *The multi-focus Simulsat antenna can view up to 35 satellites at once. Either C- or Ku-band signals are directed to up to 35 feedhorns and low noise amplifiers. (Courtesy of Antenna Technology Corporation)*

The Horn Antenna

The horn reflector, shaped like a horn, sends intercepted signals entering via a large aperture down the reflective body of the horn to a focus. Thus the whole body of the horn performs the same function as the feedhorn and dish does for a prime-focus dish or Cassegrain antenna (see Figures 2-12 and 2-13). This type of antenna is almost always installed in light-of-sight relays used by common carriers terrestrial telephone, video and data transmissions.

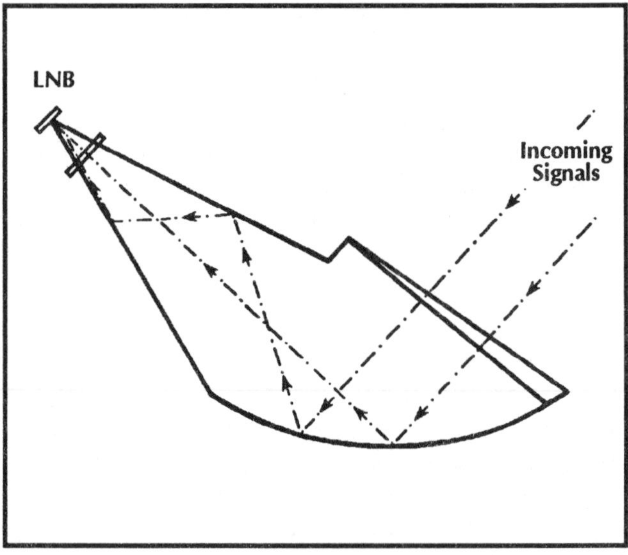

Figure 2-13. Horn Antenna. *The whole body of a horn antenna acts to capture microwave signals.*

Figure 2-12. Horn Antennas on Line-of-Sight Microwave Relay. *Before the advent of satellite broadcasting, microwaves were relayed either via cables or between line-of-sight stations like the one shown here. Such towers are still widely used. (Courtesy of Microwave Filter Company)*

Antenna Performance

Antenna performance is determined by a number of interrelated factors including gain and efficiency, beamwidth, side lobes, noise temperature and f/D ratio. Unfortunately, many dish manufacturers either do not have the facilities or the desire to accurately measure some of these performance factors. However, until low cost, independent testing organizations are established, consumers and satellite dealers can still do a reasonable job of evaluating dishes with a knowledge of the underlying theory and an understanding of how antennas are designed and manufactured.

Antenna Gain

Antenna gain, G, expresses the factor by which intercepted microwaves are concentrated when they reach the feedhorn. Gain is dependent upon three factors. First, as dish reflector surface increases, more radiation can be intercepted so gain also increases. Therefore, if the area of an antenna is doubled, the gain also increases by a factor of two. For example, a 12-foot dish has 44 percent more gain than a 10-footer because the reflective area increases with the square of its diameter (12 x 12 = 144 compared to 10 x 10 = 100).

Second, gain increases with frequency. Higher frequency microwaves do not spread out as would waves in water but are more easily focused into straight lines like beams of light. This is part of the reason that Ku-band direct broadcast systems which operate at 12 GHz instead of 4 GHz are capable of being captured by smaller dishes.

Third, gain is determined by how accurately the surface of an antenna is manufactured to exactly its designed shape. Even small distortions in the surface accuracy of a dish can cause substantial signal losses (see Table 2-1). Therefore, a dish that has ripples in its surface will behave more poorly than one that is smooth and is closer to its designed shape. This is especially the case at higher Ku-band frequencies.

Antenna Efficiency

Antenna efficiency is a measure of how much signal is actually captured by the dish and feedhorn/LNB assembly. A perfect, 100 percent efficient dish would therefore direct all the power intercepted from a broadcast satellite into the feedhorn.

Efficiency is determined by the surface accuracy of the antenna, by losses that occur when microwaves are not perfectly reflected but absorbed by its surface, by reflective losses from components sited in the path of the incoming rays such as the feedhorn and its supports, and by how much "spillover" occurs. The phenomena of spillover is examined later in the section on feedhorns.

TABLE 2-1. LOSS OF GAIN WITH SURFACE DISTORTION (C-band Microwaves)

Average Surface Distortion (inches)	Loss of Gain (percent)
0.01	2
0.05	17
0.10	44
0.25	28.8

Typical efficiencies range from lows of 40 percent for quite poorly designed systems to as high as 65 or 70 percent for high quality antennas. Offset fed antennas can have efficiencies in excess of 80 percent because there are no structures between the incoming signal and the dish surface that reflect the incoming signal energy.

Antenna gains ranging from the theoretical maximum (100 percent efficiency) to those for dishes having efficiencies as low as 50 percent are shown in Table 2-2.

TABLE 2-2. ANTENNA GAIN IN DECIBELS AT 3.95 MHZ

Antenna Diameter (feet)	Antenna Efficiency 100%	80%	70%	60%	50%
5.0	36.18	35.21	34.63	33.96	33.17
6.0	37.76	36.79	36.21	35.54	34.75
7.0	39.10	38.13	37.55	36.88	36.09
7.5	39.70	38.73	38.15	37.48	36.69
8.0	40.26	39.29	38.71	38.04	37.25
8.5	40.79	39.82	39.24	38.57	37.78
9.0	41.28	39.34	39.73	39.06	38.27
9.5	41.75	40.78	40.20	39.53	38.74
10.0	42.20	41.23	40.65	39.98	39.19
10.5	42.62	41.65	41.07	40.40	39.61
11.0	43.03	42.06	41.48	40.81	40.02
12.0	43.78	42.81	42.23	41.56	40.77

When reading this table remember that a 3 dB difference means a doubling of gain; 1 dB means a 26 percent difference; and 0.1 dB means a 2.3 percent difference. Small decibel changes mean large variations in performance. Also, the numbers for 50 percent efficient dishes are not half those for 100 percent ones because the are expressed in decibels, not in the raw signal concentration factors. So, for example, a perfectly efficient 5 foot antenna would concentrate signals by a factor of 4,150 (36.18 dB); one with a 50 percent efficiency by 2,075 times (33.17 dB, the number of decibels in 50 percent of a 4,150 signal concentration or half the power).

Beamwidth and Side Lobes

Antenna beamwidth and side lobes determine what a antenna actually "sees." If a dish were set up on a test range and a microwave source was rotated across its field of view, the captured power plotted against this angle would give all the necessary information about beamwidth and side lobes. A plot of beamwidth is shown in Figure 2-14. As expected, an antenna captures most power incoming along the antenna axis, its boresight. This beam pattern is, in essence, a fingerprint of dish/feed quality and performance.

This characteristic fingerprint depends upon a number of factors, among the most important being reflector surface accuracy. The width of the main lobe, called the beamwidth, decreases as both the frequency of the satellite signal and the antenna diameter increase. A 10-foot dish will have half the beamwidth of a comparable 5-footer. An antenna receiving a Ku-band transmission has one third the bandwidth compared to when it receives a C-band broadcast. This occurs because 12 GHz divided by 4 GHz equals three, the same as the decrease in beamwidth.

The beamwidth of a dish is a measure of how well it can target a very narrow region of space. An antenna must have a sufficiently narrow beamwidth because satellites separated by 4 degrees or less and located more than 22,300 miles away appear to be very close together and most of the power received by the dish is detected via its main lobe. The beamwidth is defined as the width

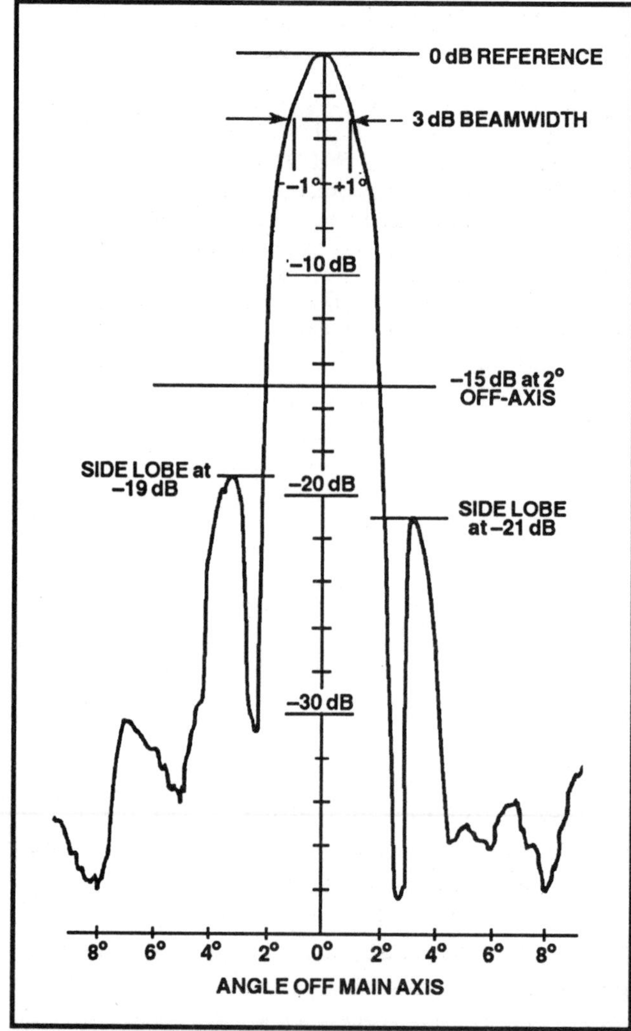

Figure 2-14. Antenna Lobe Patterns. *This graph shows what C and Ku-band power levels, expressed in relative terms, a 7.5 foot (2.3 meter) antenna detects from its boresight or main axis to its periphery. The taller and more narrow the central portion is relative to the side lobes, the better an antenna will target and pinpoint a satellite in the sky. In this case, the C-band 3 dB or half power beamwidth is 2.2° and the side lobes located at 4.5° on either side of the main lobe are reduced 25.5 dB in power. Signals arriving at 2° are reduced to 12 dB relative to the main beam at 4.2 GHz. By contrast, the 3 dB beamwidth at 12.2 GHz is 0.73° and the side lobes, that are reduced 25.5 dB in their ability to detect signals, are located at 2.1°. Signals arriving at 2° are reduced to 26 or 27 dB depending on their position in beamwidth pattern.*

of this main lobe between the "half power" or 3 dB points where power has dropped by 50 percent.

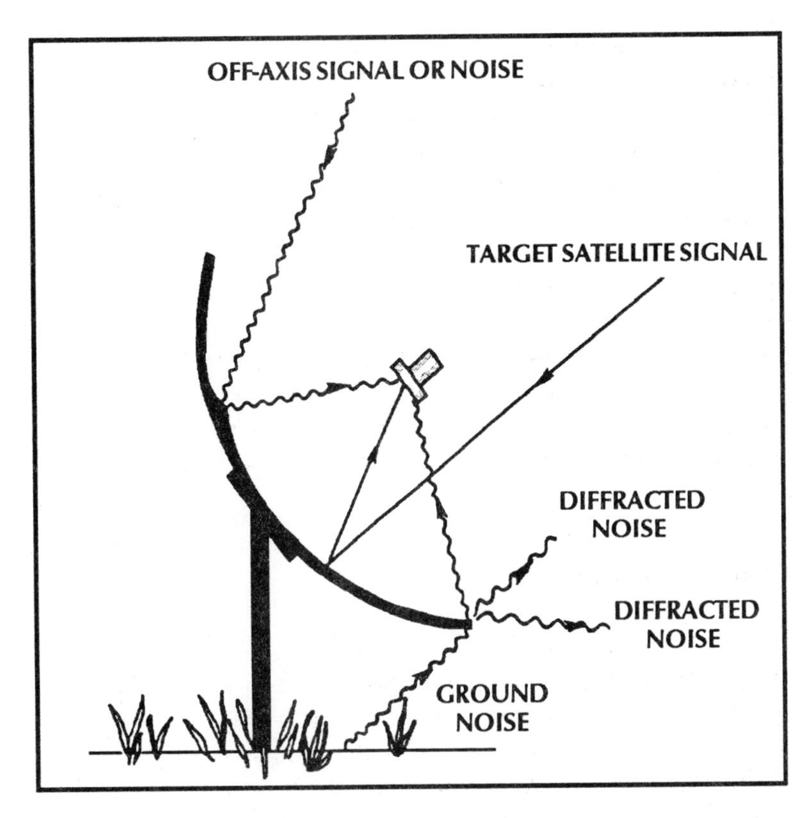

Figure 2-15. Noise Spillover and Antenna Side Lobes. *This drawing shows how an antenna can actually detect microwaves and noise from behind, 180° off its main axis. Noise from the warm ground is diffracted by the reflector edges and is scattered in all directions. Some of this noise manages to enter the feedhorn. As reflector surface inaccuracies increase, higher side lobes are formed. As a consequence, the antenna is more capable of detecting unwanted signals from the ground or from off-axis communication satellites.*

Beamwidth indicates how well a dish will detect off-axis radiation. For example, if the beamwidth is 2.4°, then signals at 1.2° off-axis will be captured at half the power of those entering the dish on the main axis. If an adjacent satellite happened to be located 1.0° away from the targeted one, this interfering signal would be received nearly as well as the desired signal and would result in an unacceptably noisy picture.

Side lobes also determine the capacity of a dish to detect radiation coming from off-axis directions. Side lobes must be much smaller than the main lobe so that off-axis signals are largely ignored. If this is the case, for example, a signal of 1 watt which might be detected at full power by the main lobe could be reduced by 20 dB when located 2° off-boresight and intercepted by a side lobe. Otherwise, a dish having a side lobe 2° away from the main lobe but reduced only slightly in power could detect an unacceptably high level of signals from another satellite spaced 2° away from the targeted vehicle.

If the power levels of the side lobes are 20 dB or more below those of the main lobe, then a dish can perform excellently. If side lobes are more than 3 or 4° away from the main lobe and reduced in power by less than 10 dB there may be trouble. Watching a channel and the messy interference between two signals coming simultaneously from two adjacent satellites would be very annoying.

Side lobes result from the inherently wave-like characteristic of microwaves and the associated "spreading" in the vision of an antenna. This "spillover" occurs when microwaves strike the edges of the rim of a dish and are bent as a result of diffraction and interference, two well-known optical effects. As a result, an antenna has the rather astonishing ability to detect radiation coming from directly behind its reflective surface (see Figure 2-15). Every dish has a theoretical lobe pattern showing how power from the full circle around the antenna will be detected. Of course, the power coming from behind a dish is detected at a much lower level than that directed along the antenna boresight.

These theoretical patterns (see Figure 2-16) are affected by other real-world design considerations. The most important practical influences on lobe patterns are unwanted reflections off of the feed and its support structure and by stray signals reflected by irregularities in the antenna surface. Offset fed dishes can potentially have lower side

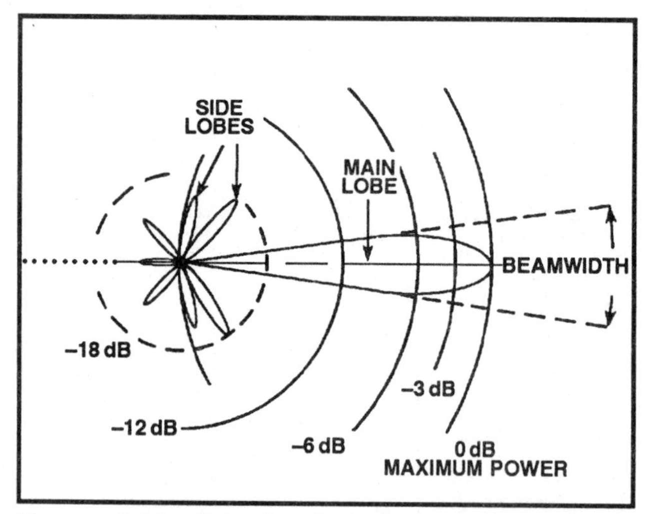

Figure 2-16. Calculated Antenna Lobe Pattern. *Although most of the radiation detected for a parabolic antenna is concentrated in the main lobe, this calculated pattern shows that some noise can be detected via side and back lobes in the full circle around an antenna.*

lobes than prime-focus parabolas because the feed assembly does not block some of the incoming signals.

The difference between an excellent and a poor-quality antenna is easily seen in their lobe patterns. Manufacturers publishing this information do a service to the industry. It is important to understand that apparently small changes in design can affect the "vision" of a dish. For example, even if square and round prime-focus antennas cut from identical larger dishes have the same surface areas and gains, they may have quite different lobe patterns.

Antenna Noise

Noise temperature is a measure of how much noise an antenna detects from both the surrounding environment and outer space. It obviously depends upon the lobe pattern which determines the ability of a dish to ignore random noise coming from the surrounding warm ground. Noise comes from both man-made sources such as fluorescent lights which emit microwave radiation and natural sources such as the surrounding terrain. In general, it is the natural, higher frequency sources of noise which are the predominant cause of antenna noise power.

Since the warm ground emits radiation, noise temperature increases as a dish is pointed at increasingly lower elevation angles (see Figure 2-17). This occurs when an antenna is targeted at both the far eastern and western portions of the geosynchronous arc and in regions further removed from the equator. A set of curves showing how much

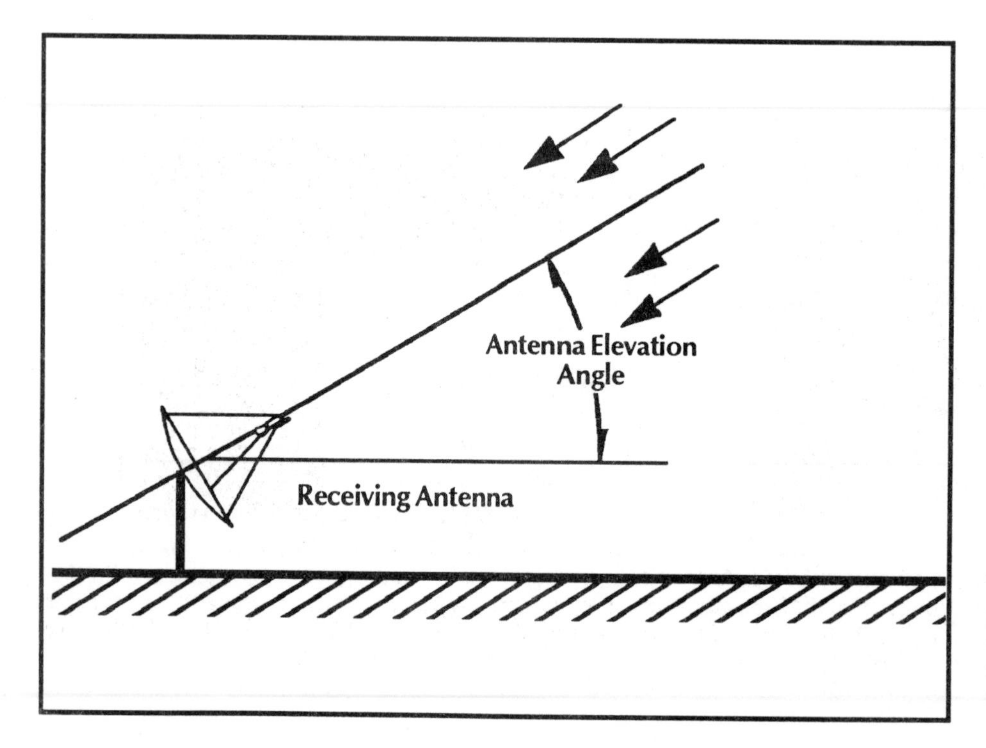

Figure 2-17. Antenna Noise Temperature. *At lower elevation angles, the more an antenna "sees" the ground and the more it is capable of detecting noise.*

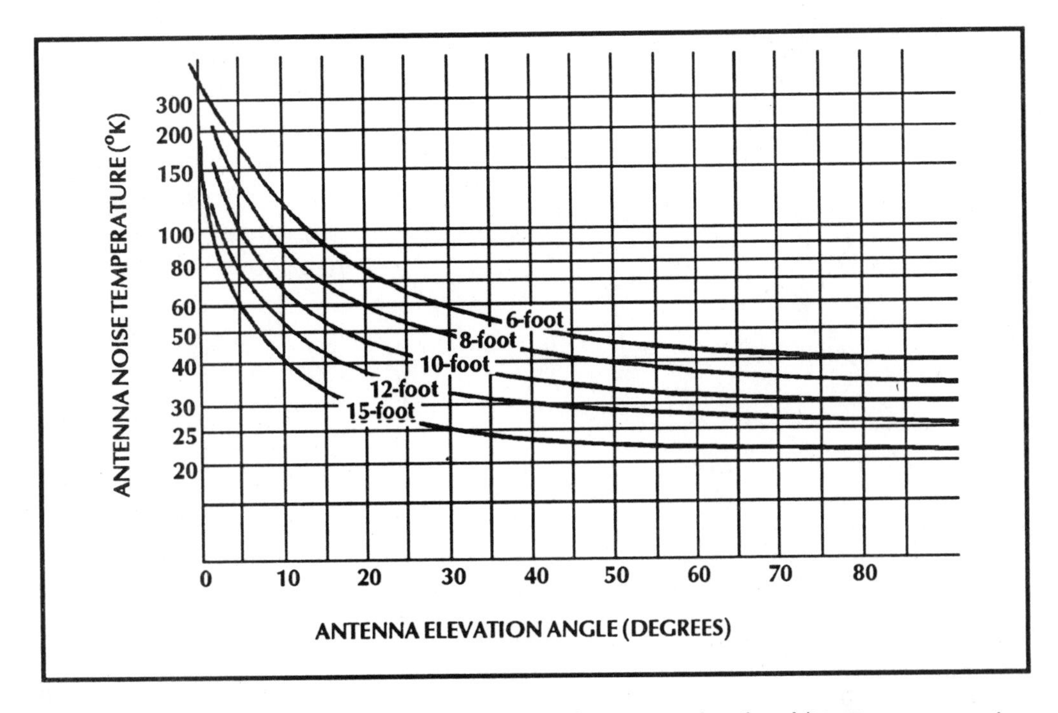

Figure 2-18. Variation of Antenna Noise Temperature with Dish Size and Antenna Elevation.

noise is added as elevation is decreased is presented in Figure 2-18. Larger antennas detect less noise because they have smaller side lobes.

Focal Length to Antenna Diameter Ratio

The focal length to antenna diameter ratio, abbreviated f/D, is another important parameter used in characterizing a dish. In general, if the f/D is lower, the side lobes will be smaller all else being equal. This is because the feed/LNB structure is mounted more closely to the reflective surface and is therefore better screened from the surrounding environment (see Figure 2-19). A similar but not equal result (because side lobes will differ) can be obtained if a reflective shroud is affixed around the rim of a dish. This arrangement is analogous to putting blinders on a horse so it will see only in one direction, straight ahead.

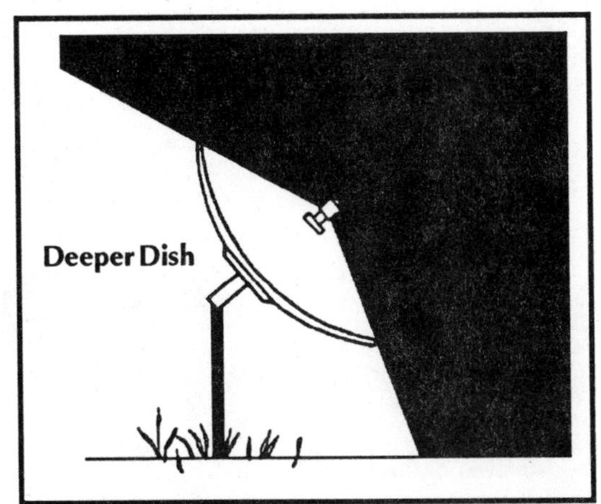

Figure 2-19. f/D Ratio and Antenna Field of View. *Antennas with a lower f/D ratio have a more narrow field of view and are therefore potentially less susceptible to noise and interference. When correctly designed, these "deep dishes" have lower side lobes. However, in general, a deep antenna performs more poorly in a dual-band offset fed configuration than a shallow one.*

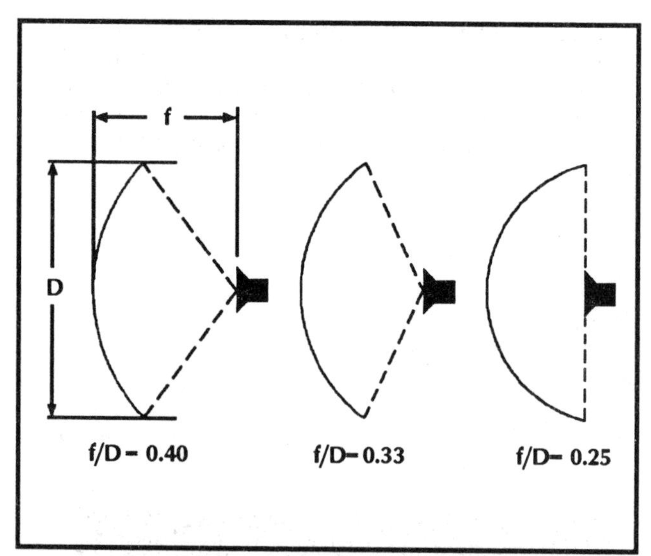

Figure 2-20. Focal Length to Antenna Diameter. *A lower f/D means that the feed assembly is located more closely to a given sized reflective surface. At a 0.25 f/D, the feed lines up with the antenna rim so that adequately illuminating the entire reflective surface becomes quite difficult.*

The f/D is used to classify dishes as either deep, average or shallow. In general, a f/D less than 0.3 defines a deep dish; one in excess of 0.45 defines a shallow dish (see Figure 2-20). Deeper dishes are generally less susceptible to environmental noise and usually have lower side lobes and lower noise temperatures than more shallow designs. Note that offset fed antennas must generally be rather shallow to have acceptable performance.

Dish Construction

Dishes can be made from a variety of materials and by numerous manufacturing processes as long as they meet basic performance requirements. They must adhere to their designed shape over a long period of time, have metal in their surface in order to reflect microwaves, be easy to assemble and be shippable at a reasonably low cost.

The details of design, materials selection and manufacturing are very important. An antenna may perform well when new but, in time, after being subjected to environmental stresses, its performance and quality may seriously degrade. Allowances must be made for contraction and expansion of different materials with changes in temperature, for wind loading, for rain, snow and hail, for the effects of intense sunlight, and for many other factors. Remember, the dish is the eyes of a satellite receiving system and a critical component needed for quality performance.

There are five major classes of dishes: spun or hydroformed aluminum or steel, stamped or drawn steel, fiberglass, mesh made from expanded metal, and, recently introduced formed plastic. Antennas in the first two categories can use either solid or perforated metals in their construction.

Spinning and Hydroforming

Hydroformed and spun aluminum or drawn steel dishes tend to have the most accurate surfaces (see Figures 2-21 and 2-22). They are manufactured by mounting aluminum or steel, typically of 0.080 inch thickness, on a die and spinning the whole assembly while pressing it into shape with either water or rollers. When done correctly, it results in a highly accurate dish with a metal surface hardened by the manufacturing process. This method

Figure 2-21. 1.8 Meter Spun Aluminum Antenna. *The manufacturer of this antenna used a spinning technique to mold an accurate reflective surface that can achieve high gain and efficiency. (Courtesy of Andrew Antenna Company)*

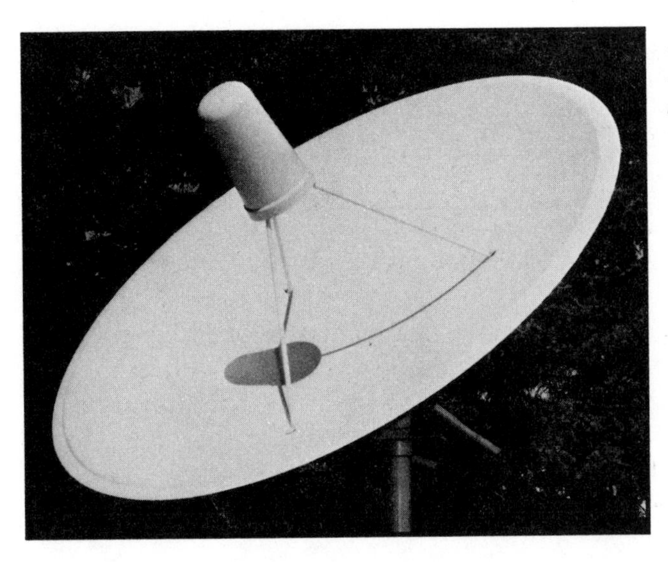

Figure 2-22. Hydroformed Antenna. *In the hydroforming process a blank aluminum or stainless steel sheet is placed onto a mold and pressed into a parabolic shape using water to aid in creating a smooth and accurate reflective surfacing. (Courtesy of Paraclipse Manufacturing, inc.)*

also has the advantage that several different size antennas can be made from one mold, each diameter having a different f/D ratio.

Hydroformed or spun dishes are manufactured in one piece and can be difficult to ship. Some are made with a perforated surface composed of metal punched with holes. In general, if the holes are less than one tenth of the wavelength used (less than 3/10ths of an inch for C-band radiation) losses through the holes are negligible.

Stamping

Stamped or drawn steel dishes are made by stamping or pressing flat sheet metal into shape (see Figure 2-23). A good quality dish can be produced by these methods at a relatively low per unit price. However, the required tooling can be expensive and only one size dish is produced for each mold.

Figure 2-23. Winegard Perforated Antenna. *This perforated antenna can be manufactured either by stamping, hydroforming or spinning. The hole diameter should be less than 3/4 inch for adequate performance at C- band. The Winegard QuadStar 3 meter antenna is constructed with 1 mm thick (0.040 inch) perforated aluminum with a hole diameter of just under 2 mm. It has an f/D of 0.278, a half power beamwidth of 1.6° and gain of 49.6 dBi at 12.7 GHz. (Courtesy of Winegard Company)*

Fiberglass

The challenge of manufacturing a fiberglass antenna is to embed wire screen, aluminized mats or other metallic materials in the fiberglass resin while retaining an accurate surface shape (see Figure 2-24). Fiberglass dishes are manufactured by three methods. In the hand layering or laminating process, a fiberglass resin called gelcoat is smeared onto fiberglass cloth laid in a mold. This gelcoat has pigment to give the dish color and inhibitors to protect against ultraviolet damage from intense sunlight. The metal is flame sprayed onto the dish resulting in a strong, lightweight but relatively expensive dish.

Figure 2-24. Miralite 3.7 Meter Commercial Antenna. *This C/Ku-band antenna has been tested for uplink operation at frequencies ranging from the lower C-band to 14.5 GHz. (Courtesy of Miralite Antennas)*

Figure 2-25. Compression Molded Reflector. *This antenna is manufactured by thermal compression injection molding whereby liquid fiberglass under pressure is injected over a reflective aluminum screen layered inside of a mold. These machines typically use one thousand tons of pressure. This technique repeatedly produces rugged, durable reflectors. A three segment, 8 foot (2.4 meter) Prodelin antenna is shown here. (Courtesy of M/A COM)*

In sheet-molding, a mixture of polyester resin, calcium carbonate and chopped fiberglass strands is pressed onto a metal screen. The resulting product is lightweight but usually brittle and easily damaged.

In thermal injection molding, the mat and screen are set into a closed mold and resin is injected under pressure (see Figure 2-25). Dishes manufactured by this process are strong, impact-resistant and usually have good cosmetics.

Wire Mesh or Expanded Metal

The fourth type of antenna is made from wire mesh or expanded metal (see Figures 2-26 and 2-27). These are held together by a rib structure which can be time consuming to assemble. The mesh can be attached to its supporting structure by a series of clips or by insertion into a channel in an extrusion. Such dishes are popular because of their aesthetic appeal, light weight and lower wind resistance but can be flimsy if not designed correctly. Today many antenna manufacturers produce this variety of antenna with four panels, known as a sectional quad dish.

Molded Plastic

A recently patented process is now used in manufacturing a molded plastic antenna (see Figure 2-27). This type of antenna can have either a clear or an opaque reflective surface. The process has a number of advantages including surface tolerances that can equal or exceed those of dishes produced by hydroforming, low weight and the potential of being produced from recycled plastic.

Figure 2-26. Paraclipse Wire Mesh Antenna. *The ribs for wire mesh antennas are manufactured either by welding pieces together or by bending an aluminum extrusion to conform to a parabolic shape. Bending is accomplished by using a hydraulic press and matching male/female dies. The expanded mesh is attached to the ribs by clips or screws, or by sliding it into grooves in the shaped extrusions. If the mesh hole size is substantially larger than 2.5 millimeters (1/10 inch) or if its surface does not adhere to an accurate parabolic shape it will not perform satisfactorily at Ku-band. (Courtesy of Paraclipse Antenna)*

Figure 2-27. Molded Plastic Antenna. *This molded plastic antenna can be either transparent or opaque. Both the feedhorn cover and the feed supports will be also made from plastic in future models. The patented forming process will permit surface accuracies equal to or exceeding those obtained by hydroforming methods.*

Wind Loading

In regions such as Wyoming, Colorado or California where winds often have speeds in excess of 60 miles per hour, wind loading is an important consideration. For example, a 10-foot dish experiences a force of approximately 2000 pounds when facing into a 65 mile per hour wind. Wire mesh dishes are subjected to smaller forces especially at velocities less than approximately 30 miles per hour. But at higher wind speeds, they tend to be more like solid surfaces even though they still experience approximately a 40 percent lower force. An approximate wind loading chart is shown in Table 2-3.

TABLE 2-3. ANTENNA WIND LOADING (Pounds of Torque)

Antenna Diameter	Wind Speed (mph) 25	50	75	100
6	40	170	400	700
8	100	400	900	1600
10	200	800	1800	3200
12	350	1400	3000	5500
14	550	2200	5000	9000
16	800	3300	7000	13000

B. MOUNTS

Purpose

The purpose of a mount is two-fold: to provide a stable support so an antenna can be accurately aimed at any chosen satellite, and to permit any other satellite in the geosynchronous arc to be targeted when desired.

Mounts must therefore be strong and rigid as well as firmly attached to both the dish and its underlying structures. They are usually set on ground poles affixed in concrete pads or other supporting structures. Securing the assembly on the ground in a protected area is a much safer bet than having a dish on a long pole or roof mount where it must be able to resist strong, uplifting winds.

Using a mount that provides stability and pointing accuracy is a critical part of designing and installing an earth station because well-designed dishes with narrow beamwidths target very small portions of the sky. For example, a 10-foot dish with a 1.7 degree beamwidth sees only one fifteen thousandth (1/15,000) of the visible sky. A 1/10th degree movement of a dish causes it to see a sweep of 46 miles at the geosynchronous arc. So the signal from a satellite would be reduced in strength if a dish moved even a relatively small distance off target. Without an adequate mount, such movement is possible when winds blow across the large sail-like, surface areas of an antenna.

Stable mounts are particularly important when larger antennas with smaller beamwidths are installed. Antennas as large as 5 meters are not uncommon in the Middle East, Africa or portions of Latin America. These require very stable mounts that can be accurately aimed. Ku-band satellite dishes which have one third the beamwidth of comparable C-band systems also require better pointing accuracy and stability.

Construction

The design of the dish/mount attachment is very important. Since the mount assembly must be securely bolted onto a dish, any mismatch can cause stresses and possibly warp an antenna. This would result in increased side lobes and poorer performance. A sufficient number of well-placed points of attachment must evenly distribute the weight in both static and wind load conditions to prevent warping of the dish when it is subjected to these loads (see Figure 2-28). Fiberglass or a thin-walled metal will bend if it is stressed when supported by a poorly attached mount. In addition,

Figure 2-28. A Rigid Antenna Mount. *The ten mount arms on this Prodelin fiberglass antenna are solidly attached to the reflector in twenty places with two 13 mm (1/2 inch) bolts per arm. This rigid design counteracts the tendency of the reflector surface to sag when the antenna is aimed at low look angles. Any flexing would result in a loss of gain and an increase in side lobes. (Photo Courtesy of Brent Gale)*

the mount must be strong enough to bear any forces encountered including potentially massive build-ups of ice and snow in some geographic areas. The method by which the mount is secured to the support pole must be adequate to prevent twisting away from the north-south setting. Such collars can be very solid structures on some well-designed mount assemblies. Finally, the points of attachment which support tracking movements must be designed with ball-bearings, bushings or other structures to prevent excessive wear over the life expectancy of the system.

Types of Mounts

There are two major classes of mounts: those that track the arc with two movements or degrees of freedom like azimuth-elevation (az-el) mounts; and those that require only one movement, namely polar mounts.

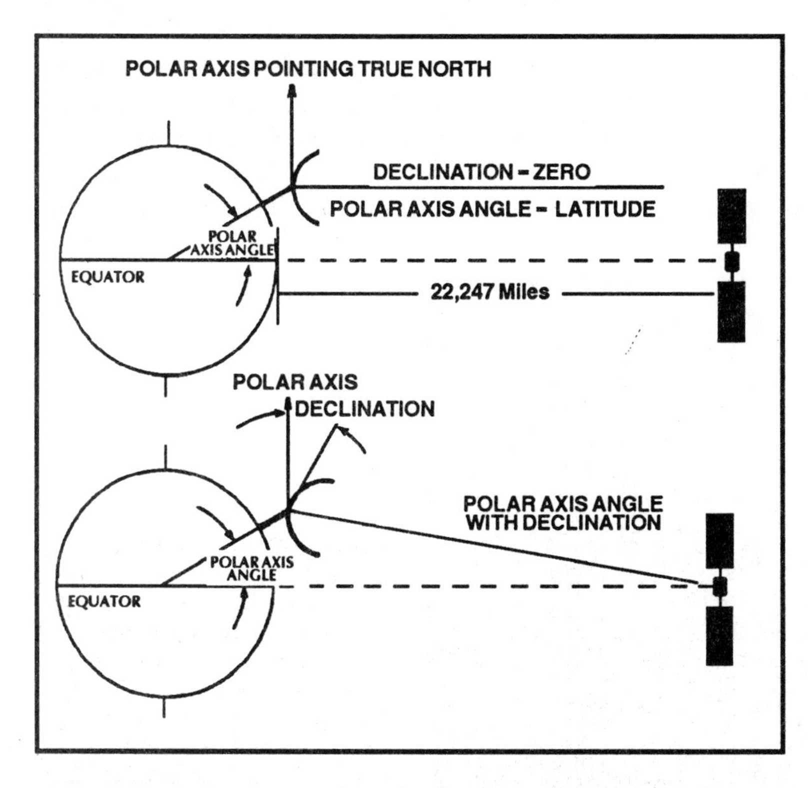

Figure 2-29. Alignment of a Polar Mount. *The polar axis angle which is set equal to the latitude points an antenna along a plane parallel to the one through the equator. Setting the declination offset angle lowers the field of view to the geosynchronous arc of satellites. Declination would not be necessary if the satellites were at an infinite distance, as the stars effectively appear to be.*

Polar Mounts

Polar mounts are designed to track the arc of satellites by rotation around a single axis, the polar axis. This movement can be driven by either linear or horizon-to-horizon actuators. These are discussed in the section on actuators.

Polar mounts were developed by astronomers who realized many years ago that it would be easier to keep a telescope sighted on a celestial body if it could be swiveled along a single axis to exactly counteract the earth's rotational motion. The polar axis is aligned parallel with a line passing through the north and south poles of the earth. The mount is adjusted by setting just two angles: the polar axis angle and the declination offset angle. The polar axis angle is set equal to the site latitude. To illustrate, in Denver which is located at 40 degrees latitude, this angle would be set at 40 degrees. At the equator, the polar axis angle would be set to zero degrees and the arc of satellites would be followed along a circle directly above. At both these locations, this setting would point the polar axis exactly along a north/south line.

The declination offset angle, which ranges from 4 to 8.5 degrees in North America, adjusts the tracking motion from a circle to a flattened ellipse. It compensates for the fact that the arc of satellites is at a finite distance; the further away this arc, the smaller would be the required declination offset (see Figures 2-29 and 2-30). Another way of visualizing this is by realizing that once the polar axis angle is set, a dish points directly along a line parallel to the plane passing through the earth's equator. The declination offset adjustment lowers the antenna's view to the arc of satellites. It should be set to the calculated value for that latitude during installation and, if held securely in place by the mount hardware, need never be touched again.

Polar mounts always have a small tracking error. If a satellite in the center of its sweep is accurately targeted then those

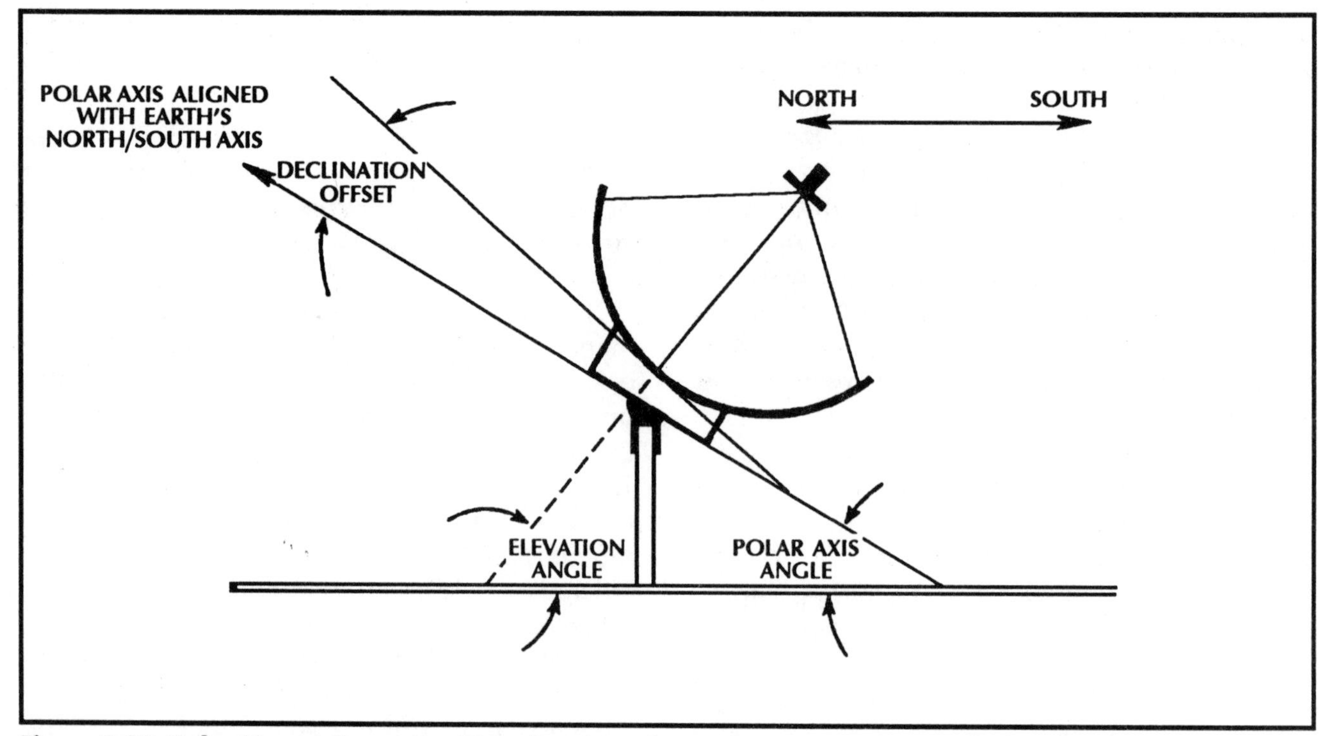

Figure 2-30. Polar Mount Geometry. *This drawing of a polar mount shows all the details of setting true north/south alignment, elevation and declination offset angles.*

at the far ends will be slightly missed. If the satellites near the horizons are accurately aligned with the antenna boresight, the central ones will be slightly off target. There is no way to avoid this inaccuracy. A properly set declination offset adjustment, however, has a tracking error of less than 0.1 degrees and allows a polar mount to sweep quite closely to all satellites from east to west. Tracking accuracy is especially critical when aiming at higher frequency, Ku-band satellites.

The Modified Polar Mount

A refinement in the geometry of polar mounts offers a way to improve antenna alignment across the entire arc of satellites. This is accomplished by fine tuning the polar axis and declination offset angles (see Figure 2-31). Tracking inaccuracies can then be reduced from 0.1 to less than 0.05 degrees.

The modified polar mount reduces tracking errors as follows. Assume that the angles have been correctly adjusted in a conventional polar mount. The polar axis is then tilted slightly forward towards the arc of satellites. Recall that this axis normally points along a true north/south line. Following this the declination offset angle is reduced by an equal amount. A satellite located due south would

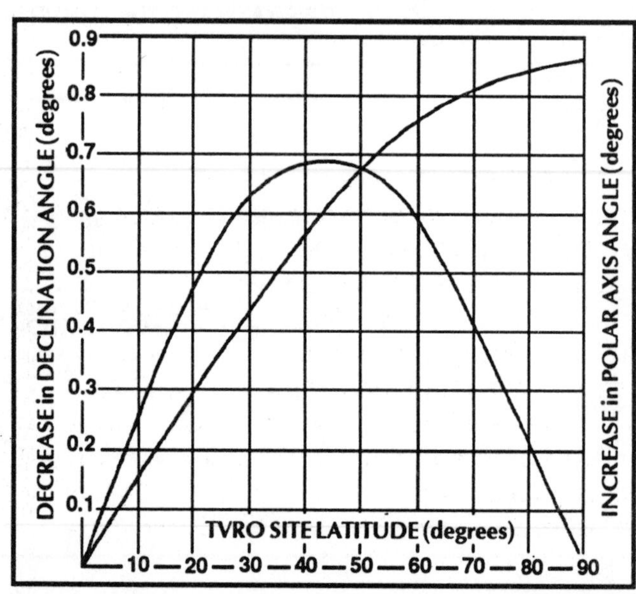

Figure 2-31. Modified Polar Mount Adjustment Angles. *In the modified polar mount, the polar axis is tilted slightly towards the arc of satellites and the declination offset angle is lowered by an equal amount. These fine adjustments allow very accurate tracking of the geostationary belt.*

be unaffected by this net zero change. However, when the antenna is rotated towards either due east or west it is inclined exactly as it would be if the polar axis had been still pointing parallel to the north/south line. This occurs because the slight inclination in the polar axis in a north/south direction has no effect when the assembly is aimed due east or due west. However, the slight decrease in declination causes the antenna to point higher than normally would be the case at all intermediate points between south and east or south and west. In effect, the most southerly satellite is targeted perfectly, while the antenna's main axis is aimed below the easterly or westerly spacecraft or any others in between substantially less than would occur with conventional polar mount adjustments. Comprehending this geometry usually takes some careful visualization.

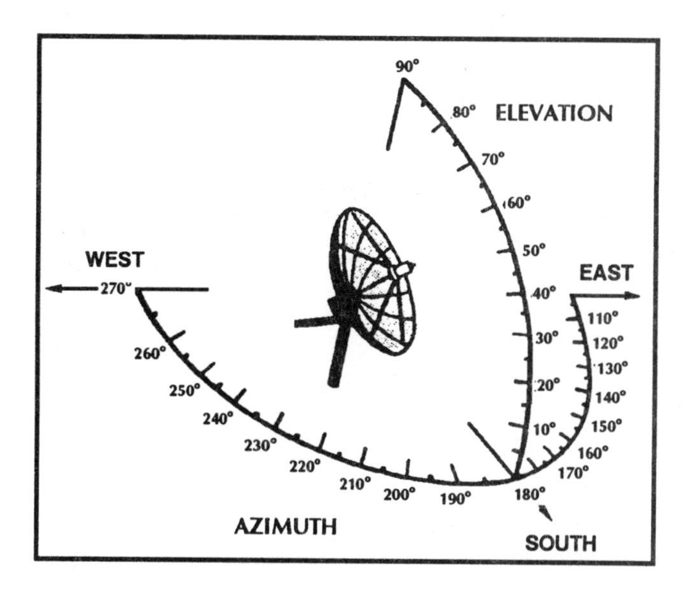

Figure 2-32. Az-El Mount Geometry. *When rotated in its azimuth, an antenna moves in a horizontal plane. Adjusting the elevation angle raises the dish above the horizon.*

Az-el Mounts

An az-el mount uses two independent adjustments to target any satellite. A satellite is located by first setting the correct azimuth angle, a movement parallel to the surface of the earth, and then by rotating up to the required elevation angle. Since two adjustments are required, it is more difficult to operate manually. However, an Az-el mount can be adapted for remote control by using two linear actuators at right angles to each other (see Figure 2-32).

Az-el mounts are commonly used in motor homes and recreation vehicles because even though the base of the mount often sits at a random angle, it has the ability to compensate by hunting in two directions. There is no tracking error so each satellite can be accurately targeted but more complex control devices are required to adjust both axes. Az-el mounts are also used in many European Ku-band installations where just one satellite such as Astra, which relays a full range of programming, is targeted.

Az-el mounts were historically used in commercial systems and in out-of-footprint locations where a much larger dish was required (see Figure 2-33 and 2-34). They were more

Figure 2-33. An Az-El Mount. *This 5-meter (16-foot) antenna is supported by a concrete pad and an az-el mount. (Courtesy of Comtech Antenna Corporation)*

Figure 2-34. The Gibraltar Az-El Mount. *This rigid, motorized az-el mount is powered by a 36 Vdc motor and is supported by a 40.5 cm (16 inch) steel pipe. Stamped aluminum dishes of 3, 3.6, 4.2 and 5 meters are available from the same company. (Courtesy of DH Antenna)*

easily installed and usually stronger than polar mounts. The process of aligning a polar mount by fine tuning the north-south orientation was quite cumbersome with large antennas. However, the technology has developed to the point that many large antennas are now controlled by polar mounts.

Tracking Non-Geosynchronous Satellites

There are numerous satellites which have been launched into elliptical orbits. While many of these non-geosynchronous satellites are designed for classified military purposes, others broadcast entertainment and information. Mounts for tracking these moving targets must have two axes of motion like the az-el devices.

The Soviet Union's Orbita system, based upon the Molniya satellites, is a case in point. Since 1965 over 100 such vehicles have been launched; 12 are currently in service providing three independent full-time programming networks. Four equally spaced vehicles serve each network. Their unusual orbits leave them over far northern Canada for six hours each day, so a different satellite must be locked onto every six hours. This tracking has been accomplished by computerized mounts but home TVRO systems have also been used by innovative operators to view this unusual television.

The Soviets have used such a system primarily because the look angle from their far northerly territories to the geosynchronous arc can be as low as 1 or 2 degrees making reception very difficult.

C. ACTUATORS

Purpose

Actuators provide the mechanical drive and control to allow an antenna to scan the arc of satellites.

Just a few years ago when there were only a limited number of broadcast satellites most dishes were either fixed on one satellite or hand-cranked to scan between satellites. Today most C-band systems have actuator-driven antennas. However, fixed antenna installations are still common in European Ku-band systems that target only one satellite that provides a comprehensive selection of programming.

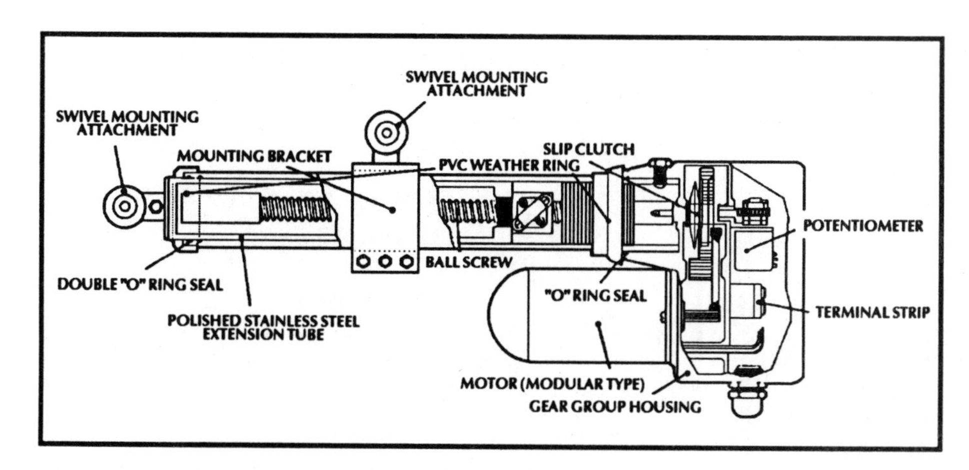

Figure 2-35. Internal Mechanism of a Linear Actuator. *The linear actuator arm bolts onto the mount at one end and onto the antenna at the other. A dc motor extends or retracts the inner arm to track any preprogrammed satellite as selected by the indoors controller. (Courtesy of Prosat)*

Actuator Construction

Most actuators in use today are either linear or horizon-to-horizon assemblies. Some actuators designed specifically for az-el mounts are occasionally used. These have two gear driven motors attached to two linear actuators for adjusting both azimuth and elevation.

Linear Actuators

Linear actuators have a telescoping arm which moves within a fixed external tube (see Figure 2-35). One end of the actuator is attached to an arm on the rear antenna surface and the other to the supporting mount. A motor-driven, worm-gear assembly drives the antenna. The gear assemblies which drive the internal arm can be an acme thread or a ball screw.

The acme jack is constructed from a threaded shaft moving in a threaded collar. It is simple and lower in cost than the ball screw mechanism (see Figure 2-36). The ball screw is similar but has ball bearings that replace the threaded collar that run in the threads. This mechanism has a much lower frictional loading than the acme so a motor of the same power will transmit more force to the dish. This increased force and the smoother ball bearing movement reduces the chance of the shaft seizing up during cold weather or after long periods of non-use.

As a general rule, a linear arm should never have to push or pull an antenna through an angle of less than 30° as measured between the arm and the back surface of a dish. The larger this angle, which depends upon the design of the mount and the length of the actuator arm, the lower the lateral force exerted on the arm during dish movement.

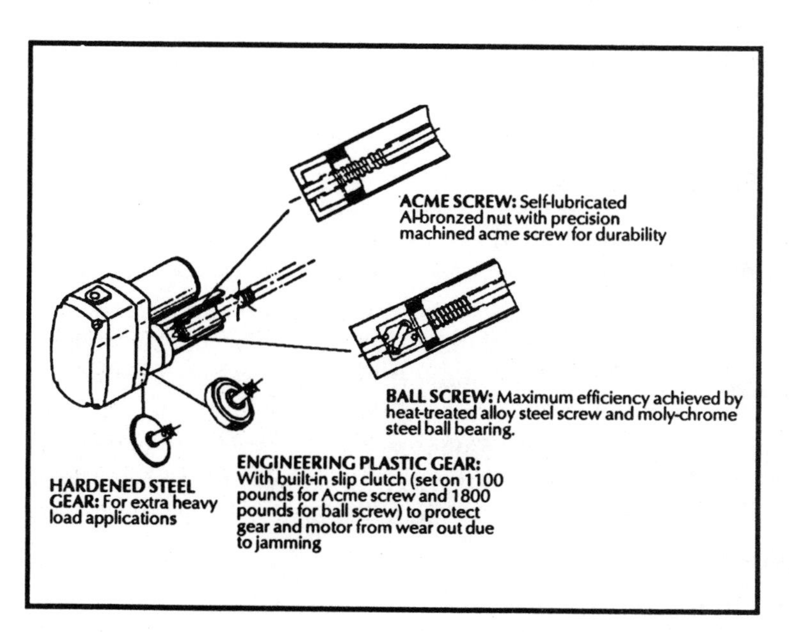

Figure 2-36. Acme and Ball Linear Jack Tubes. *An acme screwjack is constructed from a threaded screw shaft which moves in a threaded collar. This system can move up to 360 kilograms (800 pounds). Ball screwjacks are similar except ball bearings replace the threaded collar which grips the threaded shaft. This system can generate up to 700 kilograms (over 1500 pounds) of force. (Courtesy of Pro Brand)*

This lessens the chance that the internal arm could be bent and damaged (see Figure 2-37). Note that this type of actuator is limited in its motion to about a 100 degree sweep across the arc.

Horizon-to-Horizon Actuators

Horizon-to-horizon actuators are designed to move an antenna from horizon to horizon, unlike linear actuators which have more constrained movements. This ability is becoming more important in some locations where additional satellites near the far reaches of the arc are being launched into orbit.

Figure 2-37. A Linear Actuator. *This linear actuator is mounted to an 8.5 foot spun aluminum antenna. The points of attachment incorporate flexible joints to avoid potential binding.*

The gear mechanism of this actuator is encased in a housing located at the rear center of a dish (see Figure 2-38). This assembly must be solidly constructed to generate the torque necessary to rotate a dish from horizon to horizon and to allow minimal movement in winds or under other loads such as heavy snows.

Horizon-to-horizon actuators have some distinct advantages in comparison to linears. If improperly installed, a linear actuator can drive an antenna into the ground at either end of its range. When fully extended, the arm loses mechanical leverage and the internal tube can bend when driving a heavy antenna or when subjected to wind or snow loading. Actuators can seize up if water intrusion rusts threads. When this water freezes in cold climates, additional damage can result. Horizon-to-horizon actuators have the same pointing accuracy at all positions while linear actuators have decreasing accuracy closer to either end of the arc. When properly designed and manufactured, horizon-to-horizon actuators are compact and have little chance of freeze-up. They also have the ability to rotate the antenna to the point where any snow loads can be dumped.

Figure 2-38. Ajax Horizon-to-Horizon Mount. *The Ajax horizon-to-horizon mount uses a chain to drive a central gear. The antenna is firmly attached to the hemispherical gear structure.*

Monitoring Dish Position

Pioneer actuators had very simple customer interfaces. An east or west button or a switch would send a DC voltage to a motor to move the gear assembly and dish. The position of the actuator arm would be maintained by the unit's internal brake assembly. When some early customers damaged their actuators and sometimes their antennas by inadvertently driving motors past mechanical limits, programmable east and west electronic limits were introduced. However, programmable drives which sometimes "forget" their east/west limits have also damaged actuator jacks.

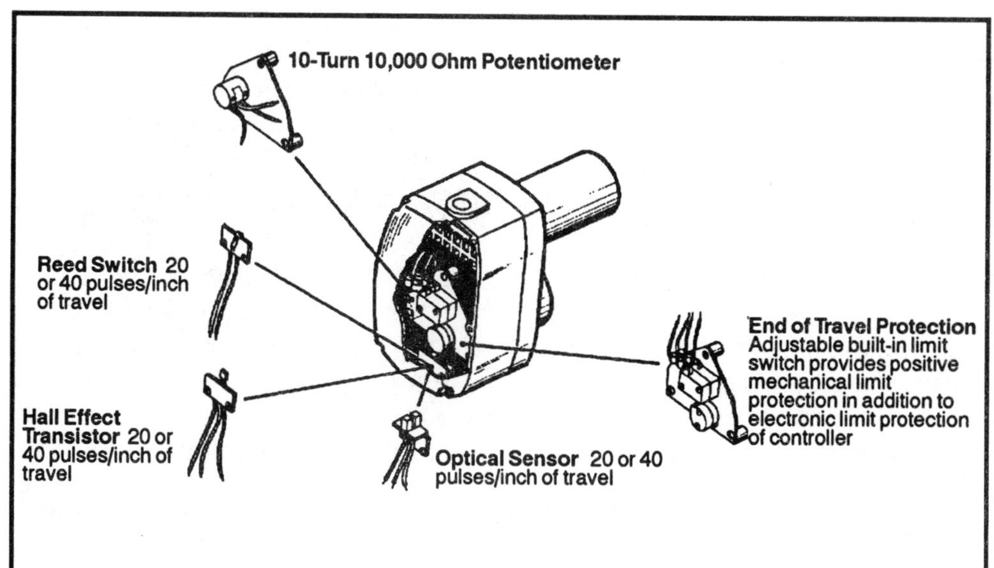

Figure 2-39. Actuator Sensor Positions. *Four types of sensors have been used to monitor antenna position. The mechanical end of travel protection illustrated here is an important design feature. (Courtesy of Pro-Brand International)*

Actuators now have built-in counters based on potentiometers, Hall effect devices, reed switches or optical couplers as well as mechanical limit mechanisms (see Figure 2-39).

Potentiometer Sensor

The potentiometer count mechanism, known as a pot sensor, uses a pair of 10 K-ohm potentiometers (variable resistors) which have a 10:1 turn ratio. (Ten turns of the shaft gives just one turn of the pot). These are stacked together on a common rotating shaft. Three wires are connected to the actuator controller. One is the constant supply voltage; a second is the common supply voltage return or ground; and the third wire is the sensor reference voltage (see Figure 2-40). Depending upon the position of the jack arm, the potentiometer returns a corresponding voltage that is used to monitor dish position.

The pot sensor is not a very common count mechanism today but can still be found in some older systems. It is generally less accurate than the three other systems which use microprocessor control of digitally counted pulses.

Optical Counters

The optical count mechanism uses a light emitting device that is aimed at a light sensitive component, usually a photo-optical transistor. As the dish moves, a plastic or metal spoked wheel is rotated through a beam of light and thus creates a pulse in the detector. This mechanism is similar to the operation of an automotive timing light or to the systems used at some markets or shops that ring a bell or sound a buzzer to signal a customer's exit or entry.

Optical counters, used in some of the earlier Amplica systems, are connected to actuator controllers by three or more wires. Any errors made

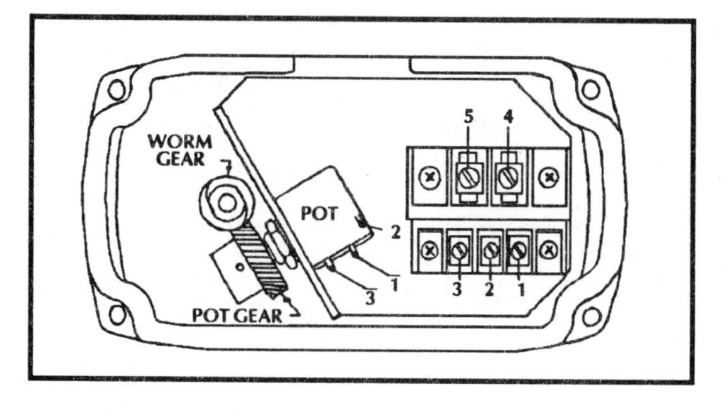

Figure 2-40. Actuator Potentiometer Sensor. *In this case, a worm gear drives the potentiometer gear and thus varies the resistance relayed to the actuator controller. Terminals 1, 2 and 3, to which the sensor wires attach, connect to points labeled with the corresponding numbers on the pot. Terminals 4 and 5 are connection points for the motor.*

during hook-up generally destroy the components. This counting mechanism is not used on systems sold today but is encountered on some older systems in the field.

Hall Effect Sensor

A Hall effect transistor is similar in operation to a reed switch. The detector, either a transistor or a pair of diodes, is positioned in close proximity to a rotor bearing four or more magnets and generates a pulse each time a magnet passes. Three wires are connected to the actuator controller: one is connected to the supply voltage, another is the common return for the supply voltage, and a third is the pulse return. These pulses serve as input to a memory chip which switches the voltage to the actuator motor off and on at appropriate times.

Hall effect transistors or reed switches generate a pulse when passed by a current carrying wire.

While earlier versions of the Hall effect count mechanism used switching transistors, diode pairs are now more common. This has eliminated the more elaborate circuit requirements of the transistor mechanisms.

Reed Switch Sensor

The reed switch mechanism, similar to the Hall effect mechanism, is by far the most commonly used today. It is a simple and economical sensor that requires only a two-wire hook-up and no power supply. The reed switch has two contacts, one of which is magnetized. It sits next to a rotor having 3 or 4 magnets or one circular magnet about an inch in diameter. Each time one of the magnets on the rotor passes by as it rotates, the magnetized reed is repelled, makes contact with the other contact and generates a pulse. The switch is designed so that the reeds separate after the magnet passes. The count can be increased by using a larger number of smaller magnets on the rotor or by increasing the rotational speed of the rotor. One of the more sensitive systems, the Houston Tracker super sensor, generates a count of 500-600 over the range of motion of most 18 to 24 inch linear actuators.

Most receivers can accommodate either a reed or Hall effect count mechanism. In some cases, for example when excessive play in the distance between the magnet and a reed switch develops, replacing the sensor with a Hall effect device is a reasonable tactic.

Mechanical Limit Mechanisms

Most actuators can be purchased with or retrofitted with mechanical limit mechanisms. Their purpose is to automatically open the motor circuit and avoid antenna or actuator damage at the far ends of travel. Two limit mechanisms are common, although other types can be found in the field.

The first utilizes four parts: two eccentric cams and two microswitches. The eccentric opens either microswitch and cuts off power to the motor at either end of the arc (see Figure 2-41). During installation, the inward position is usually set first.

The other common variety of mechanical limit consists of a threaded rod that is usually gear driven, a tapered nut and two microswitches. One of the switches, usually the inward, is fixed while the second is adjustable. The tapered nut travels along the threaded rod as the antenna moves. When the microswitch at either end is encountered, it is tripped and power to the motor is cut. The microswitch associated with the extended actuator position can be adjusted to accommodate various mount configurations.

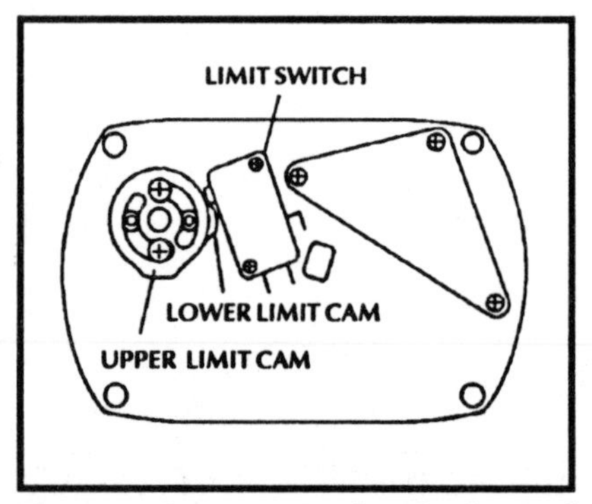

Figure 2-41. Mechanical End Limit Mechanism.

Electrical Connections

Electrical connection between the actuator motor and indoor equipment is via a cable having four or five conductors (see Figure 2-42). Two heavier gauge wires carry the 36 volt DC power required to drive the motor. Typically, for runs up to approximately 300 feet, 14 gauge wire is adequate; 12 gauge must be used for longer distances. Two lighter 20 gauge wires and often a ground are used to connect the counter to the control circuitry.

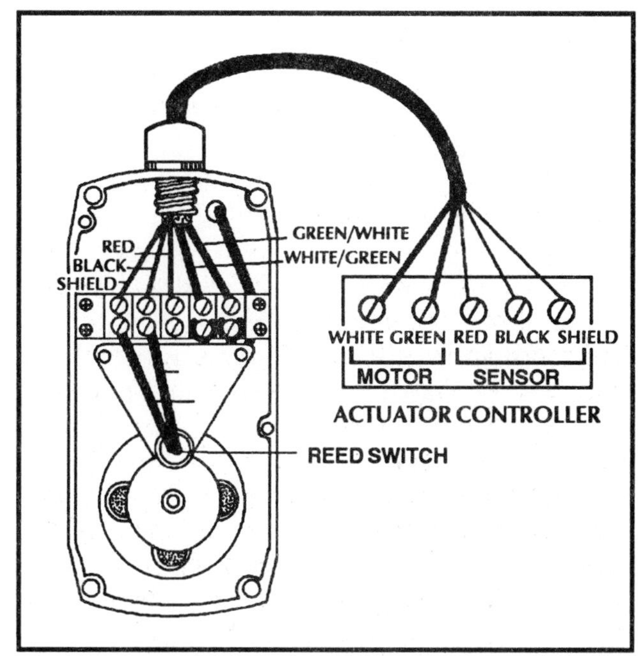

Figure 2-42. Electrical Interface Between Controller and Actuator Motor. *Five wires connect a control box with the actuator motor. Two of these send a voltage of typically 36 Vdc to the motor over a relatively heavy #14 gauge or larger cable. A three-conductor, #20 gauge shielded cable relays the counting pulses or appropriate voltage from the sensor to the indoors controller. In this case, a reed switch sensor is incorporated into the motor.*

User and Receiver Interfaces

Control boxes which operate and direct the mechanical equipment at the dish are of two basic types. The less common is the simple manual controller that features two buttons for east and west movement. The position of any satellite is identified by the count on a digital read-out on the controller display panel. More typical units have programmable memories that allow an operator to easily locate any satellite at the push of a button. Dish position is displayed by either a numerical counter or a digital read-out showing satellite code or name. Often over 100 satellite positions can be programmed into such types of actuator control memories. Both types of controllers should have electrical east and west limits that, coupled with mechanical limits, prevent possible damage to motors, arms and antennas.

Most actuator controls are now built into satellite receivers and IRDs (integrated receiver/descramblers). These feature one handheld remote to control all functions. Satellite selection is typically displayed next to channel number on the receiver or on remote TV screen read-outs that can also be monitored from other television sets.

D. FEEDHORNS

Purpose

Feedhorns have the important function of collecting microwaves reflected from the antenna surface and of ignoring noise and other signals coming from off-axis directions. This must be done with minimal signal losses and without adding significant amounts of noise. A poorly designed feedhorn assembly can add as much as 20 K noise to a home satellite system. Feedhorns also select the required signal polarity and reject or discriminate against signals of the opposite polarity.

Structure

Most feedhorns have a standard structure (see Figure 2-43). The signal reflected from a dish to its focus enters the feed assembly along its cylindrical opening called its boresight. (Boresight is also a term for the main axis of a satellite dish). A probe in the bottom of the throat is positioned to select the correct signal polarity. This probe, which is insulated from the frame by a teflon or plastic bushing, transfers the signal into a waveguide and subsequently to the low noise amplifier. A servomotor controlled by pulses generated in the satellite receiver is attached to the probe and causes it to rotate upon command. A servomotor is connected by three wires sorted according to a standard color code: red for +5 VDC power, white for pulse and black for ground.

A scalar ring attached around the front of the boresight balances the electric and magnetic fields to improve reception efficiency. The rings can be either fixed or adjustable. This ring is drilled with holes that allow it to be attached to the feed supports. The boresight has a cover, usually made of plastic, that allow signal to enter with minimal loss while keeping insects from entering its throat.

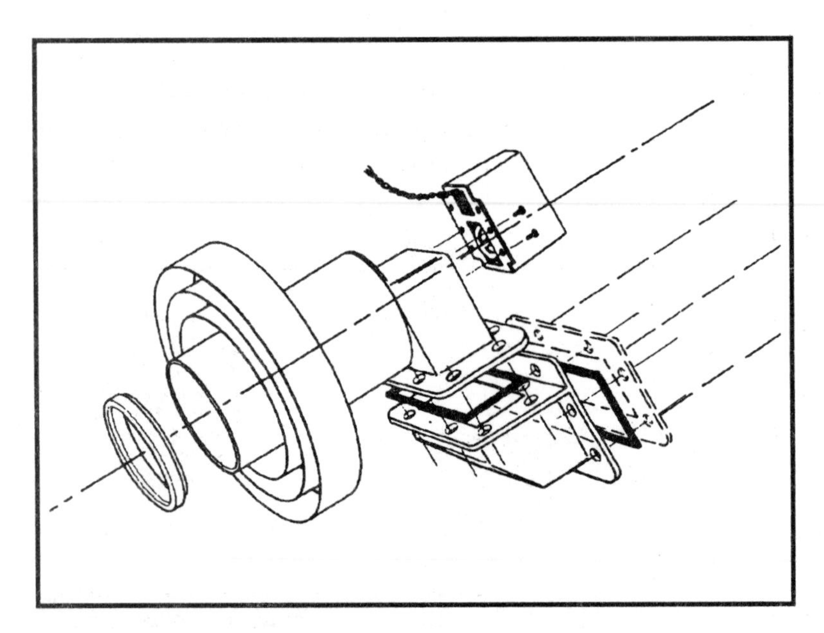

Figure 2-43. Feedhorn Assembly.

Feedhorn Operation

Illumination Patterns and f/D Ratios

The term feedhorn was derived from uplink antenna jargon. A feedhorn "sprayed" or fed microwaves onto the antenna surface below for reflection into space. Before the advent of powerful computers it was easier to design a downlink system by considering it as an uplink working in reverse. The term feedhorn has persisted as a descriptor of the collection device on a receiving antenna.

Therefore, feedhorns are said to "illuminate" a dish. The illumination pattern describes the field of view of a feedhorn. A perfectly illuminated system would detect radiation coming from nowhere but the antenna surface; it would reject radiation from any other source (see Figure 2-44).

In practice, feedhorns illuminate an antenna's central regions most strongly and are less able to detect off-axis microwaves. A feedhorn which illuminated just the central portions of an antenna would introduce little environmental noise into the system but would miss some of the signal at the dish edges and hence result in lowered gain. A feedhorn which over-illuminated a dish would take advantage of all the available gain but would introduce too much ground noise. Remember that the ground on a typical cool summer day emits noise at a "hot" 290°K.

Feedhorn designers have determined that the optimal "edge taper" for maximal collection of signal as well as noise reduction occurs when signal power at the dish edges is between 14 and 16 dB lower than that at the dish center (see Figure 2-45).

The f/D ratio determines where a feedhorn must be located. Deeper dishes require feedhorns which can see out to wider angles. Feedhorns are optimized for specific values of f/D but can be used quite effectively for a range centered on this optimal value. For example, the Polarotor™ is designed for an f/D equal to 0.375 but works over a range spanning 0.30 to 0.44. The Chaparral Gold Ring™converts this feed to one designed for an 0.30 f/D. At a f/D of 0.25, the feedhorn is located in the same plane as the edge of a dish and adequately illuminating the edges becomes more and more difficult to achieve resulting in lower gain.

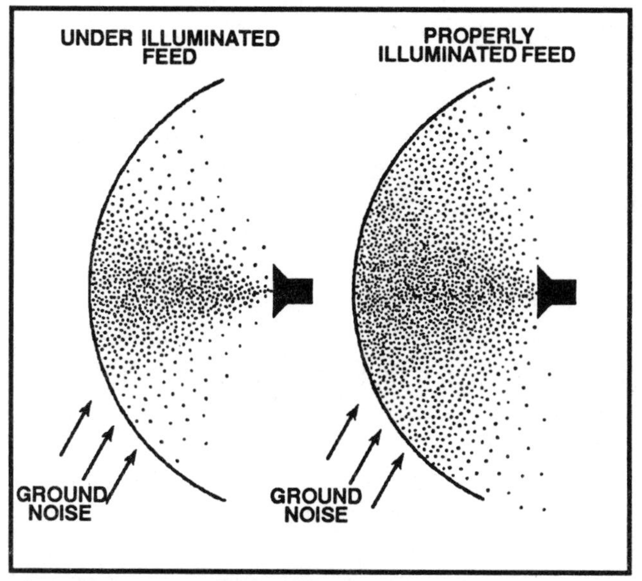

Figure 2-44. Feedhorn Illumination Patterns. *This drawing demonstrates how a reflector can be both under and over-illuminated by not selecting the correct feedhorn for a particular antenna f/D ratio. A poor match results in a lowering of signal-to-noise ratio. Under- illumination causes a loss of gain, while over-illumination causes detection of excessive noise from beyond the antenna rim.*

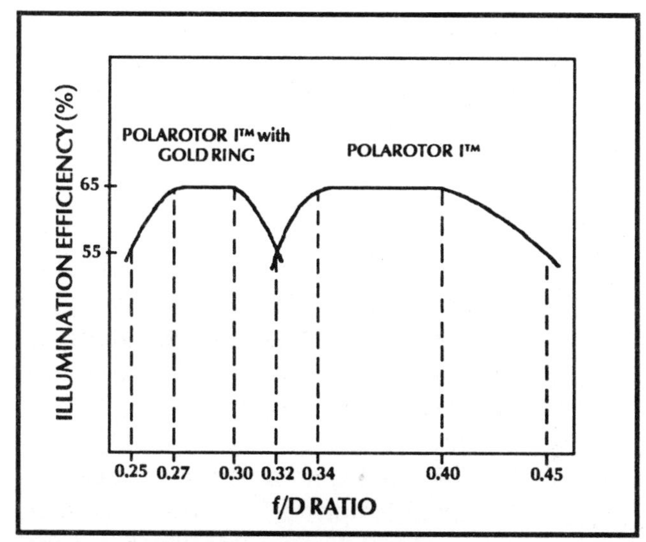

Figure 2-45. Feedhorn Illumination Taper. *This figure shows the illumination taper of a prime focus feedhorn. The design objective is to maximally illuminate the center of the reflector and to allow detected power to taper off to approximately 15 dB at its outer edge.*

Feedhorns used with offset fed dishes must be specifically designed for this configuration. As in all other systems, the offset feed should see as much of the dish surface and as little of the surrounding terrain as possible.

More on Feedhorn Design

Feedhorns detect the electric and magnetic fields that constitute microwaves that are reflected from the surface of an antenna. In order to equalize reception of both these fields and, as a result, to properly illuminate a dish, scalar rings are used. Feedhorns designed for higher frequency Ku-band broadcasts are proportionately smaller in relation to the wavelength.

Once the microwaves have been captured, they are channeled by a waveguide down the throat of the feedhorn. Waveguides are metal, hollow pipes of circular, rectangular or other cros-sectional shapes that transmit microwaves (see Figure 2-46). They can be compared to fiber-optic cables that relay light. Microwave signals cannot be transmitted through waveguides unless they have a wavelength shorter than one half of the dimensions of this waveguide. The waveguide in a feedhorn must have precise dimensions to allow transmission of as much of the received radiation as possible.

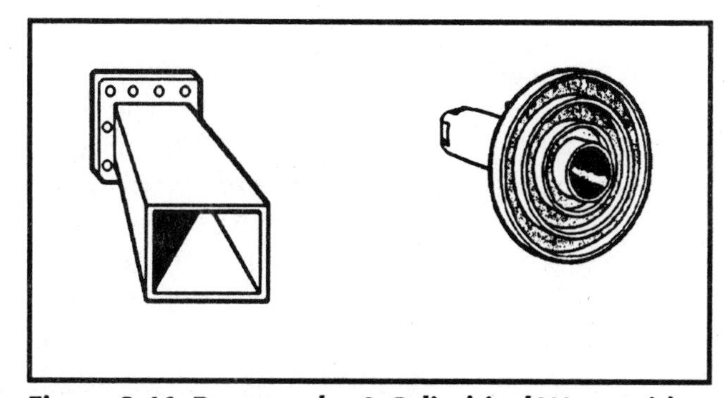

Figure 2-46. Rectangular & Cylindrical Waveguides.

A quantity called the VSWR or the voltage standing wave ratio measures how much of the incoming radiation is reflected backwards and lost (for a definition of VSWR see the Appendix). Feedhorn manufacturers must design their products to have a minimum VSWR. A perfect feedhorn would not reflect any microwaves back towards the dish and would thus have a VSWR of 1 to 1. VSWR ratios below 1.5 to 1 are acceptable but a high quality feed should have a ratio at least below 1.3 to 1. With this measurement, the lower the better.

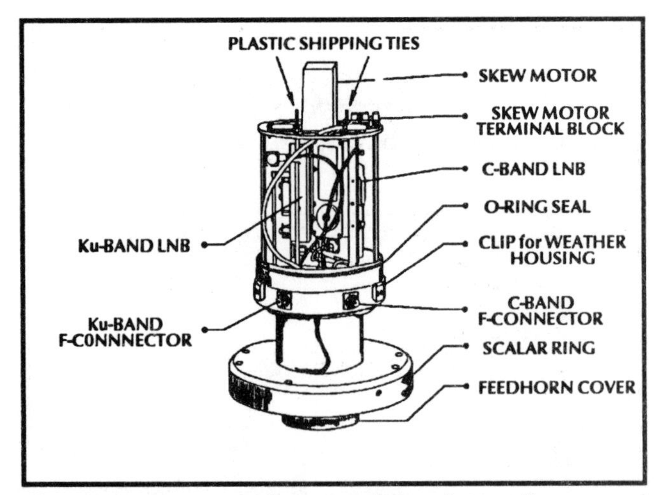

Figure 2-47. One-Shot Dual-Band Feedhorn and LNBs. *This feedhorn selects between vertically and horizontally polarized signals by rotation of the central cylinder that houses both LNBs. (Courtesy of Coast Hitech Corporation)*

Polarity Selection

Feedhorns must be capable of selecting the correct polarity format and of rejecting all others. Most broadcast satellites relay signals with either vertical or horizontal polarity. Circular polarities are used by satellites such as the Soviet Gorizont and Molniya and some Intelsat vehicles. Selection between vertical and horizontal formats is considered in this section.

Early feedhorns had a motor which simply rotated the whole body of the mechanism. This was cumbersome and subject to mechanical failure. Today a variety of more effective methods are used, although one innovative feedhorn that has recently been introduced uses a motor to rotate the probe section of the feedhorn (see Figure 2-47).

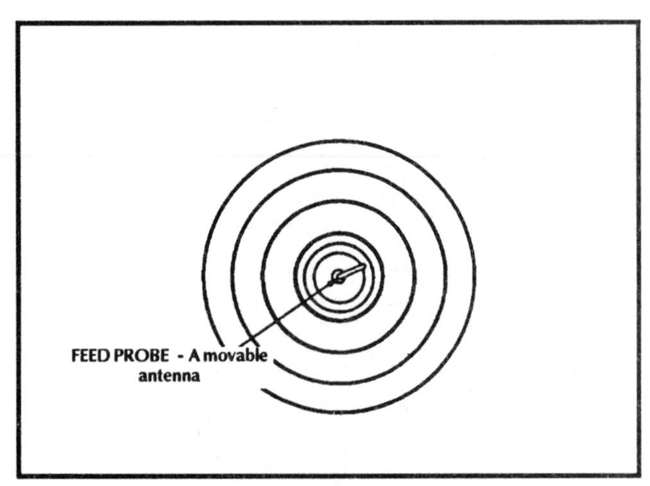

Figure 2-48. Polarity Selection. *The metal probe within most feedhorns that use servomotors rotates between vertical and horizontal positions in order to detect both linear signal polarities.*

Mechanical Rotors

In North and South America, the mechanical servomotor feedhorn is predominant. The feed probe of earlier units was rotated over a 180° range to allow reception and fine tuning of either polarity. Modern feed assemblies use servomotors that are more dependable, have a 270° range of motion, have higher voltage ratings, can be overload protected, and do not vibrate when they are being pulsed as earlier models did (see Figure 2-48). Some have built-in heating devices to allow more dependable operation in colder climates.

Some of the pioneer Channel Master satellite systems used a 12 Vdc motor that had a full 360° range of movement instead of the 5 to 8 Vdc servomptor. This motor used a two-wire connection, ground and power, not three wires because it was not pulsed like the servomotor. It rotated continuously in one direction as it switched between channels and moved more slowly than the servomotor type.

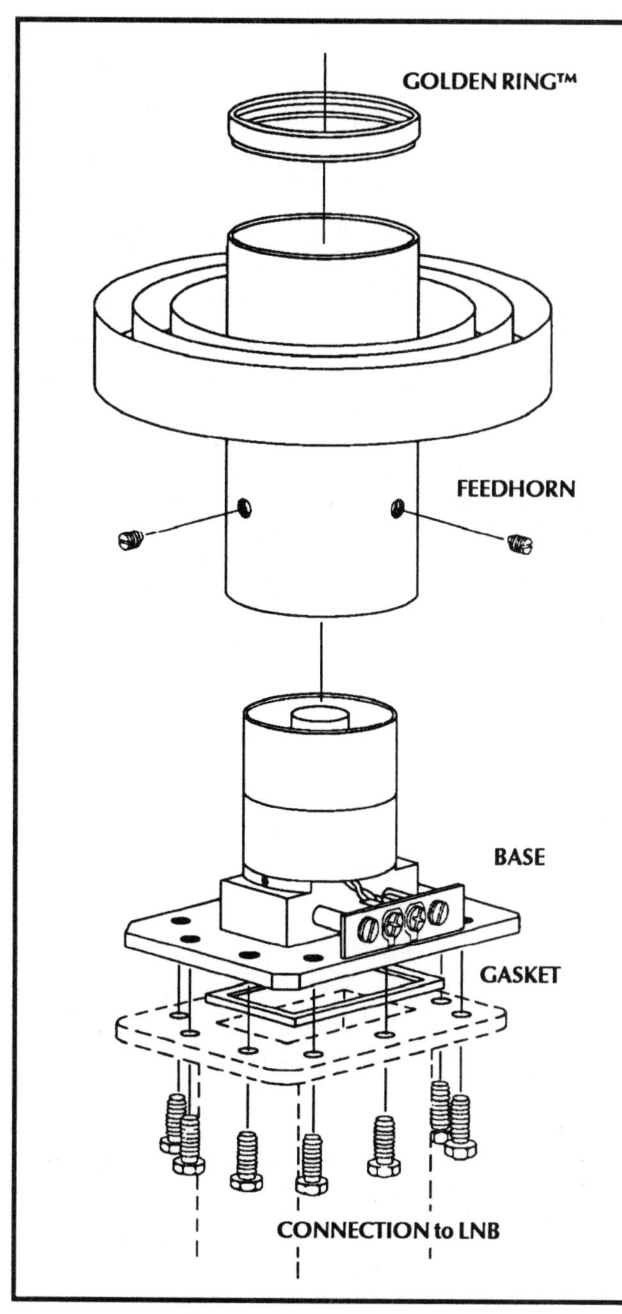

Figure 2-49. Schematic of Luly Ferrite Feed.

Ferrite Devices

Ferrite devices are solid-state mechanisms that have no moving parts. These feeds select polarity by sending current through a wire-wound piece of ferrite material. The magnetic field produced rotates the plane of polarization instantaneously (see Figure 2-49). Ferrite devices cannot have mechanical failures (except for an unlikely short circuit) and are preferable in very cold climate installations where binding could occur. Mechanically driven probes could have failures in such environments. Unlike their mechanical counterparts, ferrite polarity selection devices have some insertion loss or noise, typically about 15° to 30°K or 0.2 to 0.4 dB.

While ferrite devices are common on European Ku-band systems, they are rarely used in North America. While "international" receivers have built-in controls for both ferrite and servomotor feeds, North American receivers typically can manage only the servomotor type of polarity selection.

Ferrite feeds have one notable advantage, that of extra TI rejection. In a servomotor feed the probe detects either sense of polarity and then reradiates the signal, usually into a waveguide at right angles to the boresight. In a ferrite feed the magnetic field acts as a magnetic waveguide to discriminate between signal polarities. It is more selective than the probe mechanism and thus also more effectively rejects interfering signals that do not match the polarity of the satellite signal.

Note that some servomotor type feeds have been designed specifically to reject terrestrial interference. Both California Amplifier and Fujitsu manufacture such a feed.

Pin Diodes

Pin diode switching is the least popular since only two directions of polarity can be selected based on the orientation of two pin diodes. Polarity setting cannot be fine tuned. Pin diodes are also relatively high in loss and noise contribution. They are not used in modern satellite TV systems.

Figure 2-50. Schematic of a Dual-Band C/Ku Feed.

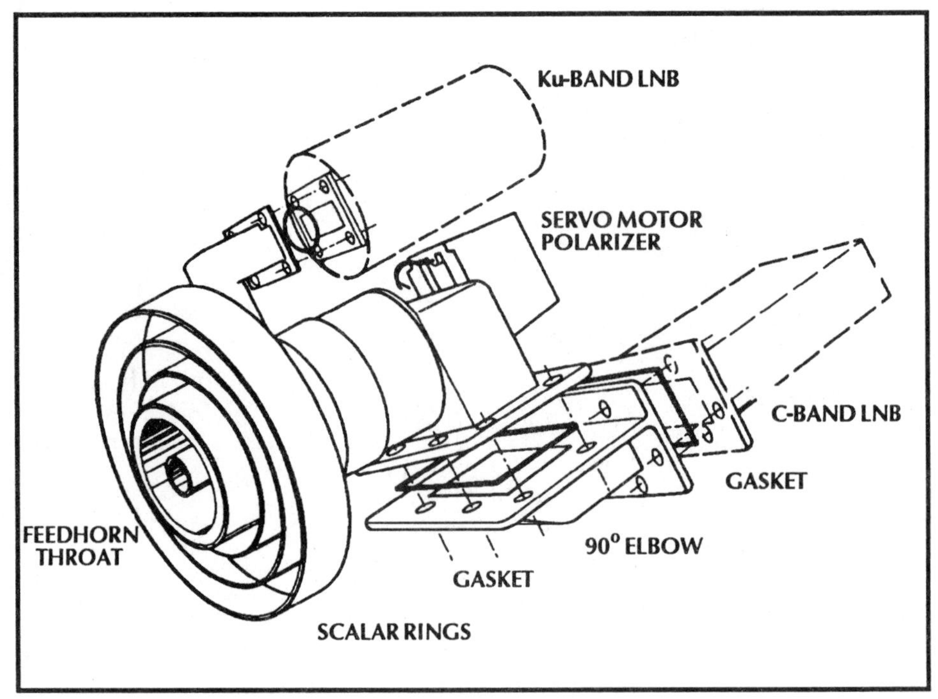

Dual-Band Feedhorns

While home satellite systems that evolved in large part in the United States were designed to receive C-band broadcasts, numerous Ku-band satellites are now in service. In fact, European satellite television is relayed predominately on the Ku-band. Feeds that were capable of simultaneously receiving both C and Ku-band signals were introduced in the late 1980s.

Dual-band feedhorns use a single motor for both frequency bands. The boresight of the C-band feed also contains a Ku-band waveguide that is positioned in the center by use of spacers. The rotor at the end of the throat has two probes, one for C-band and one for Ku-band reception, attached to a common motor. When the motor is pulsed to change polarity both probes rotate. One example of such a feed is the Chaparral Corotor™ which is an evolutionary take off from the Bullseye™ feed.

A variety of orthomode and dual-band feedhorns are available today. In addition to feedhorns dedicated to C or Ku-band alone or orthomode feeds for reception of both polarities on C or Ku-band, these include orthomode C-band with single Ku-band (see Figures 2-47 and 2-50).

Orthomode Feedhorns

Orthomode feedhorns are designed to detect both vertically and horizontally polarized signals and to permit simultaneous viewing of all 24 channels on each satellite. While such devices are common in commercial installations, they are rarely seen in home satellite TV systems.

Operation of these devices is based on the fact that waveguides reject the passage of microwaves with wavelengths longer than one half the guide's dimension. For example, if signals with a 1 centimeter wavelength reach a waveguide having a cross-section of 2 by 0.3 centimeters, only those waves polarized in the wider dimension of 2 centimeters will pass through.

A dual feedhorn uses two waveguides of the required dimensions at right angles to each other. One arm transmits horizontal polarity; the other vertical polarity. Two low noise amplifiers are used to amplify each of these signals.

Customer Interfaces

Polarity selection is controlled from the satellite receiver in three ways. First, the receiver triggers an automatic polarization change when channels are switched from an even to an odd numbered station and vice-versa. Second, once the channel has been selected, a skew adjustment allows for fine tuning of probe position. Third, since two different polarity formats are used by most broadcast satellites (either all even channels are vertically or horizontally polarized), nearly all receivers have an even/odd polarity selection control which changes probe position at the press of a button or when programmed to automatically do so.

Programmable receivers accomplish channel selection and polarity format adjustments automatically. Some sophisticated units also fine tune and optimize polarity by employing feedback circuits which maximize signal strength when a given channel is selected.

Most receivers are designed to accept the necessary wires, typically of 20 gauge, to control polarity selection and fine tuning.

Circular Polarization

Most domestic broadcast satellites relay either horizontally or vertically polarized signals. But some international satellites transmit circularly instead of linearly polarized broadcasts. Although right-hand circular polarization (RHCP) has been favored, zone beams from the Intelsat V, VA and VI series of spacecraft serving areas of Central and South America are now using left-hand circular polarization (LHCP, please see Appendix F for more details). Remember that circular polarization means that the vibrations of the electric and magnetic fields follow a circular path as they travel through space (see Figure 2-51).

Feedhorns that are designed to detect linearly polarized signals can receive circularly polarized waves, although received power is reduced by approximately 2 to 3 dB. Such a near halving of signal power is unacceptable especially when viewing the often weak global beams. In addition, standard scalar feeds cannot distinguish between LHCP and RHCP transmissions.

Fortunately, a simple alteration to standard feedhorns allows detection and discrimination of both types of linear and circular polarization. A "dielectric slab," quite simply a properly sized rectangular piece of teflon, can be inserted into the polarizer throat to accomplish this feat. A conventional feedhorn with skew adjustment is then capable of detecting and discriminating between vertical or horizontal polarities as well as between LHCP and RHCP signals. Some feedhorn manufacturers now sell this product complete with installation instructions (see Chapter 5 for more details on installing such devices).

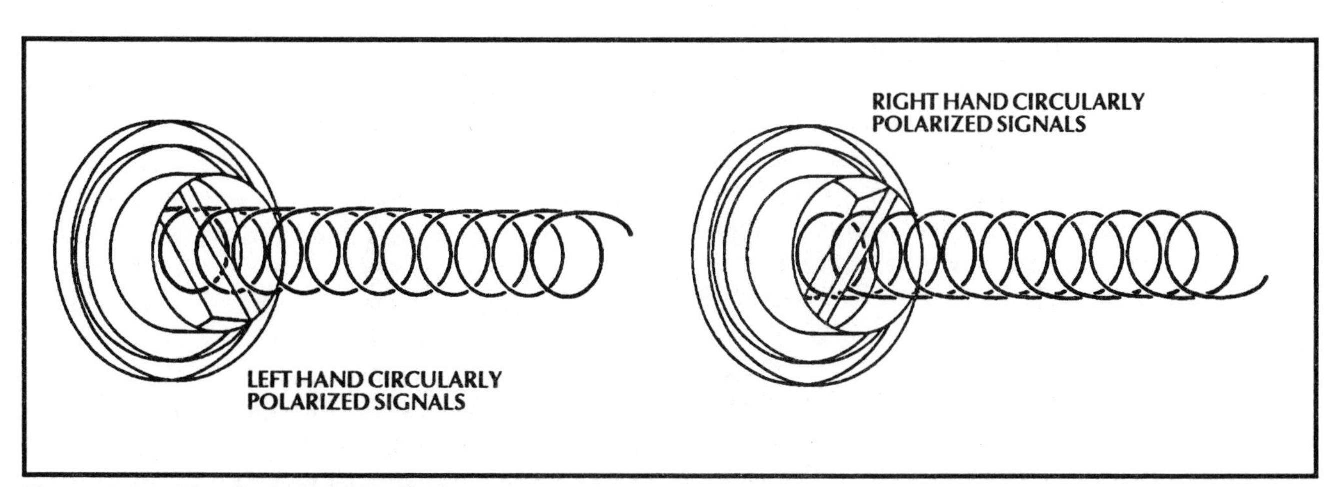

Figure 2-51. Circularly Polarized Waves.

E. LOW NOISE AMPLIFIERS

Purpose

Low noise amplifiers (LNAs) have the important function of detecting microwaves relayed from the feedhorn, converting them to electrical currents and amplifying this extremely weak signal by about 50 dB (a factor of 100,000). The antenna and LNA working in unison are the most important electronic components in determining how well an earth station functions. The LNA is the first "active" or electronic device in the chain of equipment which processes a satellite signal.

Today nearly all systems use LNBs, an abbreviation for low noise amplifier/block downconverter. LNBs combine the functions of amplification and downconversion, lowering the frequency of the signal, in one step. Downconversion is examined in the next section. The abbreviation LNB is used throughout this book except when specifically referring to LNAs.

The power reaching the input of an LNB is still at an incredibly low level of less than one hundred thousandth of a billionth of a watt. The LNB must contribute very little noise power or this signal will be drowned out in the roar of noise from the internal workings of the amplifier. This feat is made possible by recent advances in transistor technology. Without this progress satellite TV would not exist as we know it today.

The original LNAs used in radio astronomy were ordinary parametric transistor circuits which were immersed in baths of extremely cold liquid nitrogen or helium. This technique kept noise levels low by slowing down molecular motion, the source of background noise. The development of the gallium arsenide field effect transistor, known as GaAsFETs, has made the modern low noise amplifier possible. These special transistors trick the amplifier into behaving as if it were operating near absolute zero where all molecular motion ceases.

Today there are two types of low noise transistors available to LNB designers. These are HEMTs and MOSFETs, acronyms for high electron mobility transistor and metal oxide semiconductor field effect transistor, respectively. MOSFETs have been used since the early 1970s while HEMTs have been commercially available only since 1987. The basic difference between the two devices is that the HEMT performs better at higher frequencies, namely it has a better noise figure and gain than the MOSFET.

More on Noise Temperature and Noise Figure

An understanding of noise is critical in satellite broadcasting because the minute signals are just a step stronger than the ever-present noise. Noise is caused by molecular motions which generate electrical currents and, as a result, electromagnetic waves some of which are in the same microwave frequency band as satellite transmissions. The scale employed to measure noise is based on the fact that at zero degrees Kelvin, known as absolute zero (minus 273.18°C or minus 459.72°F), there is no noise. Typical C-band LNBs now have noise temperatures ranging from 35° to 70°K while Ku-band LNBs are available in the range from 70° to 120°K. Even lower noise temperatures will most likely be available within the next few years, especially for Ku-band LNBs. (Ku-band LNBs are typically rated in dBs of noise; common values available are 1.0, 1.2 and 1.4 dBs).

Interference is also considered a form of noise even though it usually originates from some other man-made source or communication device. Higher quality LNBs having lower noise temperatures will detect less random noise but are more capable of seeing other organized signals from either satellites or man-made sources.

Noise power is directly related to the signal bandwidth as well as the temperature inside electronics and in the environment. As bandwidth increases, more noise can be detected and added to any signal. So decreasing the bandwidth of signals

TABLE 2-4. EQUIVALENT NOISE FIGURES and TEMPERATURES

Noise Temperature (°K)	Noise Figure (dB)
30	0.428
40	0.561
50	0.691
60	0.819
65	0.881
70	0.942
75	1.002
80	1.061
85	1.120
90	1.177
95	1.234
100	1.291
110	1.401
120	1.508

allowed into an amplifier decreases the amount of noise and interference which can be detected.

The noise characteristics of an LNB or any other electronic device are sometimes described in terms of the noise factor. This quantity is directly related to the amount of noise added by the internal workings of such devices. The noise figure is simply the noise factor expressed in decibels. More details showing how noise temperature and noise figure are related to each other are outlined in the appendix. Some typical values of these quantities are shown in Table 2-4. How does substituting an LNB having a lower noise temperature affect system performance? The method used to calculate this change can be clearly seen from an example. If a 120°K LNB were replaced with a 60°K unit in a system whose dish added 40°K of noise, the signal to noise ratio would improve by:

$$(120 + 40)/(60 + 40) = 1.6$$

This factor of 1.6 or 60 percent performance improvement is equivalent to a 2 dB change (equal to 10 log 1.6). It is interesting to realize that an equivalent 2 dB improvement in gain can also be achieved by upgrading from an 8 to a 10 foot dish (see Table 2-2).

LNB Design and Operation

All C and Ku-band LNB designs are similar in shape because the waveguide section must have the appropriate dimensions to channel C and Ku-band microwaves, respectively (see Figure 2-52). The input flange and waveguide of a C-band LNB, known technically as a WR-229 choke with waveguide, has dimensions of 2.29 by 1.145 inches. Occasionally, fine tuning controls are built into the waveguide portion of an LNB to allow slight changes in its internal dimensions to improve VSWR. These permit the precisely manufactured LNB to be even further tweaked to maximum performance on sophisticated equipment. Such tuning minimizes the amount of signal lost in entering the LNB.

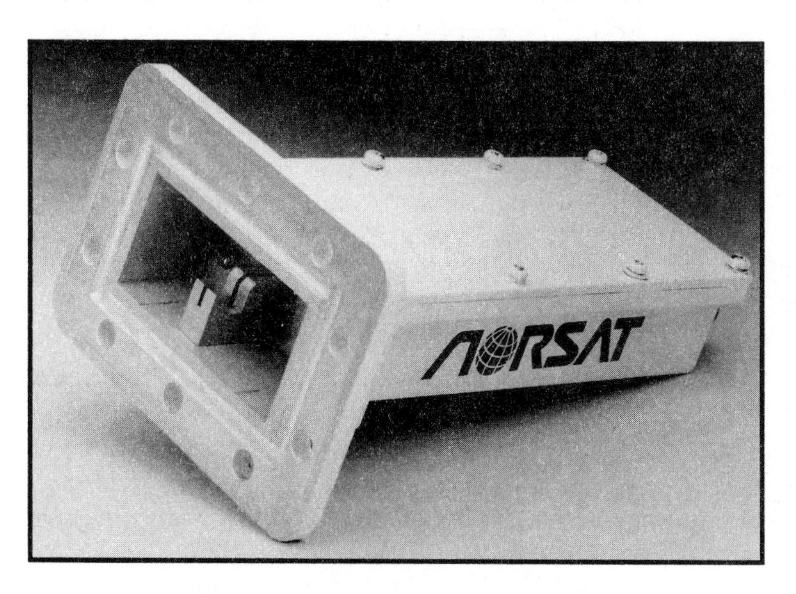

Figure 2-52. Norsat LNB. *(Courtesy of Norsat International)*

Every LNB has an internal probe which is the real microwave antenna. This small metal antenna receives microwaves and converts them into electrical currents. A DC shorted probe prevents high voltages which are caused by nearby lightning strikes from frying the internal components. Of course, a direct strike would destroy any LNB. The probe is set at precisely the correct position to maximize signal reception and should never be tampered with even if it appears to be slightly bent.

The electronic components of an LNB are enclosed in a hermetically sealed box. Water vapor has a very destructive effect on the operation of electronic components and is carefully avoided by such design.

The amplification section of an LNB consists of several GaAsFET transistor stages (usually two or three) in a cascaded arrangement followed by several conventional amplification stages. A voltage regulator is also included in the circuit. LNBs usually draw between 80 to 150 milliamps of current and operate at 15 to 24 Vdc (see Figure 2-53).

Judging LNB Performance

The major factor in judging LNB performance and quality is its noise temperature. This assumes that the gain is sufficient and that important design issues such as waterproofing and grounding have been adequately addressed.

Noise Temperature

The LNB is the "front end" of a satellite TV receiving system. The noise it adds to the incoming signal sets the noise floor and plays a large part in determining picture quality. LNBs having noise temperatures ranging from 60°K to as low as 30°K are now available at reasonable prices. This is extraordinary considering that as recently as 1981 an 85°K LNA retailed for over $5000 while today a 40° unit could be purchased for under $100 U.S.

LNB noise temperatures vary across the design frequency band. Most manufacturers will print the noise temperatures measured at 3.7, 4.0

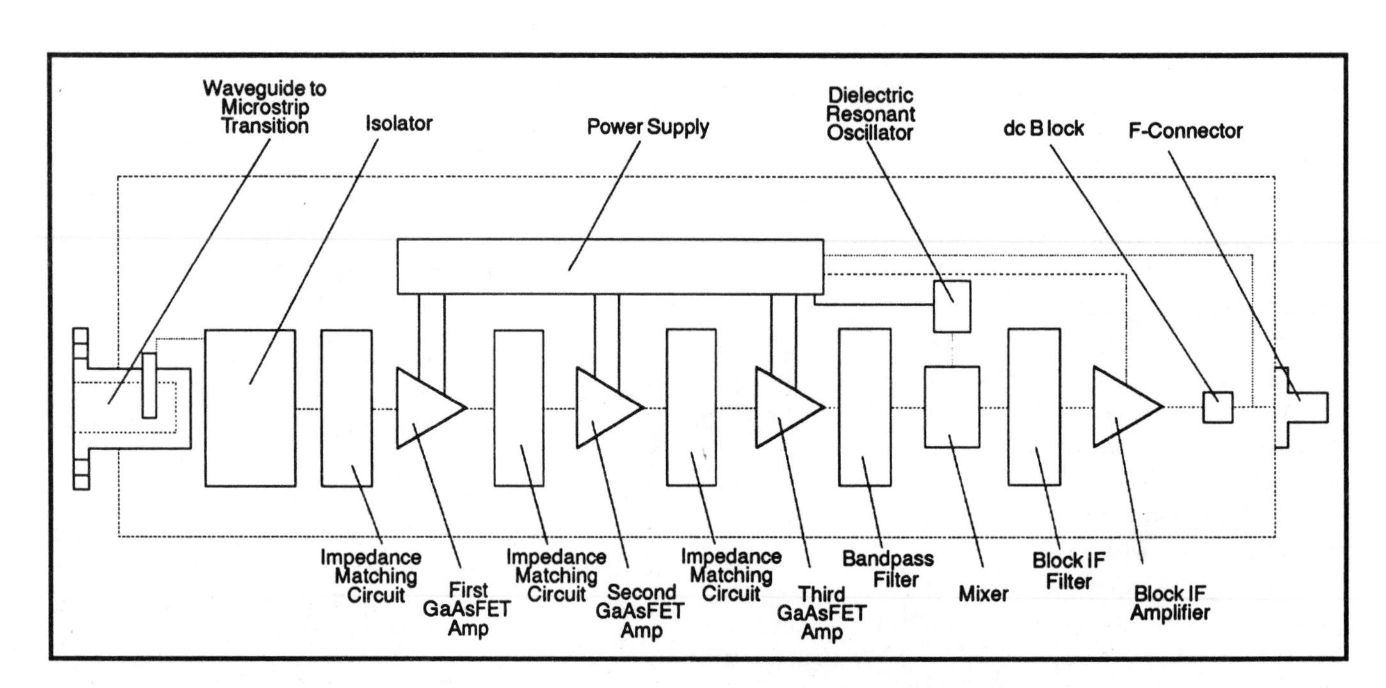

Figure 2-53. LNB Schematic. *A low noise block downconverter combines the functions of an LNA and block downconverter in one case. The LNB has a single female F-connector output which feeds directly over coax to a satellite receiver. Each stage contributes noise and gain. The GaAsFET or MOSFET stages which are responsible for the low noise temperature characteristics, contribute approximately 12 dB of gain; the two or more bipolar stages contribute a total of 30 dB additional gain.*

and 4.2 GHz on an attached plate. The overall rating should reflect the highest noise temperature in this range.

LNB performance is determined to some extent by ambient operating temperature. Thus, for example, if an LNB with a black weathercover were functioning at high noon in a desert and had its body temperature raised to 140°F, it would certainly generate more noise. The sensitivity of LNB noise temperature to ambient temperature can vary from 0.8 to 1.5°K for every 10°C (18°F) temperature rise above 25°C (77°F). To illustrate, a 90°K LNB would perform as a 92° unit if the operating temperature were raised by 20°C (36°F). Performance would improve by the same measure if the operating temperature were lower. Most manufacturers shield their products from temperature extremes by using a rubber or enamel paint.

Gain

While LNB gains have ranged from as low as 30 dB to over 50 dB, today most units have gains in the region of 50 dB. The bottom line with LNB gain is simple. Either it is enough or it is not enough. Past a certain point adding extra gain will do nothing for improving system performance. Tests have shown that this point occurs somewhere between 35 and 40 dB, but 40 dB should be considered a minimum gain especially when using an undersize or a small dish. In older systems that use a separate LNA and downconverter with 4 GHz cable runs between the LNA and downconverter exceeding approximately six feet in length, low gain can cause serious performance degradation.

Gain also varies with ambient temperature and frequency. As temperature increases, gain decreases, typically about 0.6 dB for each 10°C rise in temperaturer. This variation is of little significance except for operation in extremely hostile environments such as outer space. Small variations in gain also occur across the C-band frequency range. A perfectly "linear" LNB would amplify signals on any frequency equally but generally the variations are small enough to be negligible. This is fortunate because large deviations from linearity which could cause an LNB to become unstable and oscillate are more possible at extremely low ambient temperatures below -40°F.

A perfect LNB would amplify signals only in the designed frequency band and reject those outside. In practice, gain drops off rapidly but measurably at frequencies outside the band. Out-of-band rejection is important especially in cases when other local communicators using nearby frequencies could be inadvertently detected. The two most likely culprits would be microwave amateur or Air Force radar at 3.5 and 4.4 GHz, respectively. Out-of-band rejection is also more of a concern for block downconversion than for older 70 MHz systems (see next section) where frequencies ranging from 950 to 1450 MHz nearer to the C-band are transmitted. Some poorer quality LNBs have produced significant gain and noise in this 450 to 1450 MHz range. However, most well-made LNBs have bandpass filters, devices which reject frequencies under 3.7 GHz and over 4.2 GHz, to minimize the likelihood of such problems.

LNBs, like feedhorns, also have a rated VSWR. This voltage standing wave ratio is found by dividing the power of the input signal by that of the input signal actually entering the LNB. When no signal is reflected back towards the feedhorn in a perfect device, the ratio is 1 to 1. VSWRs are below 1.3 to 1 in higher quality LNBs.

LNAs, LNBs and LNCs – An Historical Perspective

Modern low noise amplifiers not only amplify the satellite signal but perform the functions expected of a downconverter. Downconverters do just what the name implies. They lower all or part of the frequency range from 3.7 to 4.2 GHz to either the final targeted 70 MHz or to some intermediate range. In pioneer systems, a short cable run or connector was used between an LNA and downconverter, which was mounted adjacent to the LNA or just below the dish for best results. Many such systems are still installed and working, especially in North America.

An LNC combined the functions of a downconverter and LNA in one unit. This device lowers the frequency of each satellite channel to 70 MHz in turn as required by the indoors receiver.

An LNB combines the functions of a block downconverter and LNA in one unit. This device lowers the frequency of the whole 500 MHz satellite band in one step to an intermediate range, most commonly 950 to 1450 MHz, although several other ranges have been used including 430 to 930 MHz, 900 to 1400 MHz and 930 to 1430 MHz. The lower and higher frequency blocks are referred to as low and high block conversion, respectively.

LNBs have become the standard in modern home satellite systems for two reasons (see Table 2-5). They allow flexibility in designing multiple independent receiver systems and are free from problems of channel drifting that occur when channels are selected by outdoor electronic components at a dish. LNBs are examined further in the following discussion on downconverters.

Note that it is quite simple to distinguish between an LNA and LNB by examining the output connector. The N-connector used on LNAs is nearly an inch in diameter while a F-connector is about half this size and is commonly used for television set inputs and outputs.

TABLE 2-5. COMPARING LNAs, LNBs & LNCs

	LNA	LNB	LNC
Amplification	x	x	x
Downconversion		x	x
Channel Selection			x

Electrical Connections

The electrical connections to an LNA, LNB or LNC are simple. LNAs use a female N-connector joined either directly or via a short RG-214 coaxial cable run to a downconverter. LNBs and LNCs use an F-connector and can be directly cabled into the satellite receiver. Power is relayed via the signal-carrying coax.

F. EXTERNAL DOWNCONVERTERS

Once the downlinked 500 MHz band of frequencies has been amplified by the low noise amplifier, it is downconverted either internally or, in earlier systems, relayed to an external downconverter via either coaxial cable or a male-to-male direct couple N-connector. The downconverter translates either part or all of the 3.7 to 4.2 GHz signal to a lower frequency range. This contains the identical information. The purpose of this step is to allow use of inexpensive, lower loss cable, such as RG-6 or RG-59, to send this signal in-doors.

Background and Evolution

The first satellite receivers which appeared on the market had downconverters built-in along with all the other components necessary to extract the original audio and video information. This required use of expensive, low-loss air or foam dielectric cables. These cables could not have sharp bends and required the use of difficult-to-install N-connectors. Antennas had to be mounted as closely as possible to the indoors receiver or expensive 4 GHz line amplifiers often costing thousands of dollars were required. All in all, installations had severe limitations and higher costs.

The first basic improvement was to break the satellite receiver down into two components: a

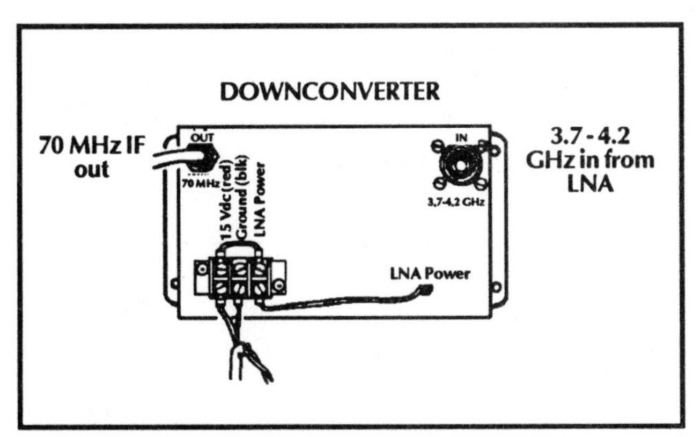

Figure 2-54. Downconverter. *A downconverter accepts a C-band input on an N-connector and feeds its IF output via an F-connector to a satellite receiver.*

downconverter and the remaining electronics (see Figure 2-54). Thus, a short length of less expensive RG-214 cable could be used between the LNA and downconverter and even lower cost RG-6 or RG-59 relayed the downconverted signal to its final destination, the satellite receiver. Cable runs even in excess of hundreds of feet could then be tolerated.

Next, the downconverter was incorporated into the LNA to create an LNC so that amplification and downconversion would be accomplished in the same component mounted at the dish focus. This eliminated a separate box, cable and two connectors.

LNCs were never well accepted by the industry for two reasons. Unlike the more conventional LNA/downconverter, if an LNC had a problem, a lower cost LNA could not simply be swapped for a working unit. Also, systems using LNCs could not be used for multiple receiver, independent channel selection since only one LNC could be used per dish.

As the industry evolved, customers began to request that satellite receivers be installed in more than one location. LNB systems, introduced in the late 1980s, are ideal for multiple receiver system designs. Such devices lower the complete 500 MHz band at once to some intermediate frequency. This result can be accomplished using a separate LNA and block downconverter or by merging these two components into one box thus creating an LNB.

Downconversion Methods

Three distinct types of downconversion schemes are in use today: single downconversion, dual downconversion and block downconversion. In order to understand how these operate it is necessary to examine the way channels are selected.

Each of the twenty four C-band downlinked channels occupies a 36 MHz bandwidth. The selection of any one of these is accomplished when its 36 MHz band is translated down to one centering on 70 MHz or another final IF. Thus, for example, if channel 10 having a center frequency of 3900 MHz and a bandwidth of 36 MHz is desired, the downconverter/receiver electronics lowers this block of frequencies to one centered on 70 MHz. If channel 22 centered on 4140 MHz is selected, the 36 MHz range around this frequency is also lowered to one centered on 70 MHz. All other channels are tuned in the same way.

Downconversion is accomplished by a process called heterodyning or mixing. To illustrate, if a 1000 MHz signal is mixed with one of 900 MHz, the resulting sum and difference frequencies of 1900 and 100 MHz are created. So when a selective filter removes the higher frequencies, the effect has been to lower the 1000 MHz signal to one still containing all the original information but centered on 100 MHz.

A satellite receiver sends a voltage to a voltage tuned oscillator (VTO) which produces a desired mixing frequency. In single or dual conversion units the VTO is in the downconverter. In block downconversion systems the downconverter oscillator is set onto a fixed frequency, usually 5150 MHz, and channels are selected by a second oscillator in the satellite receiver.

Single Downconversion

Single downconverters lower the chosen satellite channel to the 70 MHz IF in one step. A channel selection knob or button chooses a voltage which is sent to the VTO (voltage tuned oscillator). A mixing frequency 70 MHz lower or higher than the center channel frequency is produced and, after mixing, the resultant difference frequency is 70 MHz, as desired. To illustrate, if channel 10 having a center frequency of 3900 MHz is selected, the oscillator mixing frequency is set at 3830 MHz; if channel 17 having a center frequency of 4040 MHz is chosen the VTO generates a mixing frequency of 3970 MHz (see Figure 2-55).

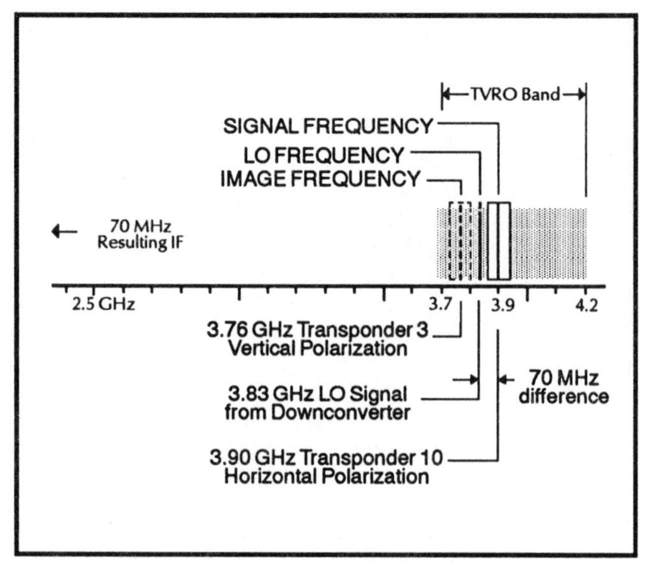

Figure 2-55. Single Downconversion. *This example, which corresponds to that of the text, shows how channel 10, which has a center frequency of 3900 MHz, is selected by a LO frequency of 3830 MHz. The resulting difference frequency is 70 MHz. However, the LO also mixes with channel 3 which has a center frequency of 3760 MHz. In this case, the difference frequency is also 70 MHz resulting in the formation of a potentially interfering image signal.*

Single downconversion receivers have a potential problem because these units not only receive the desired transponder but also an image frequency band. For example, an oscillator frequency of 3830 MHz selects transponder 10 which has a center frequency of 3900 MHz (3900 minus 3830 equals 70 MHz). However it also tunes in transponder 3 which has a center frequency of 3760 MHz (3830 minus 3760 also equals 70 MHz). Both channels end up entering the 70 MHz IF amplifier, causing interference between the two. Most single conversion receivers feature a circuit called an image reject mixer to reduce the strength of the image signal. A ferrite isolator installed on the downconverter input provides additional protection by blocking these unwanted signals from entering the system.

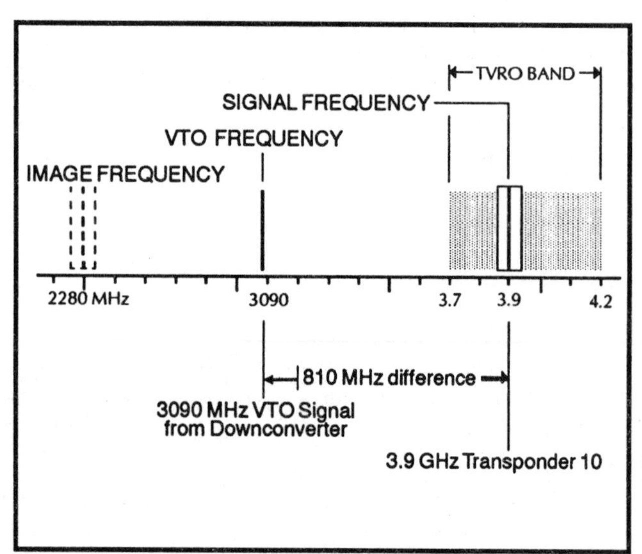

Figure 2-56. Dual Downconversion. *This example, which corresponds to that of the text, shows how the image frequency on a dual downconversion system falls below the TVRO IF band (see Figure 2-55). In this case, the VTO frequency is 3090 MHz. When mixed with channel 10 center frequency of 3900 MHz, the resultant difference frequency is 810 MHz. Then a fixed LO set at 740 MHz produces the final 70 MHz IF. However, the image frequency which also produces a 810 MHz difference signal when mixed with the VTO frequency lies far below the TVRO band at 2280 MHz.*

Dual Downconversion

Dual downconversion systems attain the final IF in two stages. An intermediate frequency, often but not always 810 MHz, is used as the target of channel selection. Following the first downconversion, a fixed oscillator called a local oscillator (LO) set at 70 MHz below the intermediate frequency is

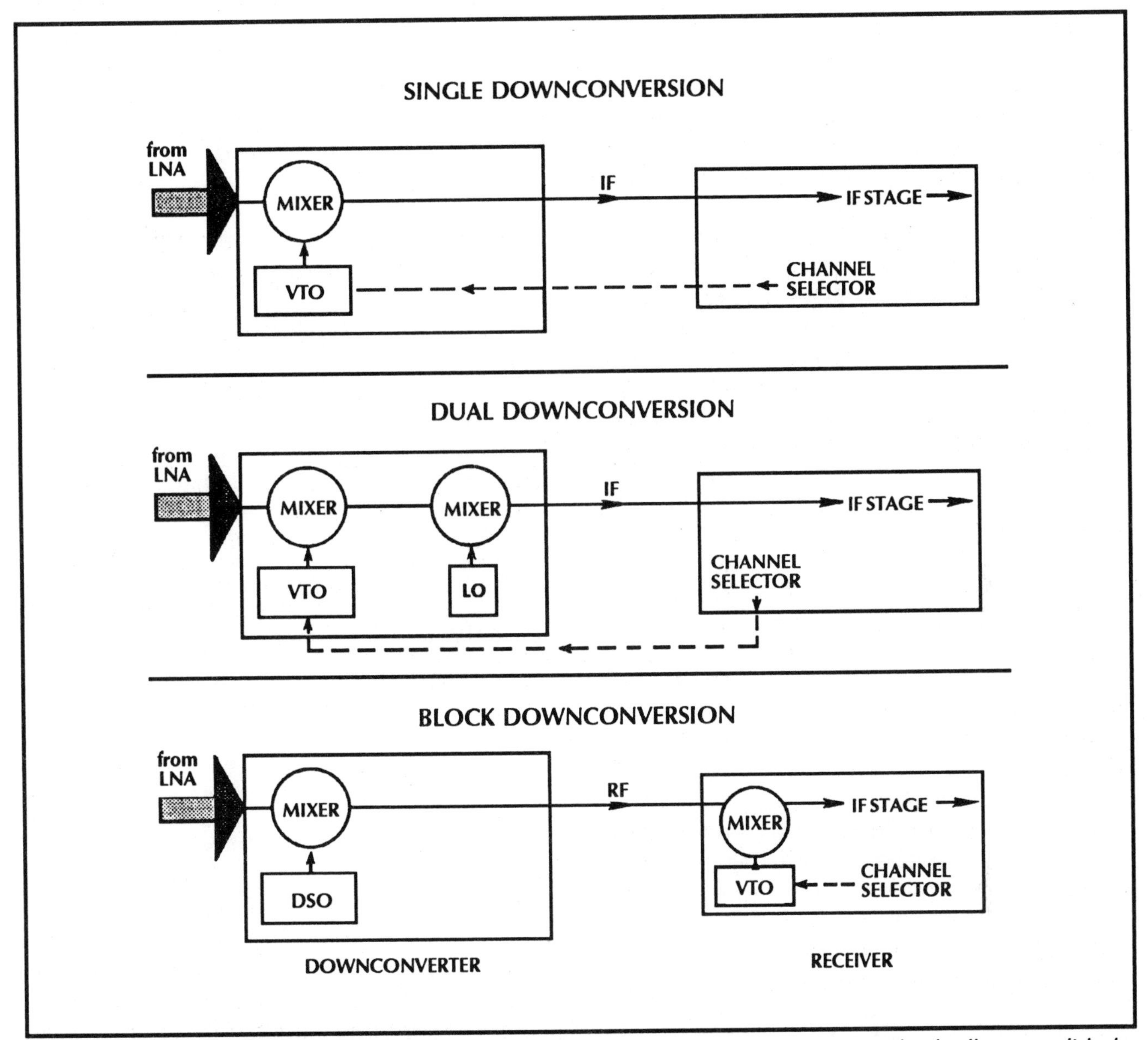

Figure 2-57. Downconversion Methods. *Single, dual and block downconversion methods all accomplish the same end result, converting the extremely high microwave signal to a lower frequency IF and selecting one of the many channels to be viewed. Note that the intermediate frequency (IF) is typically 70 MHz and RF falls in the 450 to 1450 MHz range.*

KEY: VTO--Voltage Tuned Oscillator
LO-- Fixed Local Oscillator
DSO--Fixed Dielectrically Stabilized Oscillator

mixed to produce the final 70 MHz. To illustrate, if channel 10 having a 3900 MHz center frequency is chosen, the VTO mixes it with a frequency of 3900 less 810 MHz or 3090 MHz. Then the intermediate 810 MHz is mixed with 740 MHz from the LO to lower the signal to one centered on 70 MHz. This design eliminates reception of the image signal that is present in single conversion systems (see Figure 2-56).

Block Downconversion

Block downconverters use an LO to downconvert the whole 500 MHz satellite band to an intermediate range (see Figure 2-57). Today one frequency range is favored, 950 to 1450 MHz. Some earlier receivers used 430 to 930 MHz. Both of these ranges have 500 MHz bandwidths and containing all the same information as the 3.7 to

4.2 GHz band. This block is relayed indoors to a receiver where a VTO mixes this signal with the channel selection frequency in the second downconverter to produce the final IF, typically from 70 to 612 MHz.

Block downconversion systems have one outstanding advantage. Two or more receivers can independently select channels because each one using its internal mixer and VTO can choose from the lower block frequency that contains all the information relayed from the LNB on a chosen polarity. Block downconversion satellite receivers also are less subject to drifting off channel because channel selection occurs indoors where electronic components are protected from large temperature and humidity swings.

Both of the intermediate frequency ranges have advantages and disadvantages. The 430 to 930 MHz range allowed use of off-the-shelf, low cost UHF TV circuitry, i.e. amplifiers and splitters. However, local UHF transmitters, mobile cellular radios, telephones and cable TV systems could possibly interfere with satellite reception since they all share the same frequency band. Using the higher frequency 950 to 1450 MHz range avoids all potential problems with interference but more expensive components having higher cabling losses are required.

Block downconversion can be accomplished by an LNB or an LNA/block downconverter combination. The only difference is that the LNB combines the two functions in one device.

G. COAXIAL CABLE

Purpose

The LNB and satellite receiver are connected to each other by a special configuration of conductors called coaxial cable, also known as coax. Single wires of copper or aluminum are adequate for conducting electricity at lower frequencies encountered in most familiar electrical devices. However, when higher frequency microwaves are relayed, single strands of metal behave like antennas and can radiate away most of the power. With exceptionally high frequency signals, such as those used in satellite broadcasting, specially designed cable must be used to prevent almost complete loss or attenuation of the transmitted signal.

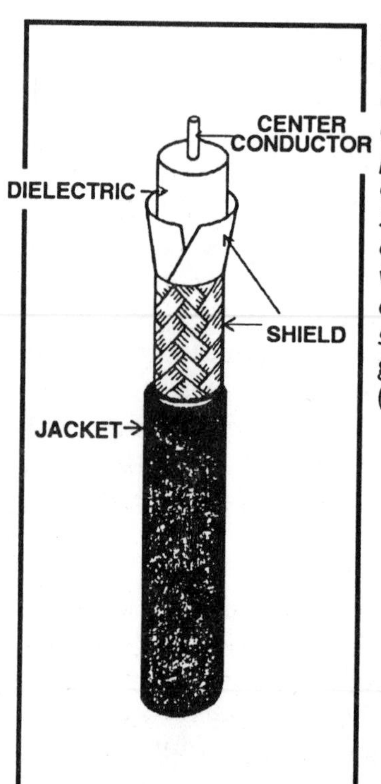

Figure 2-58. Coaxial Cable. *Coax cable is composed of two concentric conductors separated by an insulator called the dielectric. Signals are transmitted on the central conductor while the outer shield is designed to contain the signal and eliminate ingress interference. (Courtesy of M/A COM)*

Construction

Coax is composed of two concentric conductors separated by an insulating material called a dielectric. This whole assembly is sheathed in a non-conducting jacket for protection against the elements. The signal travels on the surface of the central wire. The external cylindrical conductor is grounded and greatly reduces radiative losses at high signal frequencies (see Figure 2-58).

Coaxial cables are generally sweep tested by the manufacturer to ensure that there are no breaks or discontinuities. In this test signals spanning a range of frequencies are fed into one end of the cable and the output is measured at the far end.

Types of Coax

A wide range of coaxial cables is available. Coaxial cable is usually referred to as either hard-line, coax, or foam or air dielectric coax depending upon the construction of the sheathing material, the dielectric. Coax has one or two pliable metal grounded layers wrapped around a plastic dielectric. Dielectrics used for home satellite TV coax are either polypropylene, a hard, translucent substance, or polyethylene foam, a soft, white material. The outer conductor is a braided copper or aluminum foil sheath with an aluminum braid.

The degree of cable braiding is rated by the percentage of the inner area shielded. 98 percent shielding is a typical value. Cables shielded 100 percent are recommended for installations where ingress interference caused by local communicators using frequency bands near the satellite C-band may be expected.

Foam or air dielectric coax uses either foam or compressed air as the dielectric material and is generally lower loss and more expensive than ordinary coax. Hard-line is similar in construction to either of the previous two types except it has even lower losses because it has a more rigid, metal sheath and a minimal amount of high quality dielectric (see Figure 2-59). Hard-line is usually used in the main trunk lines of cable TV or other similar communication networks such as SMATV systems.

Cable types often encountered in satellite TV installations are RG-6, RG-59 and RG-214. The latter is rated for microwave frequencies and is used, when necessary in older systems, between the LNA and downconverter. RG-6 and RG-59 transmit intermediate frequency IF signals from downconverter or LNB to a satellite receiver. Note that the nomenclature for coaxial cable is of military origin so these designations have no particular significance in the home satellite TV market.

Nearly all of the coaxial cable used in the satellite industry today is of the RG-6 type. It is usually sweep tested up to 2,000 MHz to ensure that it will conduct the 950 to 1450 MHz LNB output as rated. Such cable typically bears a stamp of having been sweep tested. Many in the satellite industry simply refer to this cable as 1 GHz coax.

Judging Cable Performance

Coaxial cable is judged according to performance criteria including characteristic impedance, loss of signal power per unit of distance and material composition. These factors determine which types are suitable for which uses.

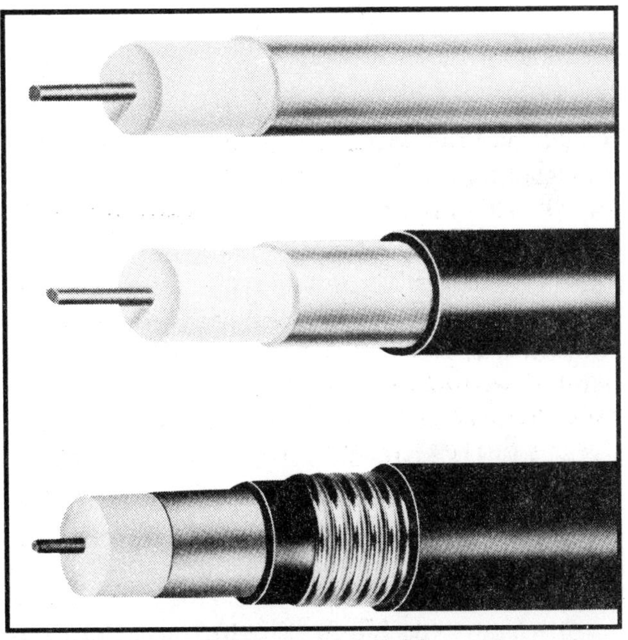

Figure 2-59. Hard-line Coaxial Cables. *These three examples of coaxial cable for high frequency signal conduction are protected from interference to increasing degrees. The middle one has a polyethylene jacket. The lower coax has two metal jackets, one being a corrugated chrome plated steel armor that prevents sharp cable bends. This or any other deformation could cause impedance mismatches and signal attenuation. (Courtesy of M/A COM, Comm/Scope Marketing)*

Characteristic Impedance

Every conductor has a certain resistance to current flow which causes some signal loss. Also, voltages on the inner and outer conductors interact with each other (known technically as capacitance and inductance). These two factors determine a value called the characteristic impedance. 75 ohm cables are typically used in TV distribution systems.

Knowing the characteristic impedance is crucial in designing electronic systems. Every electronic device also has a characteristic impedance. If the impedance of the cable does not match the device it feeds, there will be substantial reflective power loss. This mismatch would be similar to sending water from a larger diameter into a smaller diameter pipe and experiencing resistance to flow. Similarly, if the output impedance of an amplifier is not matched to the coaxial line impedance, reflections and losses will occur.

This idea is similar to the concept underlying VSWR (see page 49) that is used to characterize internal signal losses due to reflection for both feedhorns and LNBs. If the impedances are matched between the antenna, within the feedhorn components, between the feedhorn and LNB, and internally within the LNB, losses will be minimized. Any weak point in this chain will seriously degrade system performance.

Older LNBs used a component called a ferrite isolator to match impedances. This device allows signals to enter an LNB but absorbs most reflected power and therefore acts as an impedance matching device. It has been one of the more expensive components of LNBs. When the isolator is removed, LNB noise temperature could be as much as 20°K lower than an equivalent unit. Although some have argued that this advantage is lost because the impedance mismatch caused by eliminating the isolator results in additional signal attenuation and noise, recent measurements have shown that this is not the case.

TABLE 2-6. CHARACTERISTICS OF COMMONLY USED COAX

Cable Type	Signal Loss (db/100 feet) 100 MHz	1.45 GHz	4 GHz	Impedance (ohms)
RG-59	3.40	11	N/A	75
RG-6A	2.70	8.7	N/A	75
RG-11	2.30	7.0	N/A	75
RG-8A	1.90	23.0	50	
RG-213	1.90	21.5	50	
RG-214	2.30	21.5	50	
9913	N/A	11.0	50	
9914	N/A	13.0	50	

Signal Losses

Coax is rated according to decibels of loss per unit distance. These unit distance losses are frequency dependent. Some cables like RG-6 which have perfectly acceptable losses at 1 GHz are unusable at 4 GHz. RG- 214 has an attenuation of 21.5 dB per 100 feet (0.215 dB per foot) at 4 GHz. So relaying a signal of 4 GHz on RG-214 coax over a 14 foot distance results in a loss of 3 dB or a halving of signal power. When using 950 to 1450 MHz block downconversion systems RG-6 loses 8.7 dB per 100 feet at the higher frequency end, so a run of 150 feet will result in 13.1 dB attenuation. It is clear that cable runs should always be as short as possible. Runs of up to 350 feet are not uncommon when using sweep tested RG-6 cable. If this distance is exceeded the alternatives are to either install line amplifiers or RG-11 cable that has lower losses than RG-6.

Minimizing cable lengths has a number of advantages. The main benefits are lower cost, improved signal-to-noise ratio and, for ground mounted antennas, less trenching. Note that in those cases where the signal- to-noise ratio generated by the antenna/feed/LNB is inadequate for transmission over a long cable run, a line amplifier may be required to boost the IF signal power.

Values for signal loss over distance at frequencies of 100 MHz, 1450 MHz and 4 GHz in cables often used in satellite TV installations are listed in Table 2-6.

TABLE 2-7. CALCULATED MAXIMUM CABLE ATTENUATION (dB/100 feet)

Frequency (MHz)	Cable Type RG-59	RG-6	RG-7	RG-11
5	0.52	0.42	0.33	0.26
55	1.76	1.40	1.11	0.89
83	2.16	1.73	1.37	1.11
187	3.28	2.62	2.08	1.69
211	3.49	2.79	2.22	1.80
250	3.81	3.05	2.42	1.97
300	4.18	3.35	2.67	2.18
350	4.53	3.64	2.89	2.36
400	4.86	3.90	3.10	2.54
450	5.17	4.15	3.31	2.71
500	5.46	4.39	3.50	2.87
550	5.74	4.61	3.68	3.02
600	6.01	4.83	3.86	3.17
650	6.26	5.04	4.03	3.31
700	6.51	5.25	4.19	3.45
750	6.75	5.44	4.35	3.58
800	6.99	5.63	4.50	3.71
850	7.22	5.82	4.66	3.84
900	7.44	6.00	4.80	3.96
950	7.65	6.18	4.95	4.08
1000	7.87	6.35	5.09	4.20
1450	9.59	7.77	6.25	5.18

In general, the higher the frequency of the signal the greater the losses (see Tables 2-7 and 2-8 and Figure 2-60). Errors made when installing cables and connectors can cause more deterioration in performance when higher frequency signals are used. To illustrate, if a coaxial cable is not properly connected to the output of an LNA, there will be over 70 percent more losses at 12 GHz than at 4 GHz. Modern satellite receivers which downconvert at the dish have two advantages over earlier models which relayed the higher frequency 4 GHz signals in-doors: less money is spent on coax and problems caused by potential installation errors are minimized.

The losses in each cable type can also depend on construction details such as the diameter of the central conductor. For example, Alpha cables 9059 and 9803, both RG-59 designations, have different diameter central wires. As a result, they have different attenuations at 900 MHz of 10.7 and 10.2 dB per 100 feet, respectively. Table 2-9 shows some examples for cable manufactured by Belden. The attenuation values at 400 and 900 MHz are applicable to some of the older, lower frequency block downconversion systems.

TABLE 2-8. RECOMMENDED COAXIAL CABLE SIZES

Maximum Usable Frequency (no amplification)	Cable Lengths (Meters) 25	50	100
70 MHz	RG-59	RG-59	RG-6
950 MHz	RG-6	RG-6	RG-11
1,450 MHz	RG-6	RG-6	RG-11

Losses in excess of rated values can also occur in coaxial cables. If cables are bent too severely, the impedance at the sharp bend will be changed and reflective losses could occur. To illustrate, RG-8A has a foam dielectric and keeping the central conductor aligned directly down the middle of the grounded shield is difficult if a wide turning radius is not used, particularly when the cable gets hot. Misalignment causes the impedance to change at this sharp turn. A suggested minimum cable turning radius is 5 cable diameters.

Losses can also be substantial where connectors join cables. If these connectors are not attached properly, impedance mismatches and resulting losses as well as ingress interference can occur. It is important to examine the inside of each

TABLE 2-9. COAXIAL CABLE LOSSES & CONSTRUCTION

Cable Type	Model	Shield Type	Attenuation (dB/100 ft) 400 MHz	900 MHz
RG-6	8228	Foil & Wire	4.5	6.9
RG-6	9248	Foil & Copper Braid	4.5	6.9
RG-11	9230	Foil & Wire	3.2	5.2
RG-11	9292	Foil & Copper Braid	3.2	5.2
RG-59	8241	95% Copper	7.1	10.9
RG-59	9275	Foil & Wire	5.4	8.4
RG-8	9846	80% Copper	3.8	6.0
RG-8	8214	97% Copper	4.2	6.7
RG-21	38267	97% Copper	4.7	8.0
RG-21	48268	Silver Covered 98% Copper	4.7	8.0

connector before mating. This will ensure that the center pin or conductor is not broken off and that it is extended far enough out from the connector to make electrical contact but that it is not too far to short out to a chassis causing damage when attached.

In a C-band satellite system the cable run from an antenna to the indoors electronics nearly always carries frequencies in the 950 to 1450 MHz range. RG-6 should and generally is used. RG-59 coax should only be used after the satellite receiver when distributing the 70 MHz television signal. (Usually a "launch amplifier" is used to boost the signal for distribution).

To summarize, coax should be carefully selected so impedances are properly matched and so the frequency carrying ability is adequate. The distances between the LNB and receiver should be minimized. Also, connectors selected must be rated to carry the frequencies chosen and all connections must be secure and waterproof. As a rule, never make a cable connection that will be below ground level and always use silicon grease where possible.

Water and Aging

Coaxial cable can be destroyed by intrusion of water, especially salt water. Any water leaks at connectors or along the cable body can short out a signal and, perhaps, damage a receiver or downconverter by shorting the LNB power to ground.

Underground moisture that comes into contact with any components of buried cable can cause very rapid corrosion. It is reported that tubular aluminum outer conductors have been almost completely destroyed by water within 90 days. Even small pinholes in jackets can allow this chain of events to occur. Poor installation or cable handling techniques or even rodents can cause such damage.

Most direct burial cables on the market today are often sufficient for the job. But conduit should be used to protect all buried cables. Most cables that are considered to be direct burial types cannot withstand the damage caused by gophers, moles, ground squirrels or digging equipment such as shovels and forks. For even more protection, some commercial installations use, when necessary, more expensive cables having a flooding compound under the jacket to protect against water intrusion.

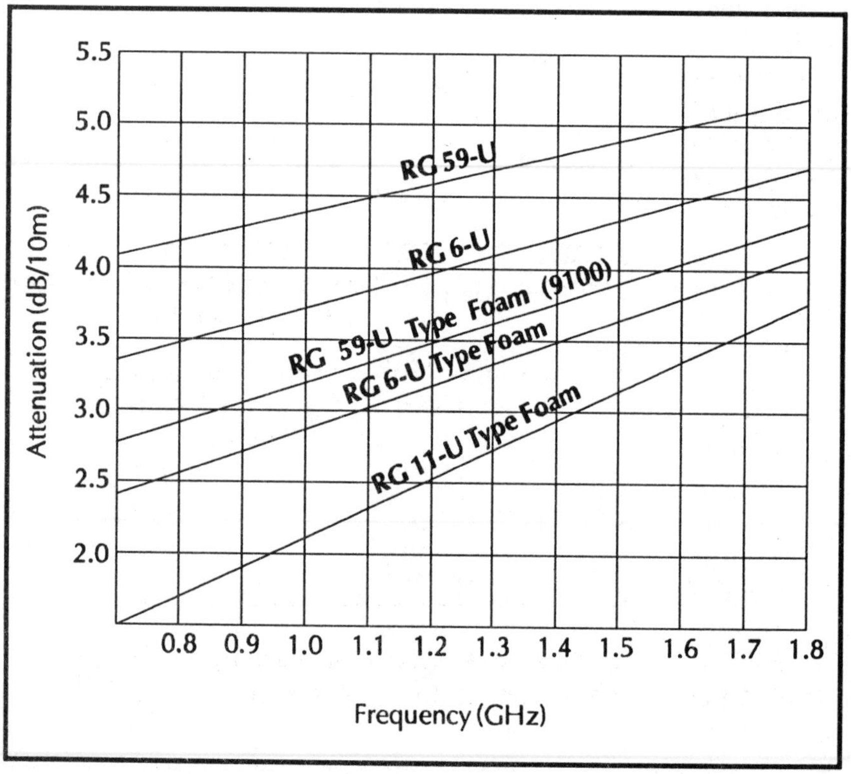

Figure 2-60. Cable Attenuation versus Frequency. *Signal losses per meter of coaxial cable are graphed here. The attenuation per meter is measured in dB per 10 meters for the range of frequencies typically encountered in home satellite systems.*

H. MULTI-RUN CABLES

Cables for satellite TV systems are generally available in a multi- stand form as a ribbon called "one run" or "dual run" (also known as direct burial cable). Single run ribbons incorporate one coaxial cable, a 3-wire polarity servo motor control run and a 5-wire actuator power/control cable all bound together. Dual run cables have two coaxial lines (see Figure 2-61).

Ribbon cables are much easier to install than individual wires. Tangles are eliminated and there need be no concern for one or more of the individual cables being too short. It is generally wise practice to use dual run cables if an upgrade to a dual band system might be a future possibility. In some cases, installing an additional 3 or 4 conductors might be considered if electronic switches, known as coaxial relays, will later be used in dual-band systems (see Chapter VII).

The individual wires in multi-run cables that relay power to drive servo or actuator motors and transmit pulses to count actuator position have improved during the past few years. In the earlier days of satellite television, solid wires (Romex) were used. However, the multi-stranded wires now incorporated into ribbon cables for satellite television use are generally less likely to break and can have lower resistance per unit length. Multi-strand wires are also better able to contain the signal within the confines of their insulating jacket than solid core conductors. As a result of these improvements, longer runs can be used.

Polarizer control wires should be contained in a grounded sheath and be of the minimum gauge as outlined in Table 2-10. In dual-band installations where the lines to two polarizers are wired in parallel, the recommended cable diameter should be increased by two gauge numbers (e.g. 20 to 18 gauge).

The actuator line consists of two #14 gauge 36 Vdc motor wires and three #20 gauge sensor lines including one for a ground connection. Such cables are usually adequate for runs up to about 60 meters. For longer distances of up to 400 feet, #12 gauge can be used to power the motor.

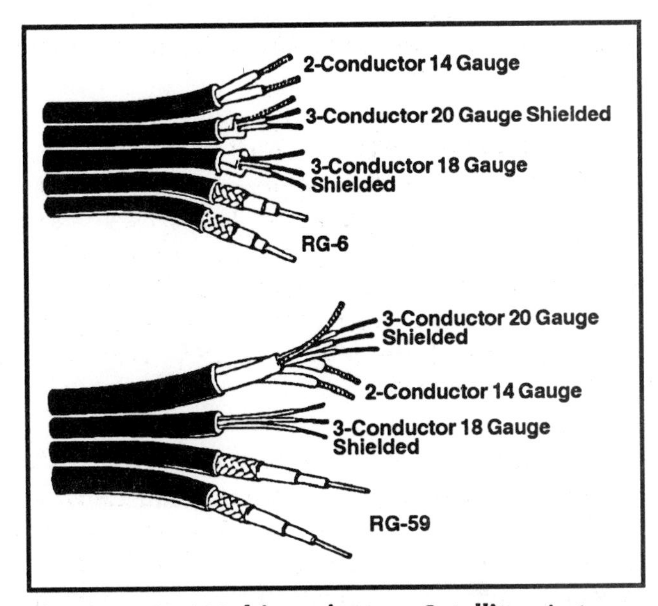

Figure 2-61 Multi-conductor Satellite Antenna Cables. *A ribbon type of flex cable is the commonly used multi-conductor cable in satellite installations. These have all the necessary coaxial cables and voltage carrying conductors to control and receive the signals to and from the satellite antenna. This cable uses dual RG-6 with 67% braid coverage with 20 gauge center conductor with 3-conductor 18-AWG, 3-conductor 20-AWG and 2-conductor 14-AWG wires. All these cables are stranded and shielded with a ground wire. (Courtesy of M/A COM)*

TABLE 2-10. WIRE SIZES RECOMMENDED for POLARIZERS

Maximum Cable Length (meters)	Wire Gauge (shielded)
25	20
50	18
50 to 100	16

Cable runs to the actuator and LNB should be minimized (see Table 2-11). When runs exceed 100 meters, heavier gauge lines must be used to prevent problems such as servo hunting or incorrect actuator counting. (These phenomena are explored in detail in Chapter VIII).

One final note. Some manufacturers supply an especially flexible "cold-weather" type of cable because working with cables can be difficult in very cold weather. More details on cold weather installations are explored in Chapter V.

TABLE 2-11. WIRE SIZES RECOMMENDED for ACTUATOR CONTROL CABLES

Maximum Cable Length (meters)	Wire Gauge Motor	Wire Gauge Shielded Sensor
25	16	20
50	14	20
100	12	20

I. SATELLITE RECEIVERS

Purpose

A satellite receiver selects a channel for viewing and processes the signal into a form acceptable by a television or TV monitor.

The modern satellite receiver, which is light, small and attractively packaged, is the control station for a home satellite reception system. Gone are the days when receivers were large, clumsy devices whose weight was almost a measure of cost.

All satellite receivers share the same basic task in preparing the signal captured by a microwave dish for viewing on a TV or listening on a stereo. However, just as one may drive to the theater in a Porsche or a Volkswagen, so consumers can select from among basic or highly sophisticated equipment.

Operation

A satellite receiver consists of a downconverter, final IF stage, discriminator, video and audio processor and, in most cases, a built-in modulator (see Figure 2-62).

Internal Downconverter

This stage lowers the frequency to a final IF, typically 70 to 612 MHz and accepts a voltage from the tuner to select channels as described in the previous sections. Receivers use either single, dual or block downconversion formats. The block downconversion methods have the advantage of allowing use of multiple receivers for simultaneously viewing of different channels on two or more televisions.

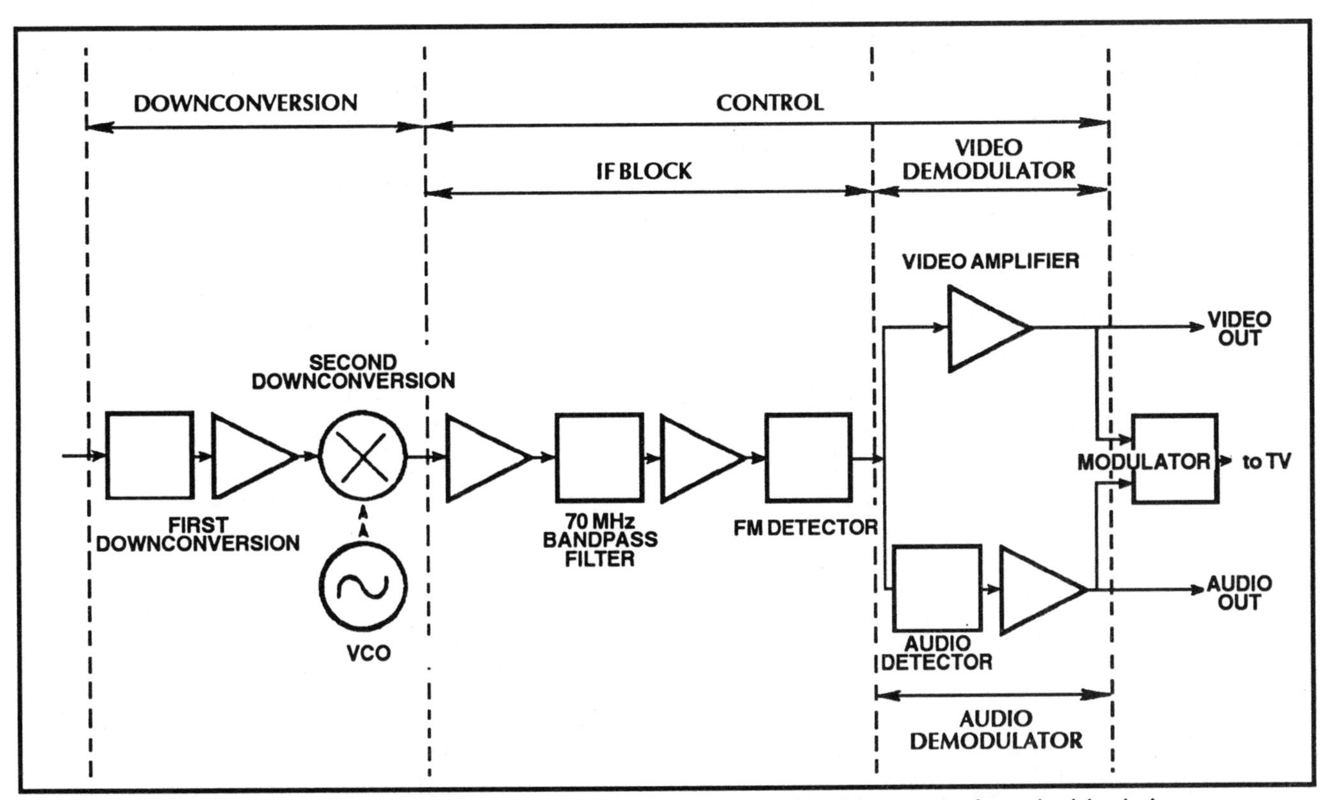

Figure 2-62. Satellite Receiver Block Diagram. *A receiver accepts the IF frequency from the block downconverter and selects one channel by downconverting one portion of the satellite bandwidth to a single IF of typically 70, 130, 134 or 612 MHz. This band centered on the IF is then amplified and fed to the detector/demodulator where the FM carrier is removed and the signal is converted to baseband. These signals are than remodulated so that AM video and FM audio can be transmitted to the TV set.*

The standard 70 MHz IF is an arbitrary choice. Another frequency could and is occasionally used. For example, DX receivers use both a 130 MHz and 134 MHz IF. Hitachi and Maspro components are designed for 612 MHz and 400 MHz IF, respectively. However, this generally accepted standard arose because when satellite receivers were being developed a circuit called a phase lock loop (PLL) detector, required for video receivers, that operated to a maximum frequency of 35 MHz was available. These electronics had been used in the telephone microwave industry. The 70 MHz satellite IF was divided by two and conventional, lower cost components were used. Today, PLL detectors operate up to as high as 612 MHz. Many receiver manufacturers have retained this relatively low 70 MHz IF because cable losses are lower than if a higher frequency PLL detector were used.

Final IF Stage

This stage, which usually operates at a final IF of 70 to 612 MHz, is composed of a bandpass filter and an amplifier. The bandpass filter sets the channel bandwidth at 36 MHz or less by selectively eliminating any out-of-band signals. The amplifier restores losses incurred during downconversion stages and strengthens the signal for the next stage.

Detector/Demodulator

The detector/demodulator circuits process the FM modulated satellite TV signal into a form called the baseband signal. This baseband signal has all the original audio and video information contained in approximately a 10 MHz bandwidth. This baseband signal is used as input to stereo processors and decoders.

Two popular types of demodulators each having advantages and disadvantages are in use today. The phase locked loop (PLL) is more capable of detecting weak signals and of discriminating between the desired signal and interference. Less expensive PLL circuits can exhibit a tendency to produce slightly fuzzy video. In more extreme cases, bright colors or high contrast scenes will tear or streak and the picture may flicker. The coaxial delay line is not as effective in discriminating between the satellite signal and interference and of detecting marginal strength signals as the PLL. However, when signal strength is adequate, this type of demodulator delivers sharp, crisp pictures with well defined colors.

Video and Audio Processors

The video processor removes a 30 Hz signal called the energy dispersal waveform and, in more advanced designs, tracks and corrects black and white levels in the video signal. The dispersal waveform was added to satellite broadcasts by order of the FCC. It gives these signals a distinguishable identity from terrestrial microwave sources sharing the same frequency band.

The video processor delivers the 0 to 4.2 MHz baseband video information to an amplifier. The audio processor selects audio information ranging in frequency from 30 to 15,000 Hz from a selected subcarrier. Satellite transmissions relay audio signals on subcarriers ranging from 5.0 to 8.0 MHz. The audio information required to supplement TV pictures is usually carried on a 6.8 MHz subcarrier which can be selected by a knob on the receiver front panel.

Internal Modulator

A modulator is required to "rebroadcast" the raw audio and video signal in an AM form which can be understood by a conventional TV (see Figure 2-63). Off-air channels, unlike FM satellite signals, are amplitude modulated. The selection of modulation frequency determines which television channel will receive the satellite programming. Channels 3 and 4 are typically selected.

The baseband signal, which contains all the video and audio information required to recreate a television picture, can be fed directly into a TV monitor without use of a modulator. However, a conventional TV, has built-in demodulation circuits designed for picking up off-air broadcasts. Unlike a TV monitor, it can accept only modulated signals. Note that higher quality reception will probably be attained with a monitor by eliminating these redundant steps. A home video cassette recorder (VCR) also processes the raw baseband signal and has its own built-in modulator driving a conventional television. VCRs can make higher quality tapes from satellite baseband signals than from demodulated and remodulated off-air broadcasts.

How Receivers Select Channels

Every satellite is capable of relaying up to 24 television channels plus many audio subcarriers per channel. A receiver selects a channel by first directing the probe rotation circuits to choose the desired polarity.

The tuner sends a voltage to a VTO (voltage tuned oscillator) either at the dish as is conventionally the case or to the correct circuit in the block downconversion receiver. This voltage causes the

Figure 2-63. Inexpensive Modulator. *Modulators that are factory installed into home satellite receivers have evolved from the same type that are built into video tape recorders. While they have adequate performance and are small enough to easily fit into home satellite receivers, they are not suitable for SMATV installations which have multiple adjacent channels. (Courtesy of Brent Gale)*

VTO to generate the proper mixing frequency for channel selection and then plucks the 36 MHz wide channel band down to the final IF. Earlier receivers and some commercial models today are set on one frequency; some semi-agile receivers permit channel selection by interchanging frequency-set crystal oscillators on their back panels. All home satellite receivers are fully agile.

Channel selection falls into two categories: detent and synthesized tuning. Detent tuners, like some of the earlier Avcom receivers, can use either a continuously variable device such as a potentiometer. Other brands use a switch which clicks between various preset resistors (eg. the Drake 324 receiver) to send the required voltage to a VTO.

Synthesized tuners store the required mixing frequency in more stable devices than resistors. For example, the Houston Tracker V has a microprocessor which generates the correct frequency. The Maspro uses crystal oscillators specifically designed to send the exact frequency to the mixing circuits. Such higher quality synthesized tuners are much less subject to drifting between satellite channels than are detent tuned receivers.

Note that the presence of a push button rather than a knob for channel selection does not necessarily mean that synthesized tuning hass been incorporated in the receiver.

Judging Receiver Performance

The quality of a receiver is ultimately judged by clarity and fidelity of the television picture and crispness of the sound. These are measured by the video bandwidth, receiver threshold and tuning method used. These, in turn, are all determined by the care taken in designing and manufacturing as well as the choice of circuit components.

Video Bandwidth

Satellite broadcasts are relayed with a full 36 MHz bandwidth. However, a watchable television picture can be reproduced using a bandwidth as low as 15 MHz. Both fidelity and clarity vary with bandwidth. Half transponder broadcasts occupying only 18 MHz are commonplace on some Intelsat and European satellites. This technique allows more channels to be squeezed onto the available satellite bandwidth.

Picture fidelity worsens as receiver bandwidth is narrowed. Picture details are slowly lost and color shades begin to vary. However, receivers having more narrow bandwidths also detect less noise power so unwanted "sparklies" are avoided. Thus clarity can be traded off for picture fidelity. It has been found that the picture quality using a 24 to 28 MHz bandwidth is virtually indistinguishable from that with a 36 MHz bandwidth. So this lower bandwidth has been adapted as an industry standard for satellite receivers. Home satellite receivers often use a bandwidth more narrow than 28 MHz to minimize noise. Some brands even come equipped with a front-panel adjustable bandwidth knob, a very desirable feature.

The judgment of what constitutes reception quality is of course very subjective. Consumer studies in which picture quality was rated according to signal to noise ratios show that although judgments vary, they are related to receiver threshold.

Receiver Threshold

The threshold of a satellite receiver determines how weak an input signal, measured by the carrier power to noise power ratio (C/R), can be before a picture is judged unacceptable. The concept of C/R can be clearly understood by imagining yourself in a room full of people. If everyone were speaking and the noise level were high, someone would have to shout to be heard. If it were so quiet that you could hear a pin drop, a whisper would suffice.

If the C/N fed into a receiver is plotted against its output signal to noise power ratio, S/N, a straight line results. This linear relationship means that for a given change in input, there is a proportional change in output. To illustrate, if an input of 1 watt results in an output of 5 watts, then an input of 2 watts creates a 10 watt output. A simple example shows why this linear relation between input and output is important. A photograph of a scene having one area twice as bright as another would not look proper if the reproduced picture had that area only 50 percent brighter, or if the reproduction was not linear.

Threshold is measured at that point where the deviation from a linear or straight line plot is one decibel (see Figure 2-64). Near or below threshold "sparklies" begin to appear. Subjective quality studies have shown judgments of picture quality as shown in Table 2-12.

Most satellite receivers have thresholds at C/Ns of about 8 decibels, which is the value chosen as an example in this table which shows that a signal of 5 dB would generate an unacceptable picture.

"Extension" techniques are available to reduce video threshold. Using the best possible circuit components that contribute the lowest amounts of noise to a signal being processed is by far the most effective method. A second method is to reduce video bandwidth. This results in some loss of picture fidelity. Another method uses a circuit that not only switches on and reduces bandwidth when the signal is too weak for clear reception, but also "tracks" and centers its reduced bandwidth around the frequency where the signal is the strongest. These bandwidth and automatic frequency centering fixes may not work in some cases. For example, if a strong interfering signal is present the circuit may track the wrong signal.

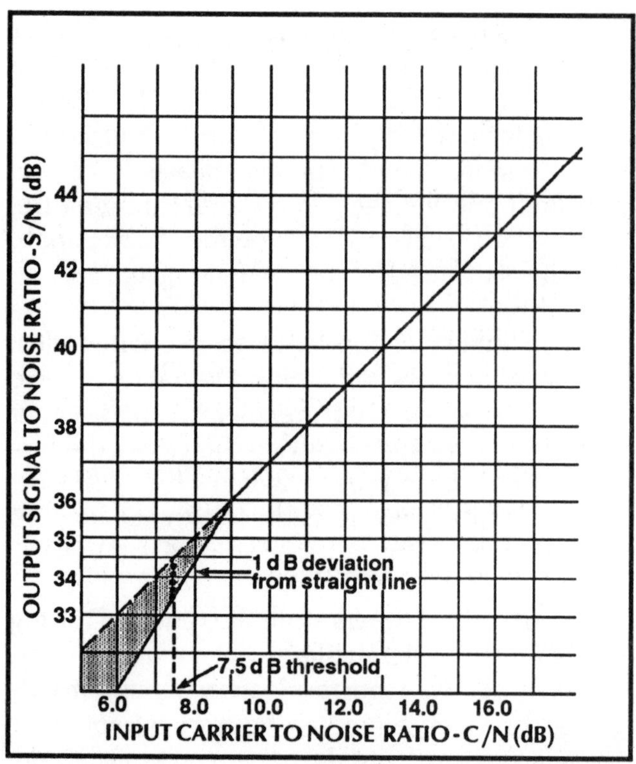

Figure 2-64. Receiver Threshold. *A TVRO should provide a sufficient carrier-to-noise ratio to exceed the receiver threshold. Threshold is measured at the point where the departure from linearity is 1 dB.*

TABLE 2-12. PICTURE QUALITY & RECEIVER THRESHOLD (Threshold chosen as 8 dB)

Signal-to-Noise Ratio (dB)	Picture Quality
5	Extremely noisy, tearing, audio noise
6	A little better, sparklies
7	Watchable but sparklies
Threshold	A few sparklies
9	Very good picture; sparklies on only saturated colors
10	Video tape quality
11	Cable TV quality

Electrical Connections

Hooking up a satellite receiver is relatively simple. All receivers have an IF input from the downconverter and an RF output to televisions. Both these terminals almost always use standard F-connectors. Three polarity selection circuit wires, ground, pulse and 5.7 Vdc, are usually attached via screw connections. A separate set of appropriate screw connectors for a DC motor polarity selection device is sometimes also provided. If the receiver has a built-in actuator, then the four or five motor and counter terminals must be attached. Downconverter power connections vary. Some receivers relay downconverter power over the IF line; some use two wire screw-on connectors.

Other output terminals are available. The unprocessed baseband signal, also known as the composite or unfiltered video, is usually accessible for relay to a stereo processor or descrambler. Some receivers with built-in stereo processors do not feature this output but should. Such receivers must have two RCA-jacks or their equivalent for connecting the left and right audio channels for relay to a stereo receiver. Often an RCA-jack output is provided to allow connection of an external signal strength meter.

User Interfaces and Adjustments

All receivers feature some standard controls including power on-off switch, channel selection, audio subcarrier selection, polarity and skew adjustment and, usually, video fine tuning. Many other controls occur on various brands. These include switches or buttons to select stereo format, channel scan mode, narrow/wideband audio filters, and video bandwidth control. Some receivers come equipped with a signal strength meter and, occasionally, a frequency centering meter.

However, the trend has been towards simpler customer interfaces. Most receivers are now available which automatically fine tune both the video and polarity and have no customer interfaces to do so. Programmable units are completely controlled via simple-to-operate, hand-held remotes as in, for example, the Chaparral Monterrey 90™, Houston Tracker VIII and many other receivers.

J. DECODERS

Scrambling has been a controversial issue in the home satellite TV industry. All agree that program producers must be paid for their efforts. It is very expensive to produce quality television programming. While consumers should pay for this service, many have argued that programming packages are not yet fairly priced.

In order for the program suppliers to control who decodes their signals, scrambling systems must be capable of "addressing" the descramblers in the field. In the simplest case where each unit in a dwelling complex is served by a separate cable, this can easily be accomplished by simply disconnecting this cable. But satellite broadcasts link the end users only by electronic signals. Therefore, each end-user such as a cable TV headend or a home satellite system must have a decoder that can be turned on or off from a central uplink. While it is a relatively easy matter to build non-addressable devices, addressable systems require a high level of sophistication.

Two scrambling systems are presently used in North America, the General Instrument VideoCipher II and II-Plus and the Oak Industries Oak Orion. Scientific Atlanta has also developed a system known as MAC (shorthand for multiple analog components) which is used in some commercial systems and in other regions of the world such as Australia and Europe. All these scrambling systems work according to a simple basic principle, the video and/or audio signal is distorted in a controlled way so as to be unintelligible except when a decoder is used to reconstruct the broadcast (see Figures 2-65 and 2-66). Those readers interested in exploring this topic in more detail can consult *Satellite and Cable TV - Scrambling and Descrambling or World Satellite TV and Scrambling Methods - The Technician's Handbook.*

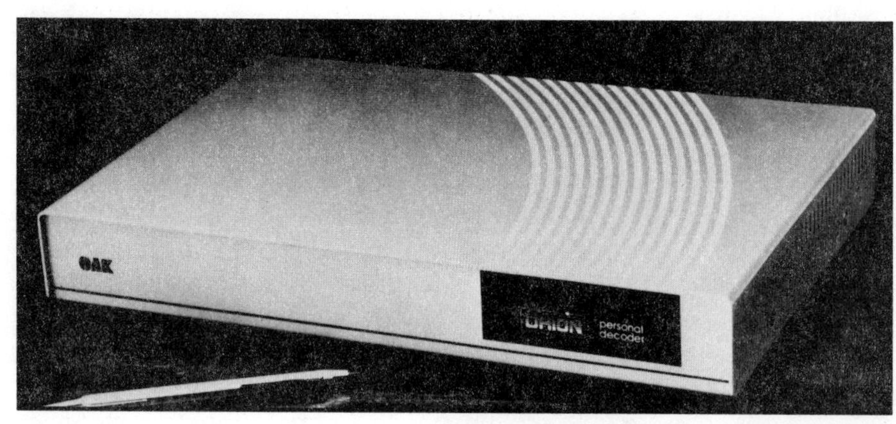

Figure 2-65. Orion-PD Personal Decoder. *This small, set-top decoder used primarily in Canadian direct-to-home applications, incorporates a channel 3/4 internal modulator. The features of this decoder match those of all other Orion descramblers. LSI microcircuits decode the digitized audio and data as well as the video information while accepting a variety of receiver video level inputs. (Courtesy of Orion Industries)*

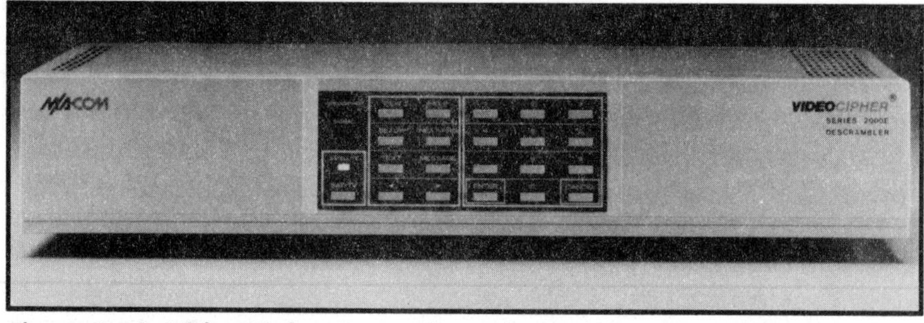

Figure 2-66. VideoCipher II. *The VideoCipher series 2000 stand-alone decoder is available as either baseband/RF model 2000 E or the baseband-only model 2000 E/B. The RF input level must fall between 55 dBmv and 0 dBmv (3 microvolts and 1 millivolt). The baseband input should be close to 1 volt. This unit has been incorporated into most North American home satellite receivers. (Courtesy of General Instrument)*

K. TELEVISIONS

The purpose of a television set is to recreate the original broadcast picture and sound as accurately as possible. A picture is created by an electron beam scanning across a screen of phosphor dots. The more intense the electron beam, the more illumination is produced for the viewer. This organized scanning occurs whether or not a video signal is present. When a video signal is received, it varies the scanning beam intensity and the resultant changes in illumination produce a picture. Without a video signal, the television screen will have a random but uniform pattern of dots ranging from black to white.

Thus a picture is "painted" line by line onto the face of a television. Information about each line's brightness (luminance) and color (chroma and hue) is translated into the video signal for interpretation by the TV. Scanning begins at the top left-hand corner of a screen. Following each horizontal scan (one line), the beam is turned off as the line is traced back to the beginning of the next one.

When the bottom of the picture is reached the beam is again turned off. These "down times" are called the horizontal and vertical blanking intervals, respectively. During these periods other information on the video signal such as teletext or captions for the hearing impaired can be relayed.

Synchronization signals are also an important component of the video signal. The horizontal sync sets the start of the horizontal scanning; vertical sync sets the timing for the beginning of vertical scanning.

Broadcast Formats

Various standards have been adapted worldwide for setting the number of scan lines and the timing of their occurrence of a TV screen. For example, the standard used in France called SECAM and the PAL format in many other countries scan 625 lines 25 times each second compared to 525 lines 30 times each second in the American format. As a result, their pictures are "cleaner" and a little closer to movie quality but tend to have an annoying flicker because of the slower scanning rate.

The various video formats are examined in more detail in the next chapter. Here we examine only the NTSC (National Television System Committee) standard used for color TV in the U.S., Canada and Japan. Note that TVs capable of switching between video formats are available especially since the introduction of sophisticated digital televisions, "TVs on a chip."

NTSC format has 525 lines scanned 30 times each second. In order to eliminate flicker, an interlaced method of scanning is used whereby two fields each having half of 525 lines, or 262.5 lines, are alternately impressed upon the screen 60 times each second. The total number of lines traced each second is 15,750 (60 times 262.5 or 30 times 525). Two sequential fields thus equals one frame.

The composite video signal is composed of the video signal, the blanking information and horizontal and vertical sync pulses. The video signal contains brightness (luminance) and color (chrominance) information.

High Definition Television

High definition television (HDTV) is a new form of high resolution TV presently under development by Japanese as well as American companies. It uses 1150 scan lines instead of the conventional 525 in the NTSC format. The first probable adapted use will be in high powered direct broadcast applications in the Ku-band where newly designed equipment will be more easily introduced.

How to Judge TV Quality

Satellite broadcasts are capable of producing near-perfect quality video on an excellent television. Picture clarity and fidelity can even be remarkably good on an older, noisy set. Probably the best method used to judge TV quality is by eye, that is by actually viewing a satellite program on the set.

A better picture will be obtained on a TV which has been design well and manufactured with the best available components, and which is adjusted properly. The most overlooked adjustment which should be done in the field is to fine tune the center frequency of the channel used for satellite modulation. This channel has often not been used so the TV set may not be tuned to channels 3 or 4.

TV Monitors vs. Conventional TVs

TV monitors are driven by the raw video, the video signal reduced to its lowest common denominator. A monitor is a conventional TV without a tuner section. The monitor usually provides a much sharper picture because these components unavoidably add some extra noise.

Tuning between Channels

Conventional TVs accept amplitude modulated (AM) video signals and then select channels by a tuner. The FCC has been careful not to allocate adjacent channels in any given broadcast area. If this were not the case, some "bleedover" between channels could occur.

When satellite signals are modulated onto channels next to off-air broadcasts, problems can occur. If the tuner is not high quality and if the satellite video signal has not been passed through a good bandpass filter, this bleedover or interference between channels characterized by "herringbone" or "venetian blind" patterns could occur. This bleedover can be overcome in a number of ways. A simple A/B switch with 60 decibels of isolation or more can be used so that only satellite or off-air television is fed into the television at one time. Or a signal combiner which has a built-in bandpass filter, usually for channels 3 or 4, can be used to mix satellite and other signals. Obviously, a color monitor dedicated to just the satellite baseband output would be the simplest and most effective method.

L. STEREO PROCESSORS

The satellite baseband signal contains raw video and audio information spanning a band from near zero to 10 MHz. Since the video signal is contained in a more limited zero to 4.2 MHz range, a great deal of audio information can be carried in the remaining bandwidth.

Audio broadcasts are transmitted in the range from approximately 5.0 to 8.5 MHz on subcarriers. For example, if the stereo receiver audio frequency control knob is set to 6.2 MHz, an audio signal centered on this frequency not necessarily related to the television picture might be heard. To illustrate, Galaxy I, transponder 3 or Satcom 3, transponder 6 are loaded with audio subcarriers (for further information see *The Hidden Signals on Satellite TV*). The sound accompanying a video broadcast on a particular transponder is usually relayed on a 6.8 MHz subcarrier. Satellite TV is more appropriately satellite radio and television since so much audio information is carried along with the video.

The satellite baseband signal relays many of these radio stations and audio for some TV programs in stereo. Some video receivers have built-in circuits to process this stereo into a form where it can be fed into a home stereo receiver. Separate stereo processors can also accept the raw, unfiltered signal from a satellite receiver and prepare it for stereo listening.

Stereo Formats

There are presently five types of stereo formats. The discrete or brute force method transmits two separate subcarriers for the left and right channel sound. These subcarriers are located in the range from 5.0 to 8.5 MHz. For example, The Disney Channel on Galaxy I uses 5.8 and 6.8 MHz subcarriers for both audio tracks. A second more sophisticated discrete technique has been developed by Wegener Engineering and is used by the Disney Channel and the Nashville Network, among others.

The matrix method also requires two separate subcarriers. One channel contains the left plus right audio intelligence (L + R) and the second contains L - R. The stereo processor then algebraically combines the two to produce a stereo output.

The multiplex stereo system, patterned after FM stereo technology is somewhat more complex. Both audio channels are transmitted by one subcarrier on the baseband signal. It uses an FM subcarrier for the L + R audio signal, and a double sideband suppressed carrier for the L - R information. A 19 KHz synchronizing signal is also relayed for stereo demodulator reference to aid in recovering the original information. This signal reproduces either a monaural or stereo broadcast.

The FCC first selected this multiplex radio stereo broadcast method in 1961. In March, 1984, the FCC authorized use of stereo audio for conventional off-air TV broadcasts and this stereo sound format was suggested as a de-facto standard. However, multiplex stereo is rarely used in satellite broadcasts.

The fifth method to broadcast stereo has been used in some of the more advanced scrambling systems such as the VideoCipher II system. The analog sound is converted to a digital signal and then actually embedded in the blanking intervals of the video signal. This method produces very clean, low noise sound.

M. MISCELLANEOUS COMPONENTS

Satellite TV systems use a variety of smaller interface components such as A/B switches, matching transformers, signal combiners, splitters, line amplifiers, pads, DC blocks and terminators. These and other related components are examined in detail in Chapter VII, Multiple Receiver Satellite TV and Distribution Systems.

III. TERRESTRIAL INTERFERENCE

Terrestrial interference, unfondly known as TI to satellite dealers, results when a satellite TV system detects unwanted earth-based communication signals. The effects on picture quality can range from a mild case of sparklies to complete picture wipe-out. Some forms of TI are easy to avoid or cure but severe interference is difficult to deal with and can completely ruin reception of a satellite broadcast.

Like an outbreak of acne, TI cannot be ignored. It has been estimated that 10 to 20 percent of all new satellite TV installations experience some form of TI, from mild sparklies to complete picture wipe-out.

The good news is that learning to understand and combat nearly any form of TI is not as difficult as first appearances might indicate. But it is also crucial to know when eliminating terrestrial interference might be nearly impossible or prohibitively expensive. Many satellite dealers have lost hundreds or even thousands of dollars on that unforgettable installation.

The most reasonable approach to combating TI follows a two stage strategy. First, TI should be avoided by correctly choosing satellite TV equipment and by properly locating the antenna. When this strategy fails, interference can be suppressed by using a combination of filters and man-made shields. These methods are covered in more detail in the Microwave Filter Company's publication *The ASTI Manual - The Avoidance/Suppression Approach to Eliminating Terrestrial Interference*. This recommended book is listed in the reference materials.

A. SOURCES OF TI

Any signals sharing frequencies encountered at any point in a satellite TV receiving system can potentially result in interference. This includes transmissions ranging from the microwave C-band and above to as low as the audio and video baseband range. Thus, for example, if an interfering 70 MHz signal leaks into a satellite receiver, pictures may deteriorate because processing occurs at the same frequency. Or if a nearby communicator is using a 980 MHz relay, the 950 to 1450 MHz block downconverter range might be affected.

The potential interference band can be broken down into two segments: the ingress interference band below 1 GHz; and the antenna interference band from 1 GHz to approximately 8 GHz (see Figure 3-1). The difference between these two bands is that while lower frequencies cannot enter via the antenna/feedhorn/LNB route, microwave signals in the 1 to 8 GHz range can do so.

Frequency (GHz)	Nature of Potential Offender
0.960-1.350	Land-based air navigation systems
1.350-1.400	Armed Forces
1.400-1.427	Radio astronomy
1.427-1.435	Land-mobile: police, fire, forestry, railway
1.429-1.435	Armed Forces
1.435-1.535	Telemetry
1.535-1.543	SAT-maritime mobile
1.605-1.800	Radio location
1.660-1.670	Radio astronomy
1.660-1.700	Meteorological radiosond
1.700-1.710	Space research
1.710-1.850	Armed forces
1.990-2.110	TV Pick-up
2.110-2.180 (1)	Public common carrier
2.130-2.150	Fixed point-to-point (non-public)
2.150-2.180	Fixed omnidirectional
2.180-2.200	Fixed, point-to-point (non-public)
2.200-2.290	Armed forces
2.290-2.300	Space research
2.450-2.500	Radio location
2.500-2.535	Fixed, SAT
2.500-2.690	Fixed point-to-point (non-public) Instructional TV
2.655-2.690	Fixed, SAT
2.690-2.700	Radio astronomy
2.700-2.900	Armed forces
2.900-3.100	Maritime radio navigation
2.900-3.700	Radio location
3.300-3.500 (3)	Amateur radio
3.700-4.200 (1)	Common carrier (telephone) Earth stations
4.200-4.400	Altimeters
4.400-4.990 (3)	Armed forces
4.990-5.000	Meteorological /radio astronomy
5.250-5.650	Radio location (coastal radar)
5.460-5.470	Radio navigation - general
5.470-5.650	Maritime radio navigation
5.600-5.650	Meteorological ground based radar
5.650-5.925	Amateur
5.800	Industrial and scientific equipment
5.925-6.425	Common carrier and fixed SAT
6.425-6.525 (2)	Common carrier
6.525-6.575	Operational land and mobile
6.575-6.87	Non-public point-to-point carrier
6.625-6.875	Fixed SAT
6.875-7.125	TV pick-up
7.125-8.400	Armed forces
8.800	Airborne Doppler radar

Interface frequencies are listed in the order of occurrence on TVROs:

1. Telephone carrier spectrum co-located with TVROs.
2. Widely distributed common microwave carriers.
3. Seldom occurring frequencies close to TVRO Band.

Figure 3-1. Potential Antenna Interference Frequencies. *The most likely sources of TI come from communicators operating in or near the 3.7 to 4.2 GHz C-band range of frequencies. (Courtesy of Microwave Filter Company)*

The Ingress Interference Band

Interfering signals below about 300 MHz in frequency usually enter through poorly grounded or improperly connected equipment. For example, a local TV station might be received along with the satellite TV signal if a ground connection is "floating" and has not been properly attached. At frequencies above 300 MHz, wavelengths are short enough that signals can leak into poorly shielded equipment cases or via breaks in cables (see Figure 3-2).

This type of interference is usually easy to cure by properly grounding cables, using wall plug filters, shielding equipment interconnects and closing unnecessary gaps in metal cases. Ingress has the recognizable effect of causing similar problems on all satellite channels seen, for example, as a characteristic picture defect or heard as an audio buzz.

The Antenna Interference Band

While ingress interference usually arises when an installation is not "tight," TI can enter via the dish in even the most professionally installed system. The antenna interference band can be subdivided into two distinct regions: in-band TI centered on 3.7 to 4.2 GHz; and out-of-band TI spanning the 1 to 8 GHz range but excluding C-band frequencies.

In-Band TI

The source of in-band interference nearly always originates from land-based, microwave repeater networks which share the C-band with satellite TV transmissions. Common carriers transmit voice, video and data traffic for a fee via such familiar microwave antennas which are usually situated on towers or tall buildings. AT&T, Sprint, MCI, and Allnet are among the better known common carriers but many regional and local companies also use C-band relays.

Figure 3-2. One Potential Source of Ingress Interference. *This 10-foot Prodelin fiberglass antenna was installed near power lines. Although these lines have little effect on the incoming satellite signal, the 60-cycle power could be a source of ingress interference. All precautions against ingress TI including proper grounding techniques should be taken to eliminate this possibility (Courtesy of Brent Gale)*

Line-of-sight C-band networks have proliferated as telephone, data and other communication needs continue to grow. Experts estimate that new license applications and requests for network modifications are being submitted at a rate of 30 to 50 every month. In North America, the growth rate has been a minimum of 10 to 20 percent yearly. Today, such networks crisscross North America like a giant spider web centered on the major metropolitan areas and spanning rural regions (see Figure 3-3). In-band TI accounts for approximately 95 percent of all interference problems.

Each antenna on a repeater station can transmit up to six different frequencies having either vertical or horizontal polarization. In addition, each tower can be fitted with multiple antennas. In some areas, only one frequency is used; in others, all 25 possible common carrier channels may create serious problems for those receiving satellite TV broadcasts (see Table 3-1).

In the United States, the FCC wisely allocated these "Ma-Bell" transmissions different center frequencies than those used by satellite TV broadcasts. The allocated frequencies lie 10 MHz above and below the center frequency used by each satellite channel. For example, channel five is centered on 3800 MHz, so common carrier relays using frequencies of 3790 and 3810 MHz can be potential sources of interference to this channel. A total of 25 frequencies are assigned to common carriers; 23 between each channel and one each above and below the satellite range.

TABLE 3-1. COMMON CARRIER CENTER FREQUENCIES

Transponder Number	Satellite Center Frequency (MHz)	Common Carrier Center Frequency (MHz)
		3710
1	3720	
		3730
2	3740	
		3750
3	3760	
		3770
4	3780	
		3790
5	3800	
		3810
6	3820	
		3830
7	3840	
		3850
8	3860	
		3870
9	3880	
		3890
10	3900	
		3910
11	3920	
		3930
12	3940	
		3950
13	3960	
		3970
14	3980	
		3990
15	4000	
		4100
16	4020	
		4130
17	4040	
		4150
18	4060	
		4170
19	4080	
		4190
20	4100	
		4210
21	4120	
		4230
22	4140	
		4250
23	4160	
		4270
24	4180	
		4290

Common carriers most often transmit signals that occupy bandwidths of 3 to 5 MHz around their center frequency. These "narrow-band" transmissions typically for telephone links can often be easily "notched out" from the signal seen by a satellite receiver. However, "wide-band" formats often used for videoconferencing links and data service networks are gradually becoming more common-

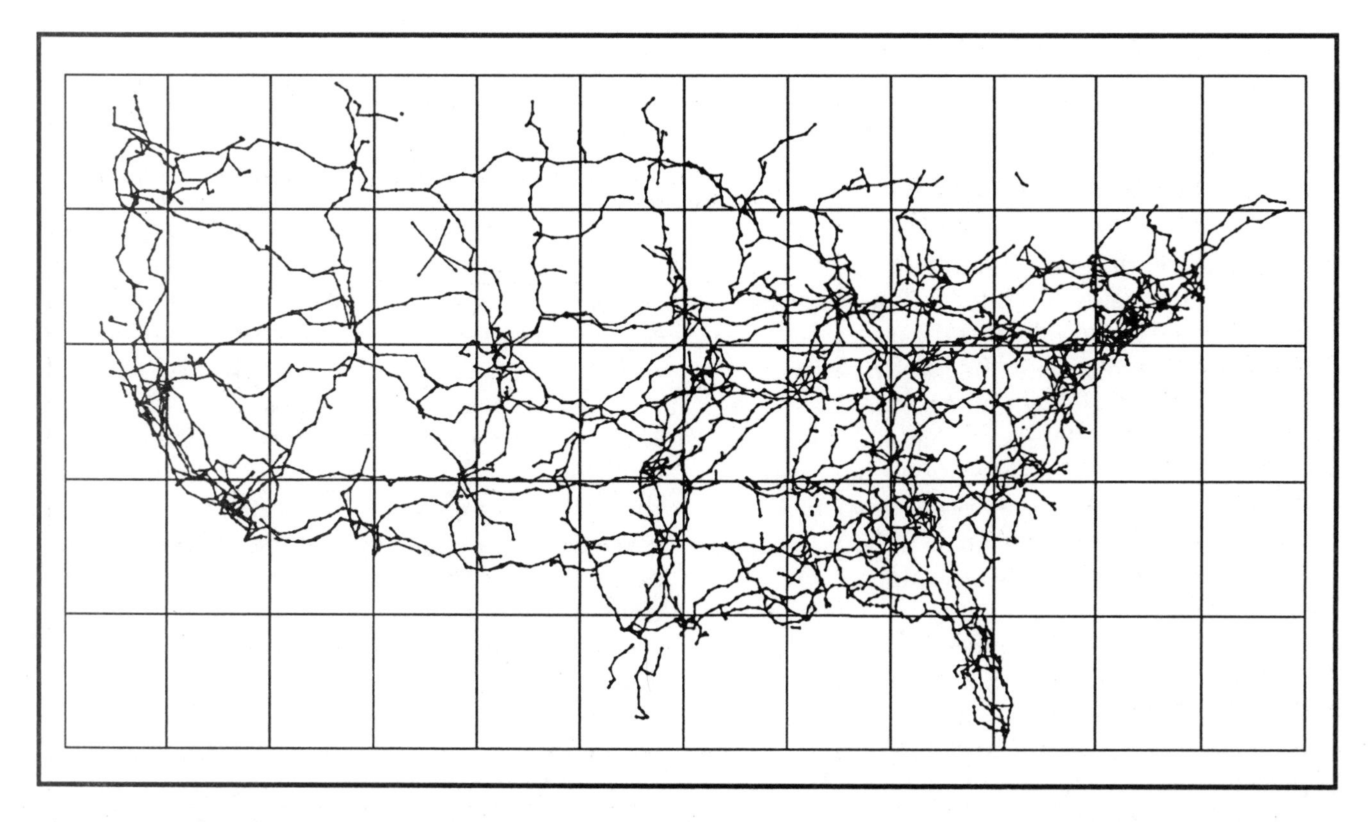

Figure 3-3. C-Band Common Carrier Network. *This map shows the locations of microwave towers and their routes. This network has more than doubled since this map was drawn. (Courtesy of Microwave Filter Company)*

place (see Figure 3-4). Such transmissions having 5 to 30 MHz bandwidths severely interfere with satellite broadcasts and cannot be eliminated with notch filters without destroying a great deal of the picture information. More expensive and sophisticated "phase cancellation" methods are required to overcome this form of TI.

TABLE 3-2 . MAJOR SOURCES OF OUT-OF-BAND TI

1.990 - 2.110 GHz	TV studios to relay towers &between metropolitan areas
2.110 - 2.180 GHz 6.425 - 6.525 GHz	Common carrier bands for relaying data
5.925 - 6.425 GHz	"Fixed SAT" for carrying remotely originating programs to uplinks for satellite broadcasts

Out-of-Band TI

Out-of-band TI has carrier frequencies that lie in the 1 to 8 GHz band. Four major communication bands account for most out-of-band TI (see Table 3-2).

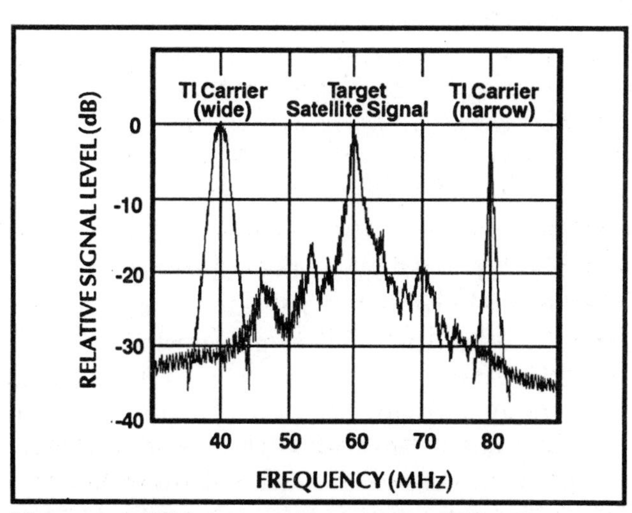

Figure 3-4. Wide versus Narrow Band TI. *The wide band interference illustrated here has a bandwidth of approximately 5 MHz.*

B. EFFECTS OF TI ON SATELLITE TV

Ingress Interference

Ingress interference characteristically affects all satellite channels with identical disturbances. It usually occurs when unwanted carriers in the 5 MHz to 1 GHz range enter the system because of poor equipment shielding, faulty connectors or improper grounding or through poorly filtered power lines. Ingress interference can usually be cured by "tightening up the system," properly grounding cables and connectors, shielding all equipment and using power line filters. The advantage of using a familiar test system, i.e. an LNA, downconverter, receiver and TV set, during installation is that it is known to be working properly so time is not wasted in a wild goose chase tracking down a bad connector or some other component. Therefore if ingress interference appears later, it can be diagnosed and easily cured. Even if ingress interference does show up during a site test, it should not be a cause for undue concern.

Ingress TI can appear when the signal from VHF TV channels 2 through 6, which occupy the 54 to 88 MHz band, leaks into a poorly shielded receiver. Co-channel interference caused by off-air channels being detected on a poorly shielded modulator-to-TV cable can also cause similar symptoms, a faint second picture in the background or "venetian blinds" across the screen. In order to avoid this problem, a shielded flat two-wire cable called a twin lead that connects the modulator in a satellite receiver to a television, should never be used. Instead, a well grounded coax should be installed.

The amateur radio band frequencies, 1.79, 3.58 and 7.15 MHz, share the same frequency region as the baseband video and the audio subcarriers. Detection of the two lower frequencies could cause herringbone patterns on the screen; interference from the latter could cause an audio buzz. As mentioned earlier, all of these irregularities would be seen on all satellite channels.

Out-of-Band TI

Out-of-band TI has characteristic effects on satellite TV reception. If its power is sufficiently high, the interference can ruin the reception of all transponders on a satellite. However, in most cases, picture deterioration caused by out-of-band TI is seen most strongly at either end of the C-band on those satellite channels closest in frequency to the interfering signal. The effect diminishes towards the opposite end of the band as the frequency separation between the TI and the transponder increases. Thus, for example, if a strong 2 GHz signal were detected by a dish and transmitted into a satellite receiver it would affect the lower channels most strongly. If this interfering signal happened to be either horizontally or vertically polarized, it would most strongly affect those channels having the same polarization. Generally, only channels having the same polarization as the TI will be affected. So a strong indicator of out-of-band TI is picture deterioration on every other channel near either edge.

The picture deterioration caused by out-of-band TI generally appears the same on all affected channels. Horizontal lines across a picture are often seen. Out-of-band TI can result when broadcasts from a local multichannel, multipoint distribution system (MMDS) are detected. MMDS systems, also known as wireless cable systems, operate at a frequency in the range of 2 GHz. Wireless cable systems broadcast signals via a local microwave transmitter to many small, "tin-can" antennas at homes of individual subscribers.

It is rare that weak to moderate out-of-band interference below 2577 MHz will disturb a satellite receiving system. The WR-229 waveguide flange on an LNB input severely attenuates signals having frequencies below 2577 MHz. However, very strong out-of-band signals can overpower the lower frequency cutoff of this waveguide and interfere with reception.

Out-of-band TI is not as rare an occurrence as ingress interference but is still quite unusual because most LNBs and LNAs have built-in bandpass filters (see Figure 3-5).

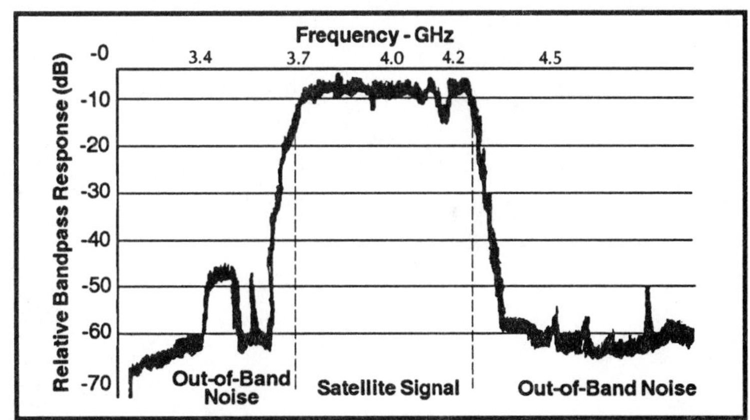

Figure 3-5. Frequency Response of an LNB with Bandpass Filter. *This drawing illustrates how frequencies outside the satellite C-band range are rejected by a bandpass filter. (Courtesy of Microwave Filter Company)*

In-Band TI

In-band TI also has characteristic signs. Since each antenna on a microwave repeater station transmits up to six horizontally or vertically polarized carriers spaced 80 MHz apart, picture distortion is often seen on every other channel. If interference from one relay were intercepted, six satellite channels spaced two channels apart could be affected. If fewer carriers are in use, as few as one channel could be disturbed. As power levels of the interfering carriers increase, channels adjacent to these can be affected.

The interference pattern across channels caused by two sets of carriers is somewhat more complex. Generally if two transmitting antennas are used, they have opposite polarity signals or widely spaced frequencies. Twelve horizontal and twelve vertical channels can be affected in these cases.

If multiple line-of-sight transmitting antennas each having different polarities and frequencies were in use, all 24 channels could be affected. Typically, patterns of good and bad channels are seen when in-band TI is present. Even in those cases where all 24 channels were affected, the degree of interference generally varies among channels.

Both wide-band and narrow-band TI have similar effects on satellite picture quality. But wide-band interference carries more power and therefore has more pronounced negative effects. For telephone relays the bandwidth is determined by how many conversations are being relayed at a time. As more people come on line the bandwidth increases until the transponder is completely wiped out. This occurs most often in the late afternoon when telephone traffic is at a peak. The varying signal strength meter often seen when TI is present reflects the changing bandwidth as thousands of people make and break telephone links.

Thus, in-band TI caused primarily by voice and data telephone relays generally appears as a pattern of disturbance across channels with alternating good and bad pictures. It can vary in intensity from mild sparklies to complete white-out or black-out. At this point a signal strength meter in a receiver would usually be driven off the top end of its scale. As the TI power level increases, so does picture deterioration. Table 3-3 illustrates the effect of interference on a satellite broadcast. TI levels are expressed relative to the satellite signal received at the downconverter. To measure these relative levels in the field a spectrum analyzer must be used.

TABLE 3-3. RELATIVE TI LEVELS AND SYMPTOMS

TI Relative to Satellite TV Signal	Symptom
Less than -18 dB	No Problems
-18 to -10 dB	None to Light Sparklies
-10 to -3 dB	Heavy Sparklies - Picture Barely Watchable
-3 to 0 dB	Lines or Random Pattern - No Picture.
Greater than 10 dB	Complete Wipe-out

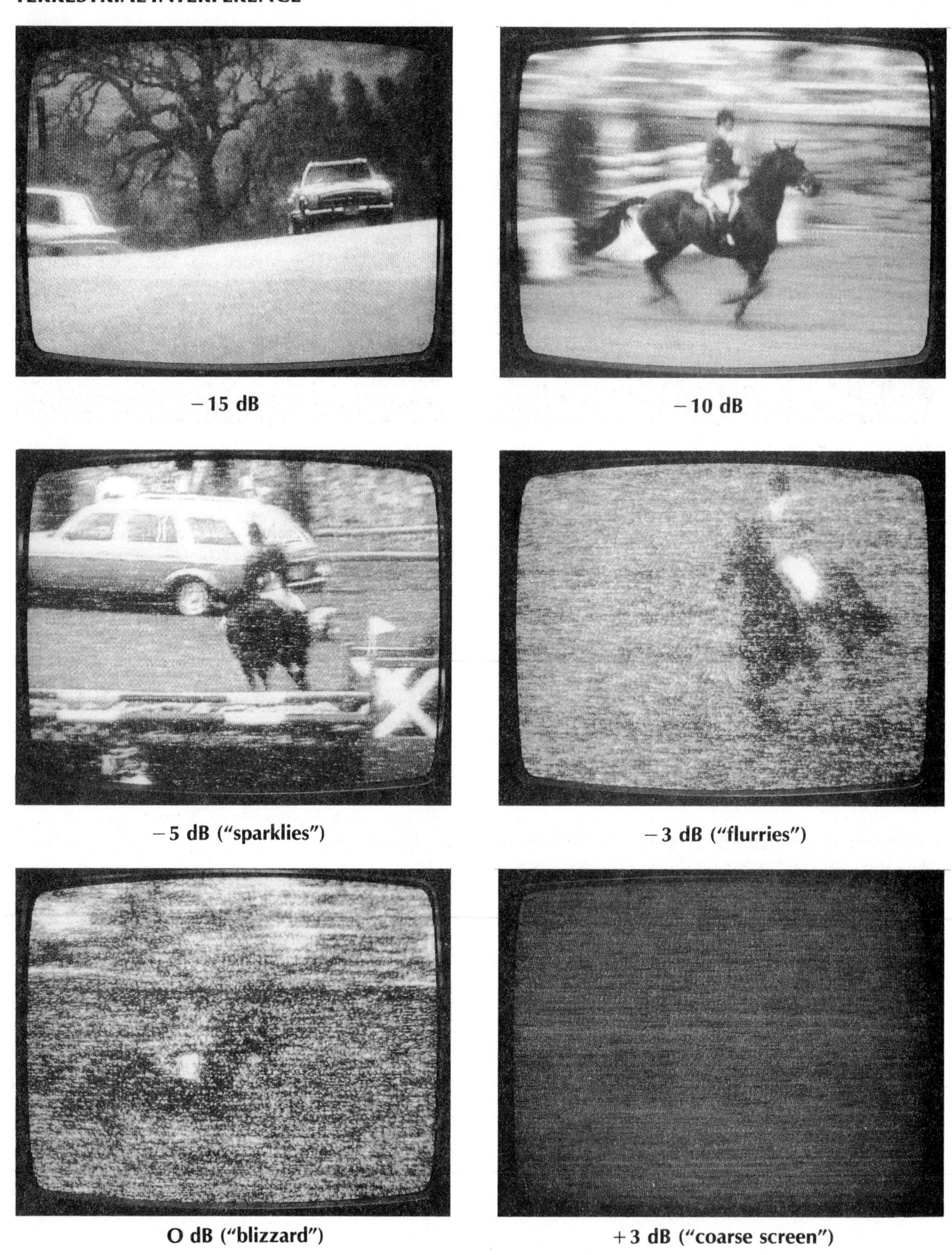

Figure 3-6. The Effects of TI on Picture Quality. *Picture quality deteriorates as the power of an interfering carrier increases relative to that of the incoming satellite signal. The interfering carrier in all these cases is at 10 MHz below the satellite center frequency. (Courtesy of Microwave Filter Company)*

+5 dB (blank, uneven screen)

+10 dB (blank screen with fine, uniform texture)

Figure 3-6. continued...

Like out-of-band interference, in-band TI may disappear when the antenna is aimed at another satellite. Microwave interference, like satellite broadcasts, can be very directional.

As mentioned in the previous paragraph, the effect that TI has on satellite television depends upon its power relative to the satellite signal. If it is more than 18 dB below the level of the desired signal, it generally is not noticeable on a TV screen (see Figure 3- 6). As TI powers rise to about -10 dB sparklies begin to appear on the screen. These increase until about -3 dB when the dots reach a "flurry" level. At 0 dB, the "blizzard" level, the picture is nearly lost. Above this power, TI begins to detune a receiver, namely, the automatic frequency control (AFC) circuit begins to track the TI instead of the satellite signal. At +3 dB the screen has no detectable picture and a coarse appearance. At +5 dB the screen is blank and uneven. Above +10 dB the screen is completely whited out and has a fine, even texture.

If the AFC circuit is disabled, usually with an external or internal receiver switch, the satellite frequency can be manually tracked. Then a watchable picture may be seen even with TI levels as high as 10 dB (see Figure 3-7). Note that if a receiver has frequency synthesized tuning it must have a very "stiff" AFC circuit which deviates less than 3 MHz from the selected frequency. If not, it may track even lower levels of TI than a receiver normally would.

The effects of in-band TI are also illustrated in Figures 3-8a and b presented on the following pages. These diagrams clearly show how in-band TI appears as patterns of "good" and "bad" channels.

+5 dB TI at −10 MHz

+7.5 dB TI at −10 MHz

+10 dB TI at −10 MHz

Figure 3-7. The Effects of TI and Shifting the Center Frequency. *If a receiver's AFC circuit is manually disabled, it no longer tracks interference. The center frequency can then be tuned away from the TI to maximize reception of a satellite signal. As a result, a TVRO can bear higher levels of TI than otherwise would be the case. (Courtesy of Microwave Filter Company)*

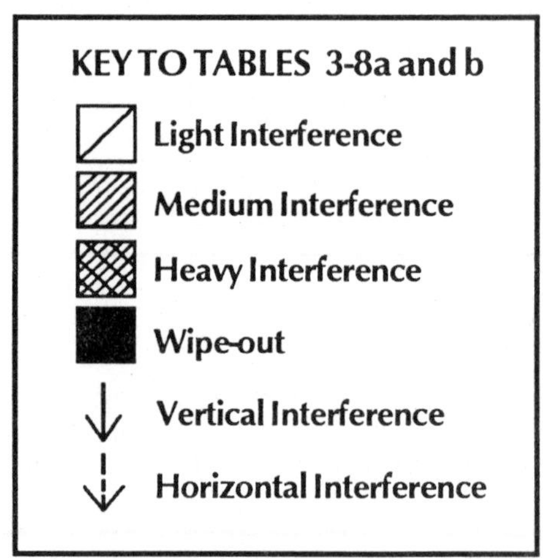

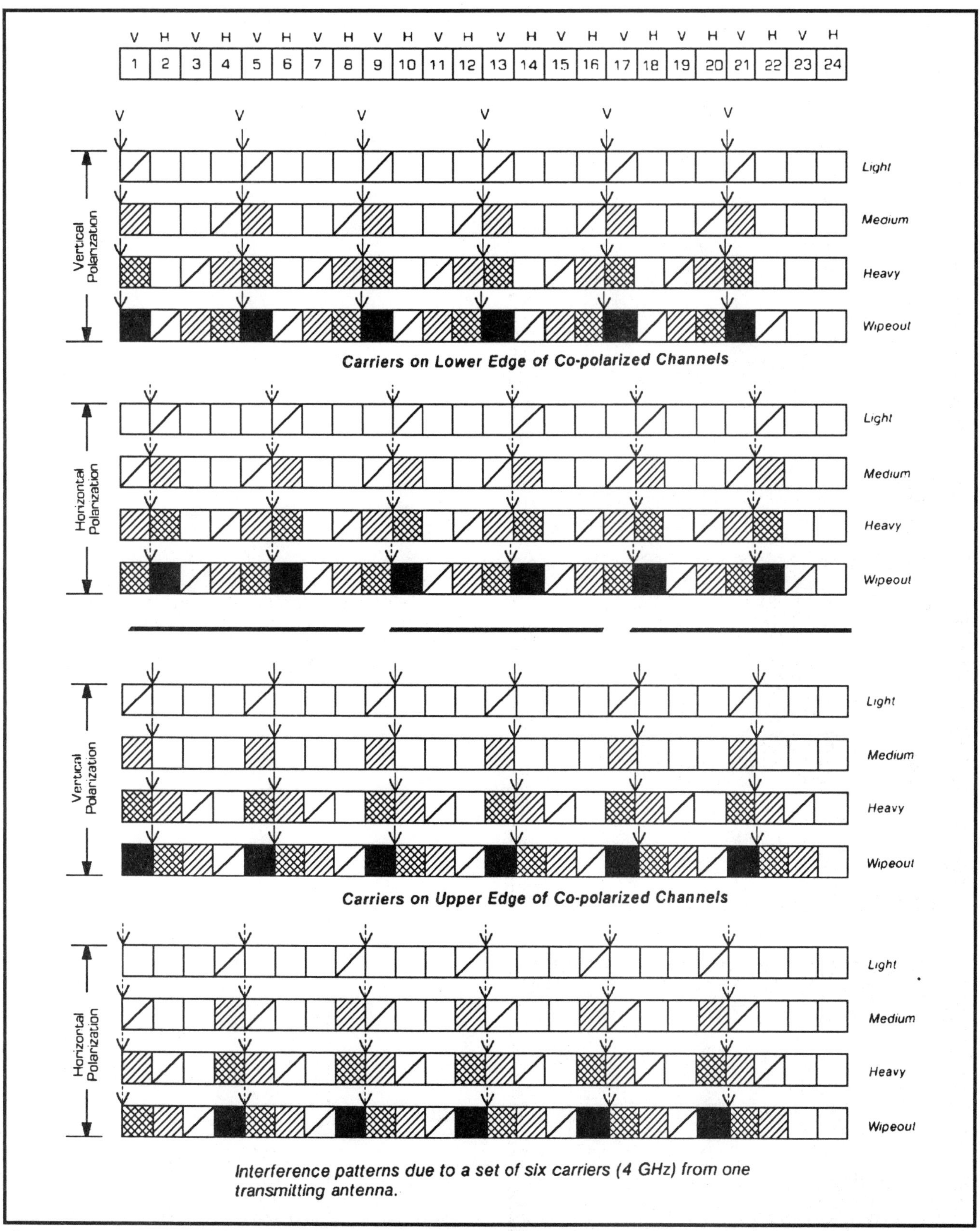

Figure 3-8a. Patterns of Interference on Satellite Channels. *Six carriers from one earth-based transmitting antenna show a characteristic regularity in patterns of channel disturbance. The key to reading both of these figures is shown on the left. (Courtesy of Microwave Filter Company)*

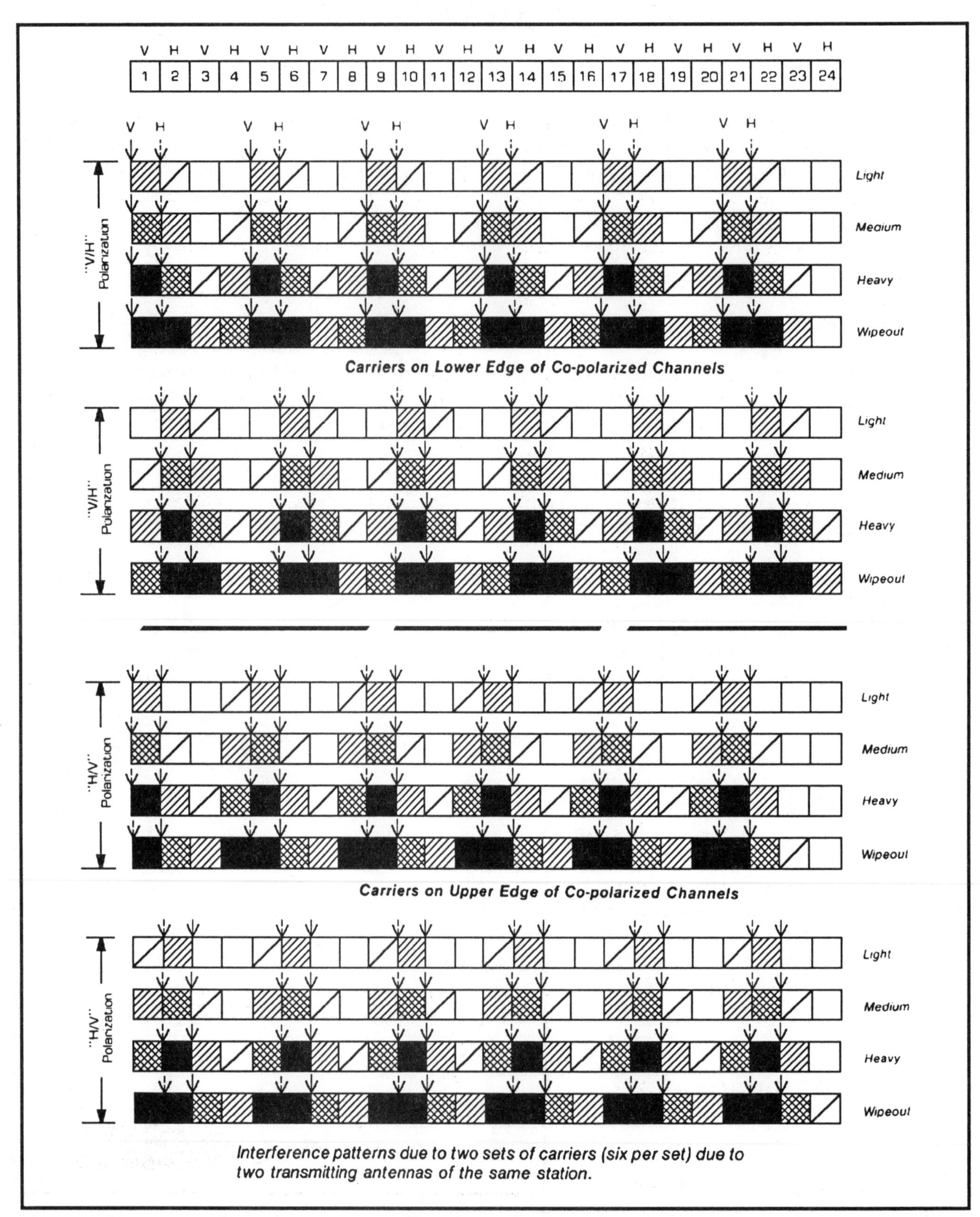

Figure 3-8b. Patterns of Interference on Satellite Channels. *Twelve carriers from two earth-based transmitting antennas show a characteristic regularity in patterns of channel disturbance. The key to reading this figure is shown on page 86. (Courtesy of Microwave Filter Company)*

C. COMBATING TI

The best strategy in combating TI is to avoid the problem at the time of installation through proper equipment selection and antenna siting. Only when this approach fails should alternative methods such as building artificial screens, installing filters or moving the antenna be attempted. Note that it is not unusual for TI to appear years after an installation when a new, local line-of-sight transmitter becomes operational.

Ingress interference is usually cured by "tightening up" the system by properly shielding and grounding all leads and ground connections. Reasonable levels of out-of-band TI can generally be eliminated by proper placement of either microwave or RF filters. In-band interference usually calls for RF notch filters or sometimes microwave traps. The most difficult situations arise when the TI level is high relative to the satellite signal or when the carrier has a wide bandwidth ranging from 5 to 30 MHz. Then either more expensive phase cancellation methods or screening methods must be used.

Terrestrial interference, whose symptoms can often mimic those caused by component failures, can be an annoying problem. However, a trained technician should be able to recognize and diagnose TI without too much difficulty. One who understands and can treat cases of TI, especially in metropolitan areas having heavy data and telephone traffic, is often in the position of a doctor who has a cure for cancer. Business will be good.

Equipment Selection and Susceptibility to TI

The receiving antenna, the eyes of an earth station, sees not only the targeted satellite but also noise and interference. Proper selection of the antenna/feedhorn/LNB assembly avoids the need for later "band-aid" remedies to combat poor performance. In fact, a system with inadequate gain and poor noise rejection will often act as if it were detecting TI. Furthermore, if TI is present or becomes a problem at some point in the future, it will be much more susceptible to its effects.

The first step in avoiding TI is to choose a dish/LNB combination that can provide a C/N in excess of the receiver threshold. If a clear picture cannot be received under normal conditions without any interference, then if even light levels are present quality will be very poor.

Optimally, an antenna having a narrow beamwidth and low side lobes should be chosen (see Figure 3-9). Interfering signals almost always

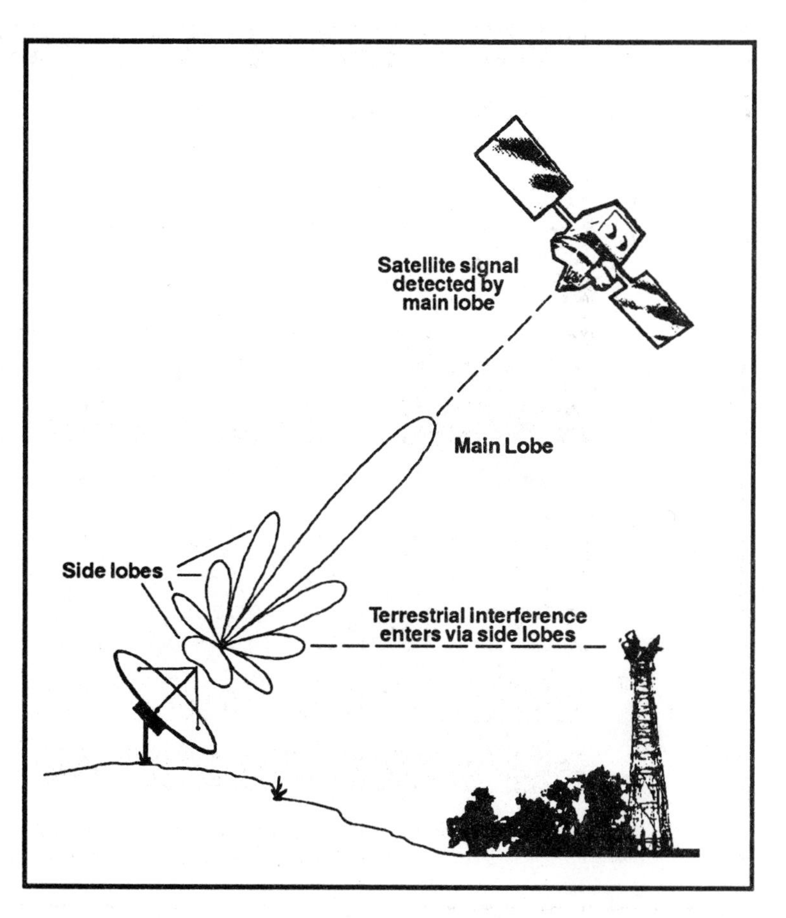

Figure 3-9. Side Lobes and TI. *An antenna usually detects interfering signals via its side lobes. If these lobes are weak relative to the main lobe, TI will have less effect on reception.*

come from directions off the main antenna axis and are detected by the side lobes. If the first side lobe is 30 dB rather than 20 dB below the main lobe, it will provide 10 dB more protection against TI. This may mean the difference between a watchable and a wiped-out picture. Antennas having higher quality reflective surfaces perform better since side lobes are substantially higher for dishes having rougher surfaces or slight warping. One simple but perhaps costly solution for light to moderate levels of TI would therefore be to install a larger dish having a better surface accuracy. This selection would increase the main lobe and decrease side lobes. Another less costly strategy would be to use a deeper dish which generally has lower side lobes.

It is important to understand that TI is detected by side lobes and, very rarely, enters via the main lobe. This explains why TI might be seen when aiming at one satellite and may disappear when targeting an adjacent one. Moving the dish even a few degrees could cause the directional source of interference to fall between two side lobes and therefore be detected at much lower strengths.

These considerations suggest that very weak levels of TI may be encountered during a site survey if a relatively small antenna is used. It would have a lower gain so the carrier to noise power would be decreased relative to the power from the interfering signal. A small site check dish can therefore be an effective diagnostic tool.

A feedhorn must be chosen to be compatible with the antenna f/D ratio for the same reason. A feedhorn which over-illuminates an antenna would cause an increase in detected side lobe power. On the other hand, one that under-illuminates while having significantly lower side lobes would not take advantage of the possible antenna gain (see Figure 3-10).

An LNB with an acceptably low noise temperature should be chosen. Most LNB manufacturers publish performance curves that graph how rapidly gain falls off outside the 3.7 to 4.2 GHz band. An LNB having a good quality, built-in bandpass filter affords protection against out-of-band interfering signals. Technicians should also realize that when using an LNB, it is impossible to insert either a microwave bandpass or notch filter between the low noise amplifier and downconverter. The alternatives in the case of severe microwave interference are either use of effective screening techniques or placement of a more expensive trap between the feedhorn and LNB.

Figure 3-10. TI and Polarizer Design. *The front end of the CalAmp "Centerline" feed (photo at right) is designed for improved rejection of terrestrial interference. Chaparral's Polarotor I features the standard configuration of scalar rings. (Photos Courtesy of California Amplifier and Chaparral Communications)*

Medium power out-of-band TI can usually be eliminated by inserting a microwave bandpass filter between the low noise amplifier and downconverter sections when a separate block downconverter has been installed. The bandpass designation means that the filter allows only those frequencies in a given band to be transmitted while rejecting all others. Note that modern LNBs are factory sealed and cannot be opened without causing potentially severe moisture problems. However, effective bandpass filters are incorporated into nearly all LNBs.

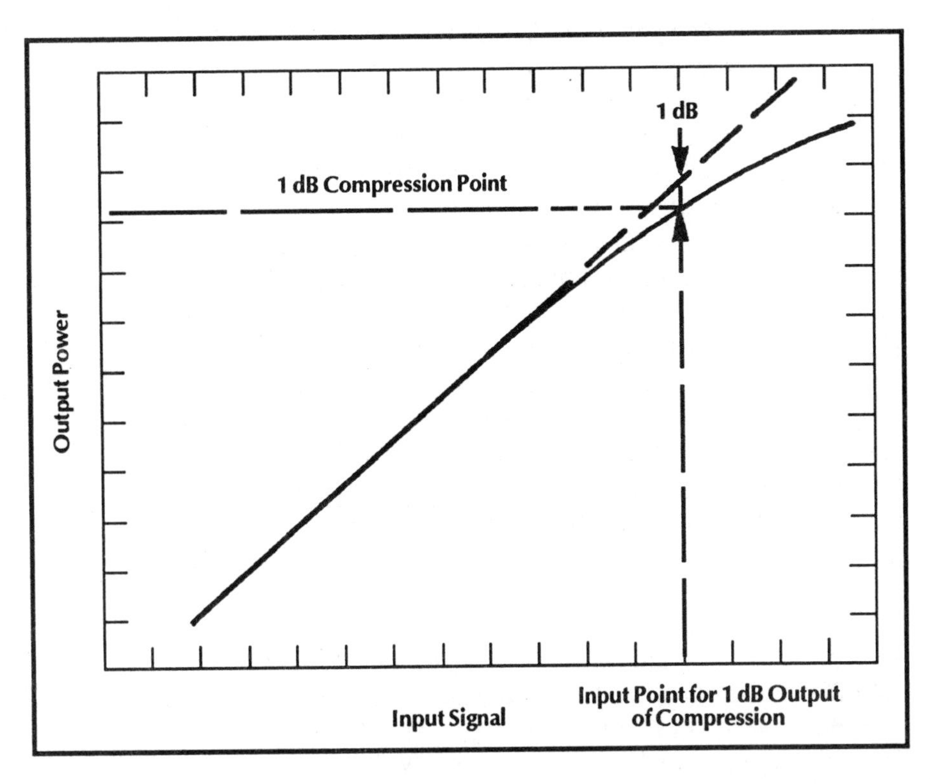

Figure 3-11. LNB Compression. *When signal levels entering an LNB are too strong, the amplifier is driven into compression. This means that it enters a non-linear response region in which its output is not directly related to the input signal. This usually occurs when a strong interfering signal is detected. The solution is then to relocate the antenna, use artificial screening techniques or to insert a filter between the feedhorn and LNB.*

In rare cases where the interfering signal is sufficiently powerful it can overdrive a low noise amplifier. This condition is known as driving a low noise amplifier into compression (see Figure 3-11). Then a more expensive microwave bandpass filter between the feedhorn and the LNA or LNB input would be required to eliminate TI.

Downconverters found in older systems are susceptible to TI. A single conversion unit can be a source of interference and should be mounted behind the dish. The oscillator mixing frequencies created for channel selection lie in the C-band range. If the downconverter were mounted directly onto the LNA, a slight leakage of signal could result in these oscillator frequencies being re-radiated to the reflective surface only to be detected and re-amplified as interference.

When selecting a satellite receiver, it is useful to have an easily accessible AFC defeat switch. In cases of light to moderately heavy interference, defeating the AFC circuit and manually tuning away from the TI onto the satellite signal can restore a picture. Similarly, a receiver with an adjustable bandwidth control can be a useful diagnostic tool during a site check or installation. If a reasonable picture can be restored by defeating the AFC or by narrowing the bandwidth, chances are good that a low-cost notch filter would overcome the problem.

Combating In-Band TI

Microwave notch filters should be used only when necessary to combat in-band TI because they eliminate some of the satellite signal and

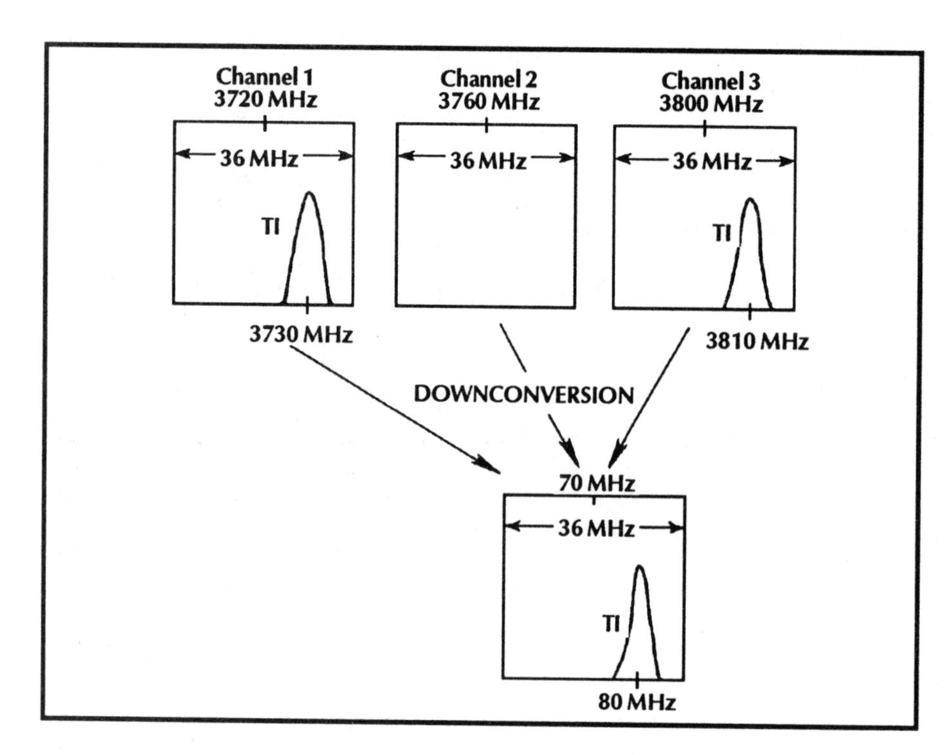

Figure 3-12. Downconverting Interfering Carriers. *In a satellite receiver, both the satellite and interfering signals are downconverted on each channel to the final IF band of frequencies. The interference can lie 10 MHz above or below the satellite IF center frequency. This drawing illustrates how TI carriers at 10 MHz above the satellite center frequency on both channels 1 and 3 are downconverted to the same 80 MHz location at the final IF. Other TI carriers on any other transponder would also be downconverted to the same frequency.*

therefore degrade picture quality. Carefully choosing equipment and siting the antenna can often reduce or eliminate interference. However the use of filters is generally preferable to relocating an existing installation or to building expensive and often unattractive screens.

The type of filter used to combat in-band TI depends upon the severity of the interference. The least expensive notch filters operate in the IF range. For low to moderate levels below -3 dB, IF notch filters, which remove a narrow band of signal centered on either 60 or 80 MHz, can restore pictures to near normal quality. These can also be adequate for levels from -3 to about 0 dB relative to the satellite signal. At 0 dB the picture is still usually visible. But IF filters are rarely effective beyond this level, since the receiver AFC begins to follow the interference. However, disabling the AFC and manually re-tuning the receiver can extend the useful range of traps and watchable pictures can result. Remember that such traps may be set for the wrong frequency range for half transponder broadcasts such as those that originate from some Intelsat communication satellites.

When using receivers having an IF different than 70 MHz, such as the DX-700 which operates at 134 MHz IF, specially made notch filters operating at 124 and 144 MHz must be used. A wide selection of 60 and 80 MHz filters and traps for more uncommon IFs such as 124 MHz or 134 MHz IF for the DX receiver are available.

In order to understand the operation of notch filters, remember that a satellite receiver lowers the center frequency of each channel to the IF frequency, usually 70 MHz (see Figure 3-12). Since TI signals are centered at either 10 MHz above or below the satellite center frequency, they are also downconverted to either 10 MHz above or below the IF frequency, usually 60 or 80 MHz. Therefore a filter that notches out 60 or 80 MHz can remove the interfering signal (see Figures 3-13 and 3-14). Just two notch filters can remove any interfering signal on any channels.

At TI levels above the -3 to 0 dB range, IF notch filters are not sufficient and microwave filters or traps must be used. One trap is required for each interfering frequency. At levels above +10 to +15 dB a special type of microwave filter, a resonator trap, must be installed between the LNA and downconverter. The complexity and cost of finding a technical solution for TI increases as TI power grows. At levels above +35 dB, very costly, specially-made traps must be placed between the feedhorn and LNA to allow the LNA to function. Note that when an LNB is used traps cannot be placed between the low noise amplifier and the internal downconverter. So if TI is severe, an LNB system may have to be replaced with an LNA and single or dual downconversion receiver. In some cases, the sensible alternative is simply not to install a system.

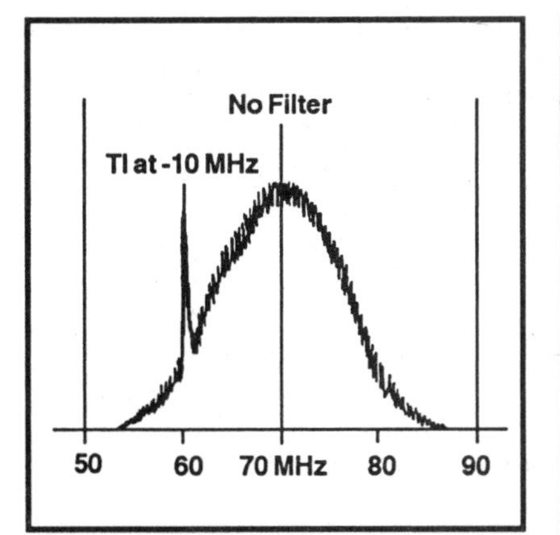

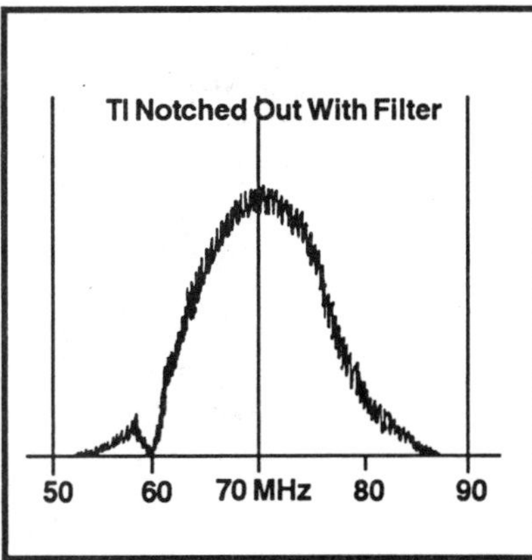

Figure 3-13. Notching Out TI. *These drawings illustrate how an interfering carrier at 60 MHz, 10 MHz below the center frequency, affects the waveform of a satellite signal (left) and how it can be "notched out" by a notch filter (right). In practice, any filter also removes a portion of the desired satellite signal when it eliminates interference.*

Filtering wide band interference is not possible with conventional notch filters which remove narrow bands of signal and interference. If the interfering signal covers the full satellite broadcast bandwidth, then notching out the TI also removes the television signal. The solution to this problem aside from moving the antenna or using artifical screens is called phase cancellation. This method is simple on paper. A sample of the TI which is taken with a second feedhorn and amplifier is subtracted from the satellite signal plus TI by shifting its phase 180 degrees and adding the signals together. This leaves just the satellite signal and eliminates the TI. Unfortunately, this method requires relatively expensive, sensitive equipment and is not one used by the average dealer on the average installation.

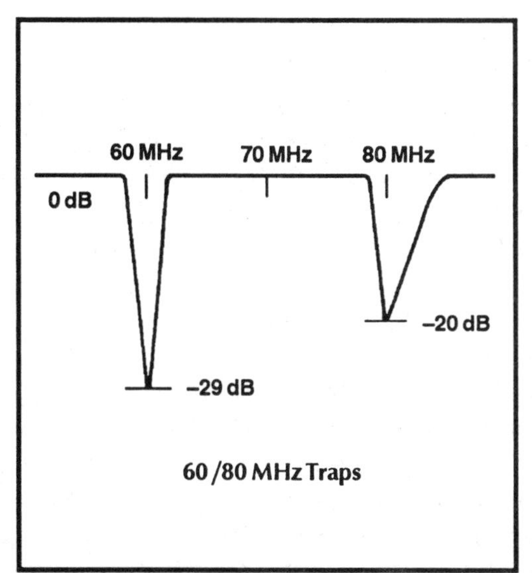

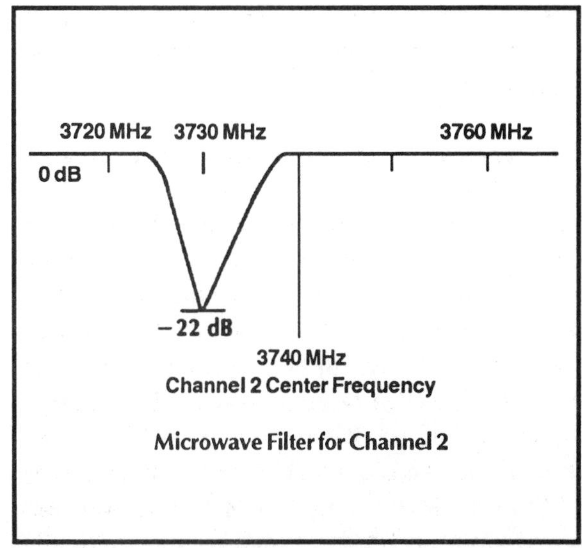

Figure 3-14. Notch Filter Characteristics. *A notch filter is characterized by the depth of the notch and its bandwidth. The left drawing shows the effects of 60 MHz and 80 MHz notch filters installed in series. The lower frequency filter has a notch depth of 29 dB; the upper of 20 dB. The right illustration displays the effects of a notch filter that operates at microwave frequencies, in this case channel 2 which is centered on 3740 MHz. This filter notches out a wider bandwidth of signals than an IF filter.*

Selection of RF and Microwave Filters

Deciding upon the brand and type of filter to employ depends upon the particulars of the TI encountered. First, any potential problem must be correctly diagnosed during the site check. TI can often be avoided. When filters are needed, the type required depends upon the source and power level of the interference.

Filters can be grouped into two basic categories: notch filters or traps, and bandpass filters. "Threshold extension" filters are a subclass of bandpass filters designed for use in the IF range. Most of these are available in either active or passive forms.

Passive filters are inexpensive and uncomplicated (see Figures 3-15 and 3-16). They are constructed from standard electrical circuit elements and do not need to be "plugged in." Since these use no power, they provide no gain as the signal passes through. In fact, they have an "insertion loss," ranging from nearly zero to substantial decreases in signal power depending upon the type used.

Active devices need power to function and usually have an integrated circuit and either a ceramic filter or a surface acoustic wave (SAW) filter to limit the range of rejected frequencies. They can be smaller than passive units and amplify as well as filter the processed signal. Their disadvantages include a need for external power and higher cost (see Figure 3-17).

All notch filters are rated by the amount of TI reduction, the depth of the notch. For example, a filter which was rated for -40 dB would reduce an 8 dB interfering signal to -34 dB.

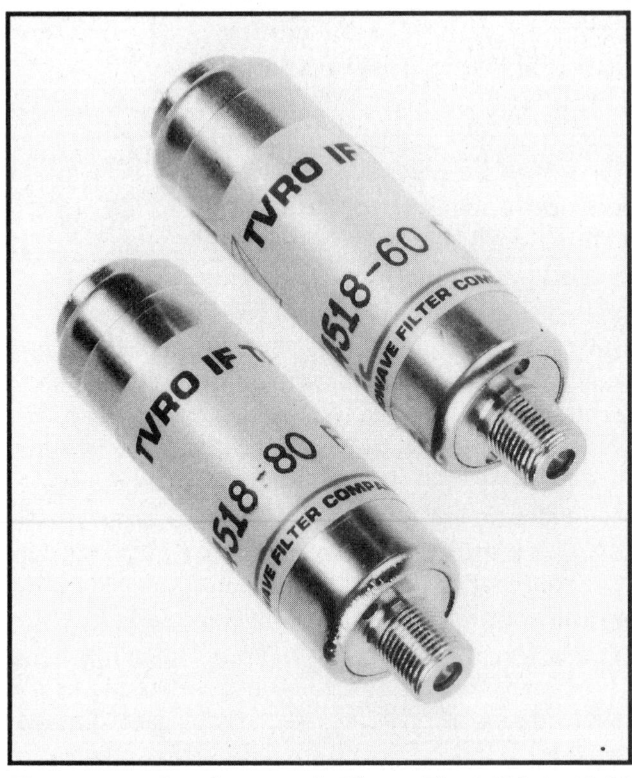

Figure 3-15. Passive Notch Filters. *The MFC 4518-60 and 80F notch filters can be used separately or in series to reduce interfering carriers in a receiver that has a 70 MHz IF loop-through. (Courtesy of Microwave Filter Company)*

Figure 3-16. Dual Switchable TI Filter. *This filter combines a 60 and 80 MHz TI notch in one compact housing. It provides effective suppression of mild to moderate levels of TI. (Courtesy of Pico Products, Inc.)*

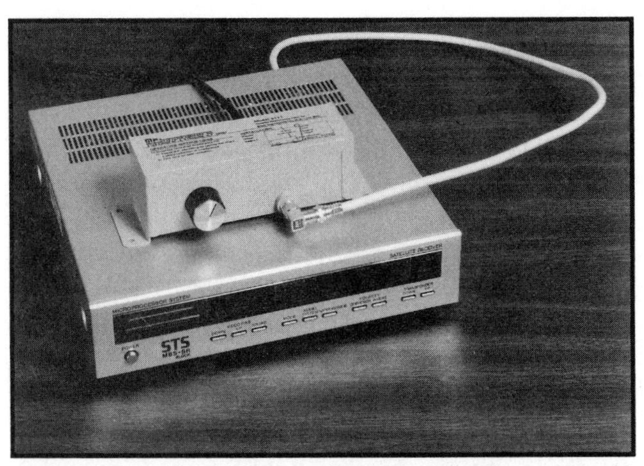

Figure 3-17. Model 5234 Block Notch Filter. *This filter inserts directly into the IF line between an LNB and receiver. The single-turn knob tunes a narrow 15 dB notch through the entire standard block downconverter band of 950 to 1450 MHz. (Courtesy of Microwave Filter Company)*

Just a few years ago about the only company catering to the home satellite market was Microwave Filter Company which manufactures a wide range of filters. Others such as A.S.P., R.L. Drake, and Phantom Engineering now offer conventional IF notch and threshold extension filters.

Notch filters, designed for use in the IF range, eliminate a portion of the frequency spectrum during the process of wiping out interfering carriers. The notches, which are centered on or close to either 60 or 80 MHz, are characterized by the relative suppression of power in this notch and by the width or range of frequencies affected. Obviously, if a telephone carrier centered on 60 MHz and having a width of 3 MHz is attacked, the ideal filter would eliminate only this band of frequencies. Eliminating additional frequencies would cause undue loss of the desired satellite signal.

Some notch filters are equipped with adjustments on one or all of depth, width and center frequency. Some can also be switched in and out of action so that when not required they do not degrade picture quality by unnecessarily removing a portion of the spectrum. For example, the Phantom Terminator offers a choice between 35 or 50 dB notches (see Figure 3-18). The MFC 3217LS-60 and 80 passive notch filters separately or in series can reduce TI carriers by about 53 dB. They are field-tuned by two screwdriver adjustments. These filters, like most other notch filters available, can easily be inserted in the IF line to the receiver by using an F-connector input and output. MFC model 4616 dual 60/80 notch filter is also field tunable, reduces carriers by about 29 dB and can be switched into and out of action without being disconnected (see Figure 3-19).

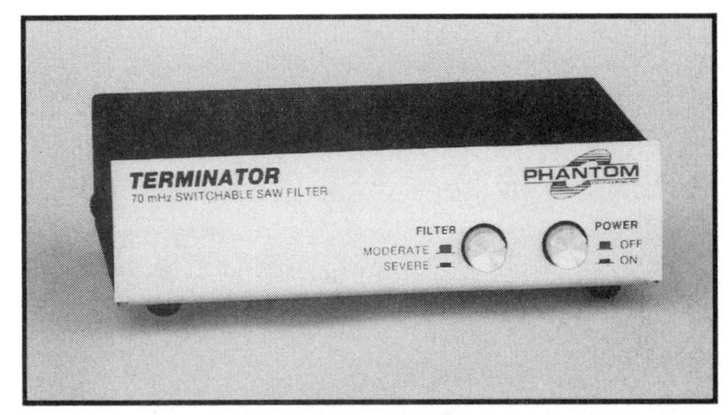

Figure 3-18. Switchable SAW Filter. *This active device has dual SAW (surface acoustic wave) filters. The notch depth can be adjusted to either 35 dB or 50 dB for dealing with moderate or severe TI, respectively. (Courtesy of Phantom Engineering)*

Figure 3-19. Switchable IF Trap Model 4616-70F. *This trap can be used with receivers having final IFs of 385, 400, 403, 510, 520, 560, 600, 610, 700 and some other special frequencies. The notch depth is 25 dB and the 3 dB bandwidth is 4 MHz. Input and output are via F-connectors. (Courtesy of Microwave Filter Company)*

Threshold extension filters are similar to those used by some manufacturers to lower the apparent receiver threshold. They restrict the bandwidth of the signal entering a receiver and eliminate noise, some satellite signal and hopefully, in the process, the interfering carrier. For example, the E.S.P. PFG-20 and 50 active devices use a SAW to produce a steep-skirted frequency response and add about 4 dB of additional gain. ESP's PGF-50 provides about 53 dB reduction in carrier power and a 3 dB (half power) bandwidth of only 13 MHz.

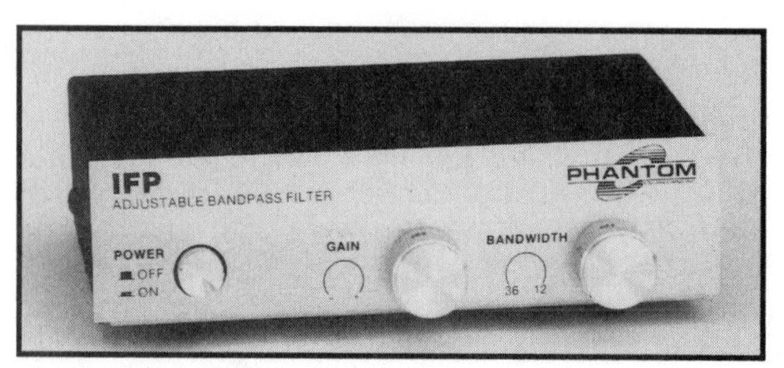

Figure 3-20. Adjustable Bandpass Filter. *This active filter features an adjustable 6 dB bandwidth that can be tuned from 32 to 12 MHz, as well as a variable gain for improving TI rejection. Maximum gain is 10 dB and maximum attenuation is 3 dB. (Courtesy of Phantom Engineering)*

The Phantom IFP adjustable bandpass filter accomplishes a similar task (see Figure 3-20).

An adjustable bandpass filter can be used, if necessary, to combat standard TI carriers by narrowing the signal bandwidth to a range less than 20 MHz. Figure 2-21 illustrates a case whereby the effects of both a wideband and a narrow band TI carrier can be eliminated. This is certainly accompanied with a reduction in picture fidelity.

Some notch filters can have effects similar to receivers which reduce threshold by narrowing bandwidth. If the width of two filters located at 60 and 80 MHz is increased, the band between them becomes squeezed down. For example, if the bandwidth of each trap were 4 MHz then the middle bandpass region would range from 64 to 76 MHz, a bandwidth of only 12 MHz.

Many brands of receivers, especially higher end units, now incorporate built-in threshold extension as well as IF notch filters. These filters can be automatically switched in and out as channels are changed. In fact, only a minority of receivers that can be purchased today have IF loops for insertion of notch filters.

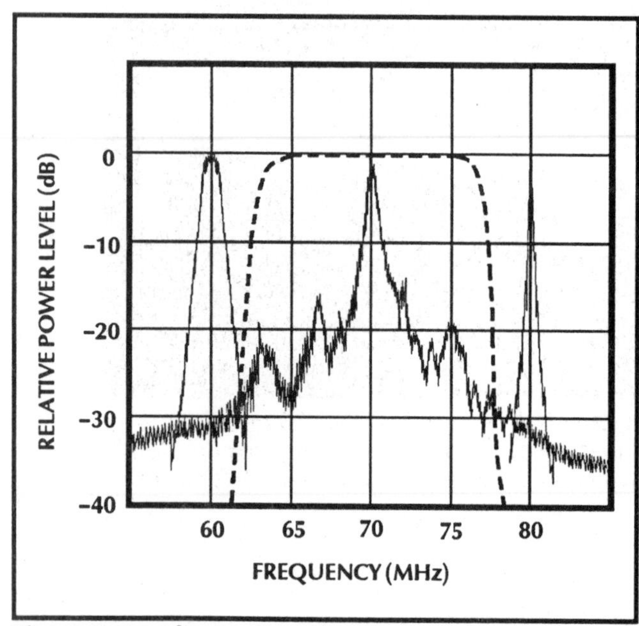

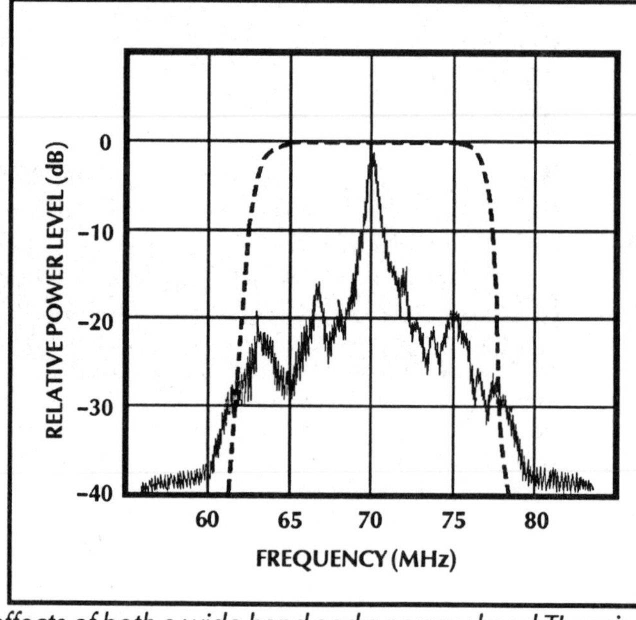

Figure 3-21. Eliminating TI with a Bandpass Filter. *The effects of both a wide band and a narrow band TI carrier may be eliminated by using a bandpass filter as shown here. In this case, the bandwidth here is reduced to about 17 MHz. Picture fidelity is degraded by reducing signal bandwidth so severely.*

Other Filters

A variety of filters are available for use in combating TI at microwave frequencies (see Figures 3-22 to 3-25). For example, MFC's 4352 bandpass filter (see Figure 3-22) passes only 3.7 to 4.2 GHz. It directly connects into the line between an LNA and downconverter with male/female N-connectors so no jumpers are necessary. The 3966A can be adjusted to pass only one of any selected transponder frequency in the C-band. However, remember that most good quality LNBs built today feature built-in bandpass filters.

Figure 3-22. Microwave Bandpass Filter. *The 4532 bandpass filter is designed to pass only those frequencies in the C-band range. It directly connects into the 4 GHz line between an LNA and downconverter so that no extra cable or connectors are necessary. This filter also passes DC power. (Courtesy of the Microwave Filter Company)*

Figure 3-23. Tunable Microwave Bandpass Filter. *This bandpass filter can be tuned to any particular transponder frequency in the C-band. It has a 3 dB bandwidth of 40 MHz with under 1.5 dB insertion loss. If several are to be used in series, a 50 dB LNA should be installed. (Courtesy of Microwave Filter Company)*

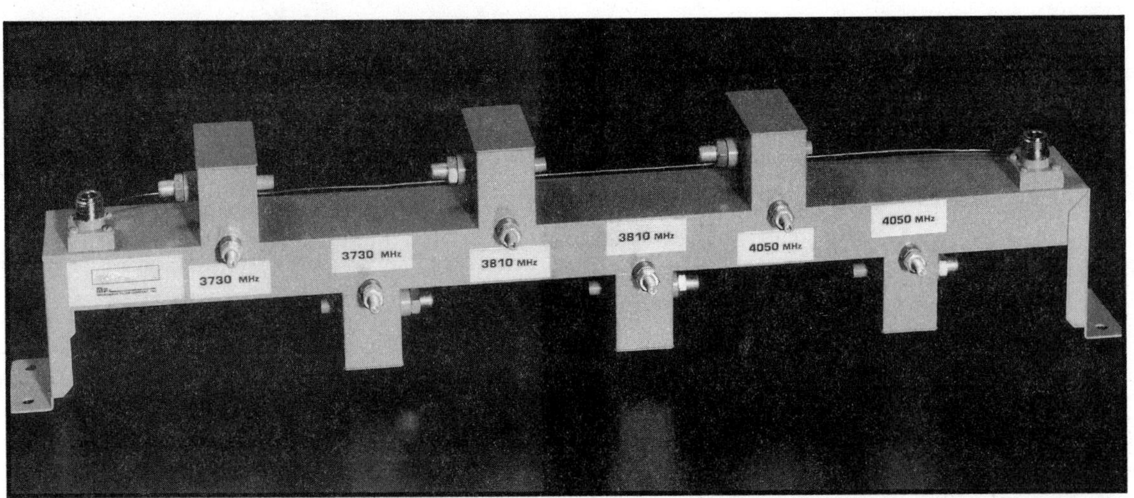

Figure 3-24. Microwave Notch Filter for 4 GHz Interface. *The model 5045-A 30 dB microwave notch filter suppresses 4 GHz interfering carriers. It is used to combat TI strong enough to completely wipe out a television picture. In these cases, the interference levels are usually too high for IF traps. (Courtesy of Microwave Filter Company)*

Figure 3-25. Block Downconversion Microwave Trap. *This filter notches out six frequencies in the 950 to 1450 MHz band. Each notch can be adjusted in depth to approximately 25 dB and the center frequency can be tuned over six 80 MHz adjacent bands to cover the entire 500 MHz bandwidth. (Courtesy of Microwave Filter Company)*

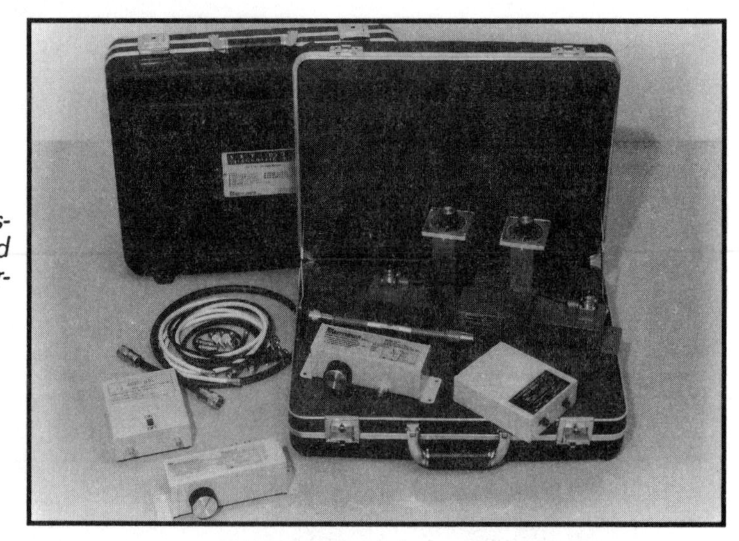

Figure 3-26. Sky Doc Kit. *This kit aids in diagnosing and curing terrestrial interference in block and single or double downconversion systems. (Courtesy of Microwave Filter Company)*

In addition, Microwave Filter Company produces a diagnostic kit to aid in diagnosing TI at both microwave and IF frequencies (see Figure 3-26).

Natural and Artificial Screening Methods

Terrestrial interference can often be overcome by a combination of natural and artificial screening methods and, if necessary, appropriate use of filters. Natural screens include trees, buildings, mounds of earth or any other structures already on the site. The best and generally least costly strategy is to make effective use of natural screens before resorting to filters or artificial screens. Natural screens would include using houses, garages or trees as protection (see Figures 3-27 and 3-28).

Materials have various abilities to absorb or reflect microwaves and to serve as either artificial or natural screens (see Table 3-4). Metal is by far the best reflector of microwaves. Most non-metallic structures are generally not very reflective but have varying degrees of absorbing abilities. Wood, for example, is a poor reflector with limited ability to absorb microwaves. However, an existing wooden structure such as a garage or barn can be made into a reflector of microwaves by lining it with a metal screen mesh. This can usually be accomplished for a relatively low cost. Bricks, cinder blocks, concrete and other masonry products are also poor reflectors but reasonably good absorb-

Figure 3-27. Using a Building to Screen Antennas from TI. *These antennas are being shielded from interference by a brick building. (Courtesy of Brent Gale)*

Figure 3-28. Natural TI Screening. *Placing a dish in a protected area can shield it from wind as well as from interfering signals. Note that if deciduous trees are used, when leaves disappear in the winter, TI may return.*

TABLE 3-4. MICROWAVE ABSORPTION ABILITIES of VARIOUS MATERIALS

Material	Absorption (dB)
Wooden Wall	5
Brick Wall	10
Concrete Wall (15")	30
Cinderblock Wall	15
4 foot Thick Earth	30
Evergreen Tree Cluster	25
Tree Cluster - no leaves	10

ers of microwaves. The moisture content of green leaves causes them to be excellent attenuators of TI. But beware using deciduous trees as shields, because when they lose their foliage in winter so goes their value as protection against interfering microwaves.

In general, the higher a dish is installed above the ground the more susceptible it is to interference. In contrast, placing an antenna in a depression or a valley is usually a safe bet in reducing or even eliminating interference. There is an SMATV installation in a motel in California that was so plagued by TI that the least costly solution was to install the 15 foot antenna at the bottom of an empty swimming pool. A new pool was con-

Figure 3-29. Using Wire Mesh Screens to Eliminate TI. *Appropriately placed mesh reflectors can protect a dish from very strong levels of terrestrial interference. (Courtesy of Microwave Filter Company)*

structed as a lower cost option than curing the interference by another method!

Some materials are optimally designed for absorbing C-band radiation. An example is space cloth, a canvas-like fabric impregnated with carbon to make it resistive to current flow. A similar material coated with metal on one side could serve a dual purpose as an absorber in one direction and reflector in the other. Bulk absorbers can be attached to existing structures to prevent reflection of TI into a dish. Another example of such a material is a carbon-impregnated foam rubber having an array of small pointed pyramids on one side. (It is interesting to note that similar techniques and materials are used by acoustic engineers to absorb sound).

Metal screens like builders' "hardware cloth" are adequate reflectors of microwaves. As long as the mesh openings are less than approximately one tenth of the wavelength of the radiation, the same amount of energy is reflected as it would be by a solid metal surface (see Figure 3-29). Since C-band signals have wavelengths of about 3 inches, openings less than 3/10ths of an inch are adequate. (This is why wire mesh dishes can reflect microwaves as well as solid antennas).

The first step in finding a screened location for a dish where TI is present is to draw a scale map of the site and note the composition of all natural and pre-existing structures. If the direction of the TI source is known, an ideal site may often be easily determined. Occasionally, extra man-made screens are necessary. The least expensive solution is to attach wire mesh to an existing structure but be careful not to cause additional problems by bouncing more interference into the dish (see Figure 3-30). Alternatively, lightweight screens mounted on wooden frames can be constructed. The screen height should be at least three feet above the dish as well as wider than the dish. The ping-pong effect which causes interference to be reflected into the dish instead of away from it should be avoided by tilting these screens at a minimum of 30° or more away from the antenna. These man-made structures must be fastened down securely to prevent them from catching a

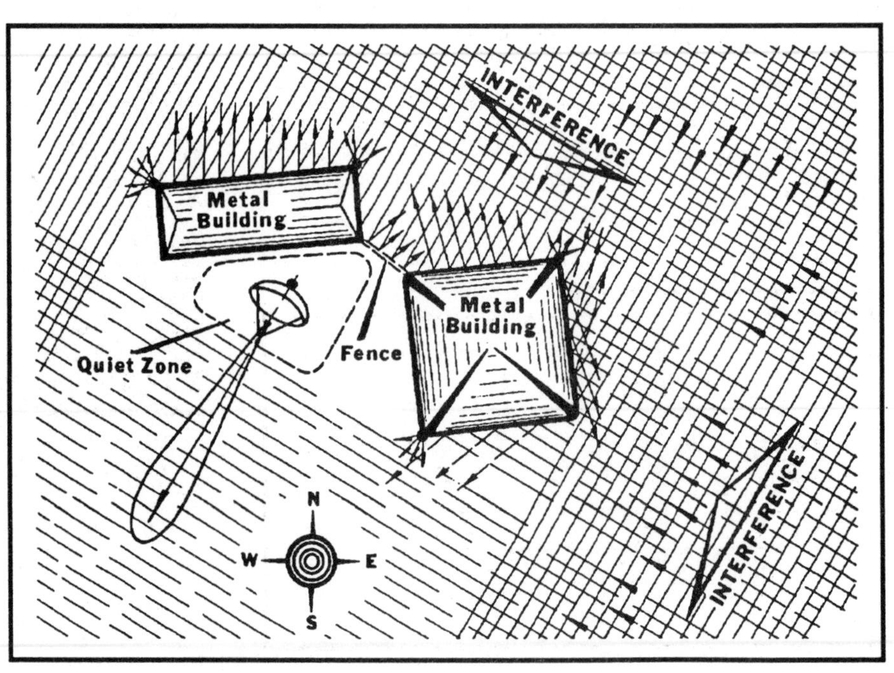

Figure 3-30. Scale Map of Installation Site. *This scale map shows interference originating from two directions. However, if the antenna would be located behind the two metal buildings and the fence a quiet zone would allow TI-free reception of satellite signals. (Courtesy of Microwave Filter Company)*

Figure 3-31a. Correct and Incorrect Use of Screens. *A screen should be tilted away from an antenna by a least 30 to 45 degrees to avoid the "ping-pong" effect which can cause even more TI to be intercepted.*

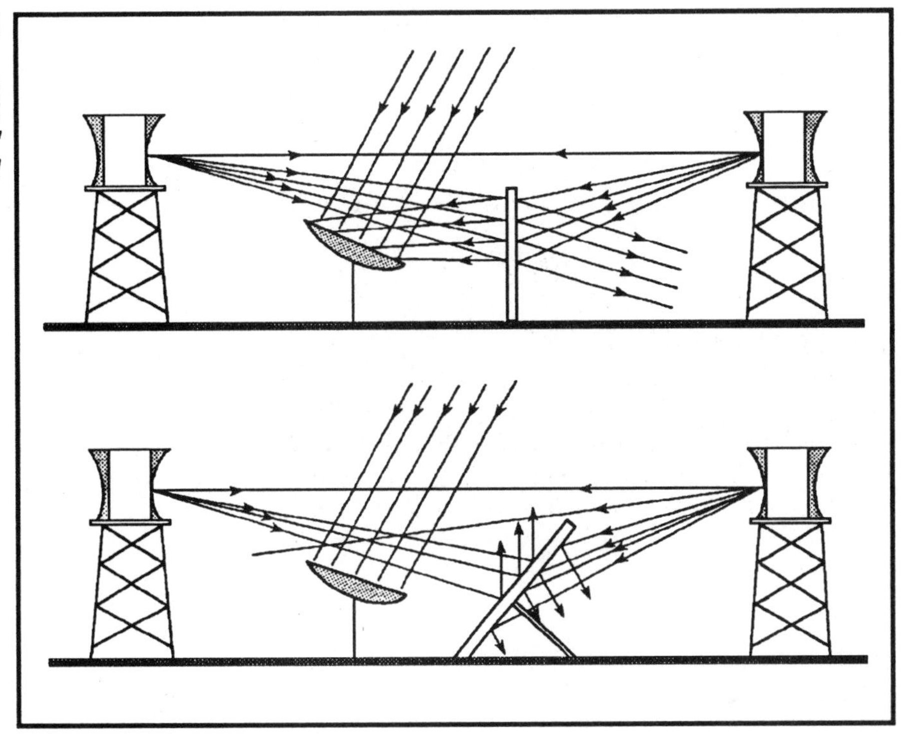

strong wind and being blown into the antenna or causing other damages (see Figures 3-31a and 3-31b).

All reflecting barriers should be terminated at the edges with at least a six inch roll of the material used. This prevents creating a sharp edge from which microwaves could be diffracted and bent into the antenna and feedhorn.

Similar effects can be achieved with edge absorbers attached to the rim of a dish. These are made from absorbing materials and reduce antenna side lobes by intercepting some of the "spillover" energy from the surrounding terrain and interfering carriers (see Figure 3-32). Edge absorb-

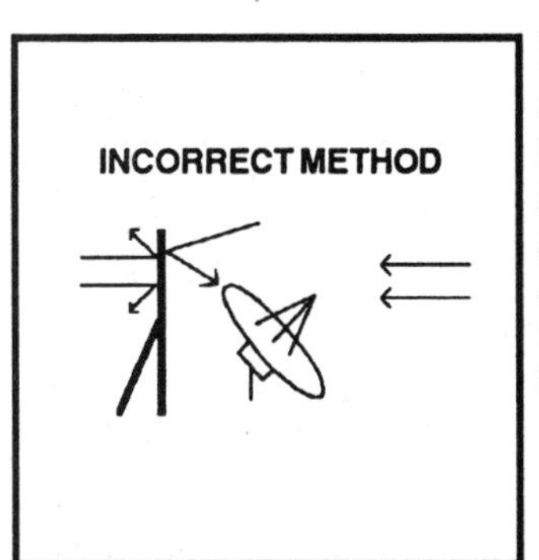

Figure 3-31b. Correct and Incorrect Use of Screens. *When correctly installed, metal screens are rolled at the top to avoid channeling TI into the dish via the edge effect.*

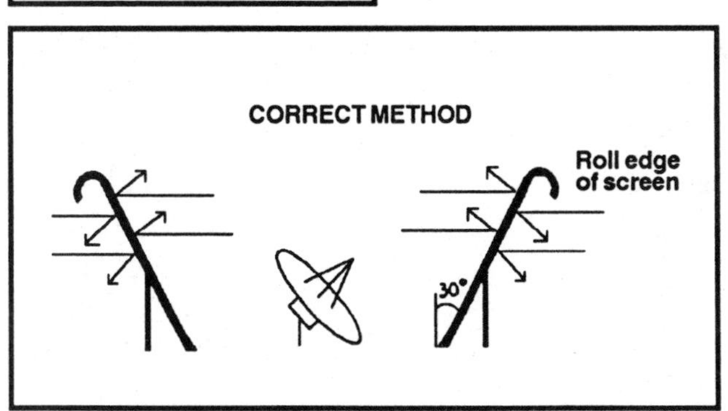

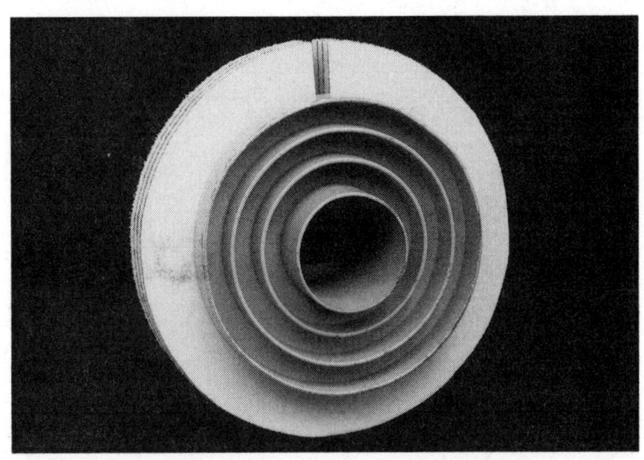

Figure 3-32. Microwave Absorbing Material. *A ring of microwave absorbing material shields the feedhorn from interfering signals. (Courtesy of Microwave Filter Company)*

ers differ from antenna extenders or skirts which increase the reflective surface area of an antenna. Since antenna extenders reflect microwaves, they must carefully follow the parabolic contour to keep side lobes low while increasing gain. They are a "fix" that allows building a smaller into a larger dish.

Occasionally terrestrial common carriers relay full video signals occupying the entire 30 to 36 MHz band from sources such as a local TV station. These will completely wipe out satellite transmissions sharing a similar frequency band. The only completely effective method of shielding against such interference is to screen out these signals with artifical or man-made barriers before they enter the feedhorn.

It should be noted that artificial shielding techniques, while generally more expensive than using filters, do not degrade the satellite signal. Whenever notch filters are inserted into the electronic system, some portion of the bandwidth is removed. This causes a loss of video information often resulting in noticeable picture degradation.

When traps in the microwave range are required, costs can escalate rapidly. If the interfering signal level exceeds 15 dB, a filter between the LNA and downconverter is required. At higher levels of TI, specially ordered, very expensive filters are required. One filter which may cost many hundred or thousands of dollars is then necessary for each affected transponder and must be inserted between the feedhorn and LNA.

Charting Problem Sites

A simple method to combat TI even before visiting the site is by charting microwave routes on a local map. The required background information includes all present and planned locations of terrestrial 4 GHz microwave relay stations in the area as well as their transmission routes and frequencies. This information can be purchased from any one of the professional frequency coordination firms, usually for less than $350, and needs to be compiled at most once a year. Or it can be obtained with some more work from either the FCC or local telephone companies.

A map of the region can then be marked with the location of each microwave repeater station. Straight lines are drawn between the sending and receiving sites. Any planned installation which falls too close to any one of these communication routes could be susceptible to TI. Since future sites of relay stations are also marked and since the chart will indicate even those potential sources of interference which might be intermittent, this method can be a very important complement to an on-the-site test.

Similar results can be obtained at a reasonable cost from professional frequency coordination companies. For a fee, these firms will provide a computer printout showing the detailed TI environment of a satellite TV installation site as well as planned sources. This includes the expected worst-case levels of interference, their operating frequencies and polarities and their beam directions. Depending upon the services offered, prices range from $200 to $1000. For example, Comsearch charges $325 for a map covering an area of 1° latitude by 2° longitude. Some scaled-down information can also be purchased at even lower costs. For example, in the United States Comsat General Corporation has offered the basic information for approximately $100 U.S.

Other packages of information including a route map and an intensity map are also available for a nominal fee. The route map, a semi-transparent overlay used with a topographical map, identifies and links each relay tower by a line. The intensity map is similar but also includes the expected power of these common carrier relays.

Another option is to register a planned receive-only earth station site for a fee with the FCC. This process may be necessary, for example, when an expensive SMATV (satellite master antenna TV) installation is planned for a condo, apartment complex or hotel. The registration protects this system from future terrestrial relays because it must be factored into their planning. Chances are low that the FCC would allow a common carrier to design a new relay which sends a beam of microwaves directly across an expensive, registered SMATV or cable TV installation. Frequency coordination firms also offer this licensing service and provide what they call "frequency protection" which warns both the licensed earth station and planned land-based communicator of a potential for interference.

It is important to realize that nothing can replace a site survey. An installation which may be far from the route of a land-based microwave communication path could easily be troubled by a signal reflected off a building or a metal billboard. Or a site directly between two relay towers may have just the necessary natural screening to avoid any difficulties. Any installation could also easily be susceptible to out-of-band or even ingress interference.

Frequency Coordination as Future Insurance

Until receive-only earth stations were deregulated in 1979, a license had to be obtained prior to installation. Today, the FCC still regulates uplinks. These require careful planning and a process called frequency coordination.

The frequency coordination begins with a study of other nearby communicators sharing the same frequency band who might interfere with or be affected by the planned installation. A detailed map, usually created by computer from an extensive data base, shows where the potential sources of interference may lie (see Table 3-5). If all seem to be right, the license is granted. Any future sites which may affect this new one must likewise register with the FCC. In this fashion, all are protected from each other.

Few receive-only stations bother to follow the frequency coordination or licensing route. Some of the larger SMATV (satellite master antenna TV) systems serving hotels or apartment complexes often take this safer route. Home satellite dealers should also be aware of these services.

TABLE 3-5. A PARTIAL LIST OF FREQUENCY COORDINATION FIRMS

COMSAT GENERAL CORPORATION
950 L'Enfant Plaza, SW
Washington, DC 20024
(202) 863-6010

COMSEARCH, INC.
11720 Sunrise Valley Drive
Reston, VA 22091
(703) 620-6300

MULTICOMM SERVICES INTERNATIONAL
266 West Main Street
Denville, NJ 07834
(201) 627-7400

IV. SELECTING NEW SATELLITE TV EQUIPMENT

The satellite TV business has evolved at an astonishing pace during its brief lifetime. Today, selecting equipment for personal or customer use can become a matter of choosing from among hundreds of antennas, receivers and other components. In 1980, only a few brands of equipment were available and antenna actuators were virtually unknown.

In order to serve customers properly, a dealer must have a thorough understanding of the basics of satellite TV operation and should also keep track of new developments and equipment. He or she must know what systems are least susceptible to terrestrial interference, how to properly design an earth station having an acceptable performance margin, the effects reduced spacing between satellites and weather will have on picture quality, as well as a host of other issues. Learning all this information is simply a matter of consistent research. But by far the most important and enjoyable task is actually installing a system.

This manual is not written to advertise various brands of equipment or to list everything on the market. The trade journals do that task perfectly well. Our intent is to provide the ammunition to allow a dealer or consumer to read and understand any trade journal or set of instructions in order to make well-informed judgments about the quality of equipment.

A. COMPONENTS SELECTION

Among the criteria used in choosing components for satellite TV systems are performance, cost, features, durability and aesthetic appeal. Each of these must be weighed in selecting equipment. Some owners will be satisfied with a $1000 system having acceptable quality pictures on most but not all transponders; others will pay $4000 for an earth station having all the bells and whistles and will demand perfect video on all channels.

Mounts

The mount, antenna, feedhorn and LNB are the most important parts of the satellite system and should be properly chosen. It all begins here. Spending extra money in the future on receivers or other electronics will usually not improve picture quality if the outdoors components have not been properly selected and installed.

A dish must be mounted on a stable foundation with a reliable mechanism for aiming the dish at the arc of satellites. A flimsy mount will allow an antenna to sway in the wind and may even collapse under the weight of a heavy snow. The dish must have a secure attachment to the supporting pole and mount, and a method for setting and maintaining declination and polar axis adjustments. The bearings which allow tracking should be well designed to move freely and to resist excessive wearing. The points of attachment to the dish must be strong but should not cause the antenna to warp or allow twisting or excessive movement under loads (see Figures 4-1 to 4-4).

By far the best method of evaluating a mount is to examine its structure in detail. Push the dish from its outer rim to see if there is any play or if the various adjustments may be upset too easily. A comparison with other mounts is usually quite instructive. There are large differences between various brands.

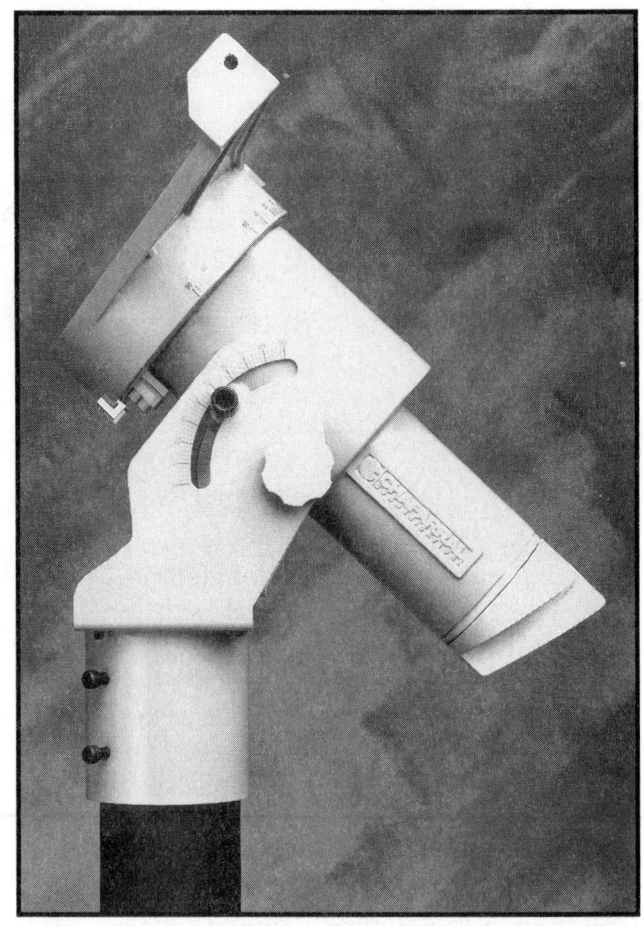

Figure 4-1. Polarmotor™ Horizon-to-Horizon Mount. *The Polarmotor™ is designed for prime focus or offset fed parabolic dishes. It tracks the full 180° arc in 50 seconds, has a pointing accuracy of 0.2° based on 5 counts per degree of rotation. It uses an optical encoder for counting that is compatible with either reed switch or Hall effect circuitry. (Courtesy of Chaparral Communications)*

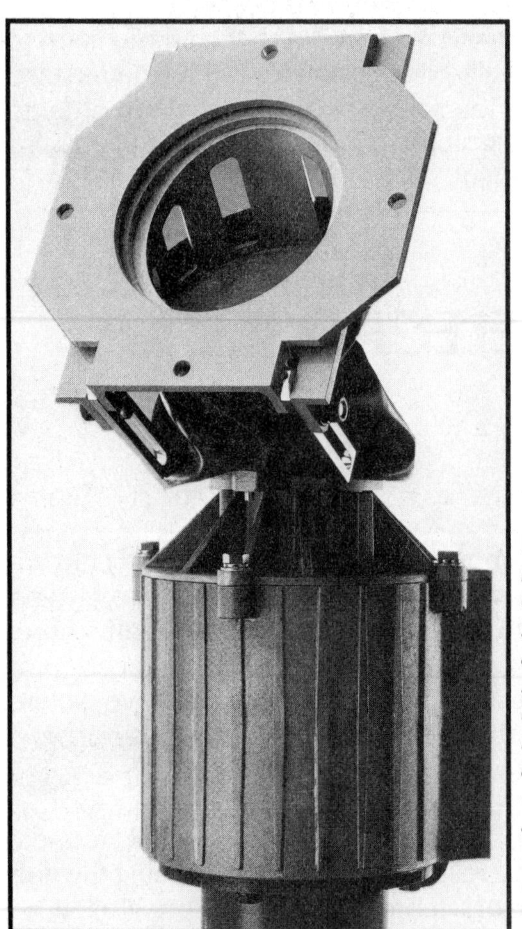

Figure 4-2. Az-El Robot Arm. *Even the most accurate satellite polar mount has some tracking error. Nearly perfect accuracy can be better obtained by using azimuth-elevation tracking. However, this type of system requires multiple motors and sensors so that off-the-shelf actuator controllers cannot be used. A complete, self-contained system having sensors, controls and motors to drive an antenna in both directions is necessary. An older variety of motorized az-el mount, the NITEC robot positioner, targets a satellite with a 0.2° accuracy. The maximum allowable antenna weight is approximately 68 kilograms (150 pounds) without the use of counter balances. The azimuth travel range is 180° and the elevation can be adjusted from 10° to 50° above the horizon. The programmable controller selects from a menu of up to 32 satellites. (Courtesy of NITEC Advanced Technologies)*

Figure 4-3. Polar Mount. *The top bar in this photo is the polar axis bar. The angular difference between the top bar and the one just below sets the declination angle. The angle of the second bar from the horizon is the elevation. The north-south orientation is adjusted by rotating the mount on the supporting pole. A linear jack tube is installed here (Courtesy of Brent Gale)*

Figure 4-4. Horizon-to-Horizon Mount. *The Ajax horizon-to-horizon mount is being readied for installation here. The entire mounting and tracking assembly is located at the base of the antenna. (Courtesy of Ajax Manufacturing)*

Actuators

Actuator arms and horizon-to-horizon assemblies are subjected to mechanical stresses. Jack tubes and motors are especially prone to problems and failures. However, satellite TV technology has developed during the past decade to the point where reliable units are readily available.

Make certain that the actuator jack does not have excessive play. This could cause an antenna to flop around in the wind. A longer arm is need for a larger dish so that adequate leverage and force is maintained throughout its sweep across the entire arc. Seals against water entry should be thorough; drain holes should be drilled at the lowest point on the motor housing. Actuator arm units should be shipped with ball joints for attachment to the dish and mount. Also make sure that the hardware is properly protected against the weather; galvanized, stainless or zinc-plated steel is the best.

Actuators are controlled from indoors by signals from either potentiometers, reed switches, Hall effect transistors or optical sensors, each type having somewhat different characteristics. Some actuators take an undue amount of time to send a dish from one end of the arc to the other. Be aware of the programming requirements for the actuator controller: some brands are easy to install and program; others are more difficult and can be unreliable. Do not purchase an actuator which does not have mechanical limits or the ability to protect itself with devices such as slip clutches in the motor should some over-ambitious user or confused microprocessor try to drive it far beyond the end of the arc.

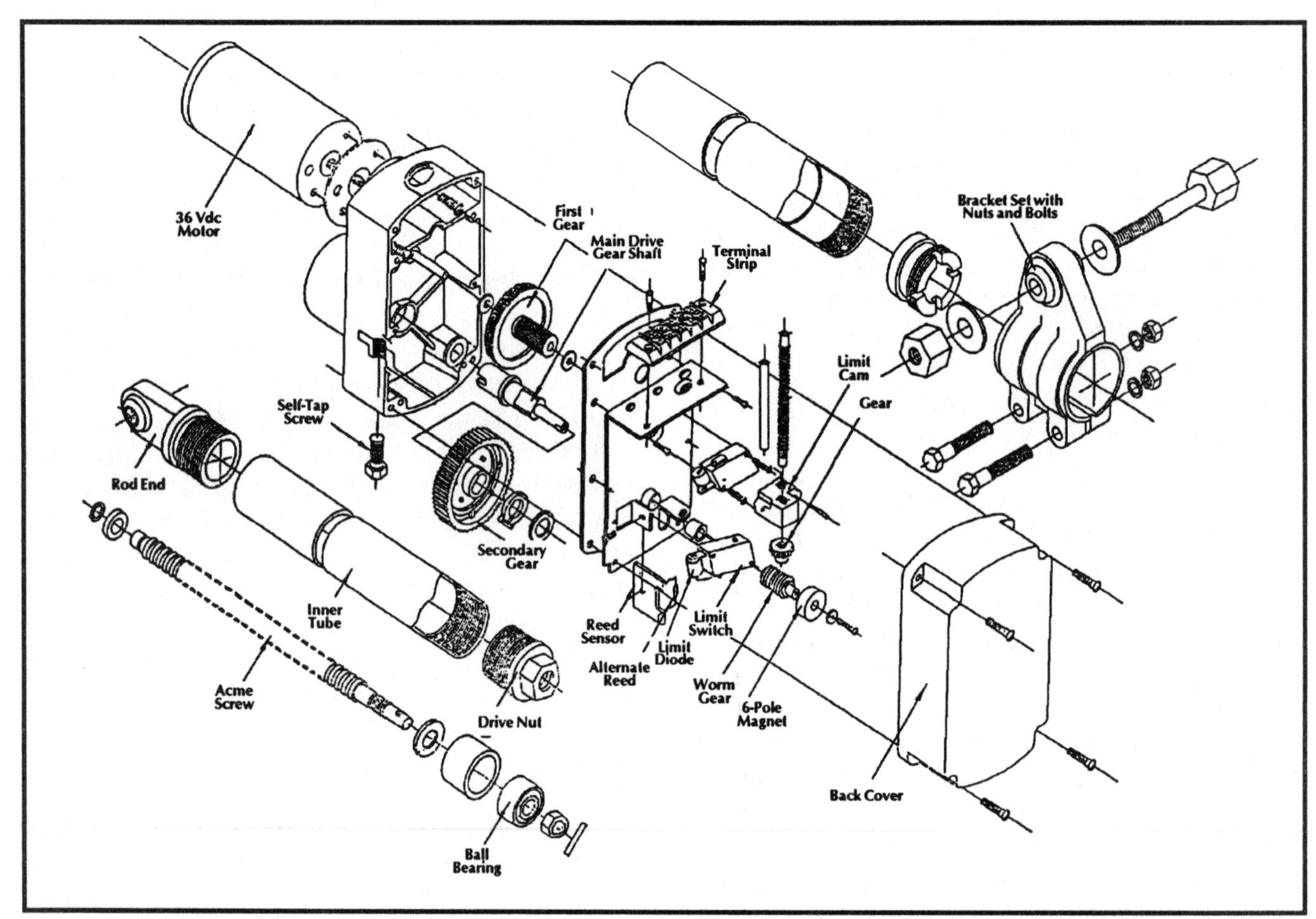

Figure 4-5. SuperJack II+ Construction. *This complete assembly of the SuperJack II+ is diagrammed here. (Courtesy of Pro Brand International, Inc.)*

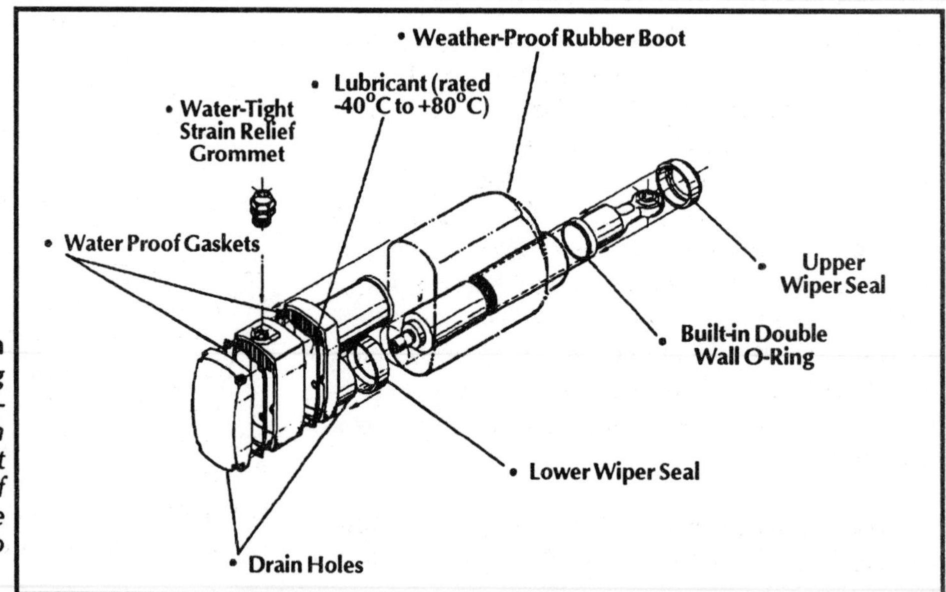

Figure 4-6. Waterproofing an Actuator Arm. *This drawing shows the points where an actuator can be protected from water entry. Drain holes must be installed and weatherproof covers as well as shaft wipes are necessary. (Courtesy of Pro Brand International, Inc.)*

Antennas

An antenna is the eyes of a satellite system. It must accurately concentrate and reflect the satellite signal as well as reject interference and noise. This is a tall order considering that a dish sits outdoors and is subjected to all the elements including corrosion, wind, ice and snow loading, hail, heat, cold and children. Of course, an antenna dish that may be perfectly adequate in Florida may not stand the test of time in Wyoming where snow and winds are often excessive.

Judging which type of dish to use is based upon many factors including aesthetics, performance, weight, ease of assembly, wind loading, durability, ease of shipping and cost. For example, spun aluminum dishes can have excellent surface accuracies but come in one piece and are not easily shipped. Wire mesh dishes which many find very attractive can be broken down and shipped in small boxes but may conform to a parabolic shape less accurately than other types if the panels are not held securely. Stamped aluminum dishes are very durable if their wall thickness is sufficient but are more easily buffeted by winds than a wire mesh antenna.

A simple physical inspection of a dish can be very revealing. First, when sighting along one rim, does the other side line up perfectly parallel? If not, the dish is warped. Second, does it feel smooth to the touch or is the surface rough or bumpy? If the surface has visible waves or has panels or mesh sections that do not line up accurately, beware. Third, how is the hub or central portion of the dish attached to the mount? Check that the mount is adequately strong and well attached to the dish to support its full weight without allowing the antenna to sag and warp in time. Some dishes are secured with only a few screws or bolts which may twist as well as rock when it is under stress. These may not withstand the test of time in very windy or snowy environments. Any rocking or twisting movement caused by wind will eventually produce metal fatigue and lead to failure and poor video quality.

The design and method of supporting the feed assembly are very important. Check how much weight is hanging out in front of the antenna. Some designs resemble a bowling ball on a bamboo pole. They certainly should be supported by guy wires attached to the outer antenna rim. The feed should block a minimum amount of the dish aperture or side lobes may be unnecessarily high.

The designs and materials of a dish must be adequate so expansion and contraction caused by temperature changes will not ruin surface accuracy or structural integrity. If a panel is damaged it should be inexpensive and simple to replace. If the dish is located near an ocean, the materials must be treated to resist the highly corrosive effects of salt spray. When a dish will be installed on a roof mount, it makes sense to use a lightweight mesh or spun aluminum reflector.

Dishes can be grouped into two major categories: mesh and solid (for more details see Chapter II). The solid antennas are either fiberglass, or stamped or spun metal. Brands from each category can be evaluated by all the above criteria as well as by some which are specific to a given type of dish.

Mesh antennas can be made from expanded or stamped steel or aluminum. These materials are protected by painting, powder-coating or anodizing. Judging which methods are best for the given climate depends upon where the dish will be installed. A painted mesh dish may be perfectly durable in a desert environment but a disaster on the coast of Oregon because of salt corrosion. It is important that a mesh dish feel strong to the touch. A flimsy mesh will probably not withstand a hail storm without dimpling , deforming or having some panels badly damaged or destroyed.

How the shape of a mesh dish is maintained is crucial to its performance and is often related to the ease of assembly. As few as eight or as many as 24 supporting struts and panels of mesh may be used. Struts are usually manufactured from tubular steel, extruded or angular aluminum for structural strength in order to maintain their shape when under wind or snow loading. In some mesh dishes the actuator arm attaches to just one rib which certainly must be strong in order to maintain a true shape. If not, the antenna could be warped from one side. Mesh dishes often have one outer circular ring by which these supports are connected to the middle hub; others have one or more middle supporting circular rings as well. Some designs and materials yield an accurate and stable surface; others are poorly designed and manufactured. Common sense judgments of these designs are usually surprisingly accurate.

Solid antennas are usually stronger and often can be more durable than mesh dishes. They can be assembled more quickly. Snow and ice tend to slide off instead of accumulating as they do on some mesh antennas. But solid dishes usually weigh more and are more subject to wind loading and sagging under weight.

Spun aluminum or steel dishes do have the most accurate surface but are one piece and rather cumbersome to ship and to move to a site. This type of dish as well as the stamped or hydroformed variety must have their surfaces protected from rusting or oxidation. Of course, those that are made from stainless steel are free from this concern. But unpainted stainless steel dishes can have problems with heat build-up. In some brands, the mount is bolted to a back plate that is attached to the reflector surface. Others have mounts which are bolted directly to the back of the reflector and, if not carefully assembled, can distort the dish surface as the bolts are tightened. This is especially so if the mount does not properly match the reflector shape and if rubber washers, which can act as shock absorbers, are not provided for insertion between the mount and reflector. There can also be an electrolysis action between the different metals which can cause severe corrosion.

The quality of a fiberglass dish depends upon which manufacturing method was used (see Chapter II for more details). If they are cured and molded properly, fiberglass antennas can have a very accurate, smooth surface. Improper techniques can result in warping, shrinking, surface deterioration and even separation of the protective gel coating from the underlying structure. In the worst cases, the embedded reflective material could delaminate. Fiberglass dishes can be very heavy and must have solid mounts attached securely at numerous points (see Figure 2-28). Like other types of antennas, sections should have compound curves and rib structures to form a true parabola and these should fit snugly together. The surface should also feel smooth to the touch.

Fiberglass antennas have the advantages of being corrosion resistant and durable. The best test of any brand of fiberglass dish is probably how it will maintain its accuracy and structural integrity over a period of time in the field.

Remember that microwave antennas are also capable of performing as solar concentrators. So the choice of colors and surfaces can be critical. A brightly painted dish or a smooth metal surface can reflect enough solar energy to melt polarizer caps or fry cables (see Figure 4-7). Even rolled mesh and fiberglass surfaces are smooth enough to cause a problem unless they are properly painted. When painting an antenna, an optically rough paint must be used so that the sun's rays are scattered not focused to generate heat. It is also better to use paints which are non-metallic since rough spots or bumps in the coat could cause reflection errors.

Feed Supports

Prime focus parabolic antennas have three varieties of feed supports: buttonhook, tripod and quad (see Figures 4-8, 4-9, 4-10 and 4-11).

Figure 4-7. The Effects of Solar Energy. *A highly reflective antenna can literally melt a plastic feedhorn cover, especially at solar outage times when the sun is aimed directly down the antenna boresight. (Courtesy of WV Publications, London)*

Whatever type is used, the feed must be stable under wind or snow loads and should not move when the dish tracks across the arc. Make sure that the attachments between the dish and support structure and the support and feedhorn are strong. A feedhorn, LNB and block downconverter mounted at the dish focus can easily stress their mounting hardware and cause the feedhorn to move from the desired "sweet spot" at the exact focal point of the reflector.

Some dishes have feeds which simply pop into a preset position so that no adjustment of the focal length is possible. This is both an advantage and a limitation depending upon the attention an installer wishes to give to tweaking the feed position. The adjustable varieties are more difficult to work with but allow additional fine tuning which can optimize the picture quality received.

Figure 4-8. QuadPod Feed Supports. *This photo illustrates the additional feed/LNB support that can be achieved compared to a buttonhook or tripod design. However, the extra support structures block additional incoming signal and can result in an increase in side lobes.(Courtesy of Brent Gale)*

Figure 4-9. Buttonhook Feed Support. *The Winegard 1.2 meter perforated aluminum antenna, model CK-4014, has the advantages of lower wind resistance and see-through appearance while retaining an accurate reflector surface. (Courtesy of Winegard Satellite Systems)*

Figure 4-10. Tripod Feed Support. *This spun antenna supports the feed and LNB at its focus by a tripod. Both the feedhorn weathercover and the feedhorn bolt to all three support struts.*

Figure 4-11. Tripod Support on Small Plastic Antenna. *The feed structure is dramatically visible against this transparent plastic dish, especially since its diameter is just 34 inches.*

Feedhorns

Feedhorns have the important job of properly illuminating a dish and of capturing as much of the incoming signal as possible (see Figures 4-11, 4-12 and 4-13). It is critical that the proper feed be chosen for each f/D ratio. Deep dishes (f/D from 0.32 to 0.25) will use different feed configurations than shallow ones.

Well designed and manufactured feeds have very low insertion losses, i.e. they transmit most of the signal into the LNB. The selection of a feedhorn can be an important determinant of picture quality. The cost is a small enough portion of the entire system expense to warrant purchasing top of the line equipment.

Polarity selection is generally accomplished by servomotors of ferrite devices. Each type has its strengths and weaknesses. For example, a ferrite device will probably perform more reliably in very cold climates where motors may seize. While ferrite polarizers are the most common variety found in European installations, they are rarely used in North America where servomotors predominate.

Both Ku-band and dual-band feedhorns are available. C/Ku-band operation can be efficiently accomplished with some dual-band feeds that are installed and function as well or nearly as well as single band C or Ku feedhorns (see Figures 4-14 and 4-15).

Figure 4-12. Corotor II™. *The Corotor II™ dual-band feed is designed to simultaneously receive both C- and Ku-band satellite broadcasts. Both the C- and Ku-band waveguides which are set at 180° from each other share the same probe. This feed has a maximum VSWR of 1.4:1, a cross-polarity isolation of greater than 30 dB and an f/D ratio adjustable from 0.28 to 0.42. (Courtesy of Chaparral Communications)*

Figure 4-13. Bullseye III™. *This orthodmode feed is designed to simultaneously receive both C-band polarities as well as higher frequency Ku-band broadcasts. Both feeds are located at the prime focal point of the reflector. (Courtesy of Chaparral Communications)*

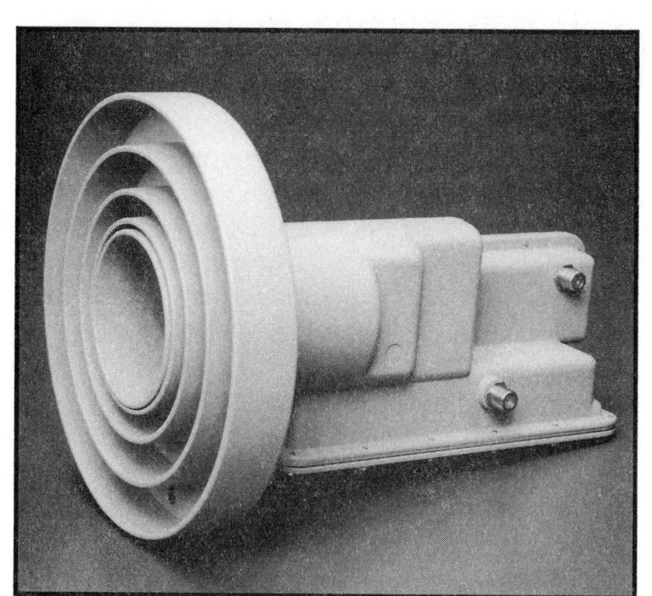

Figure 4-14. Masterfeed™ Orthomode Feedhorn. *This component combines two LNBs and the feedhorn into one compact housing. It allows reception of both horizontal and vertical channels for simultaneous viewing of any of 24 channels from two receivers. It therefore eliminates the need for an orthomode feedhorn and two separate LNBs. (Courtesy of California Amplifier)*

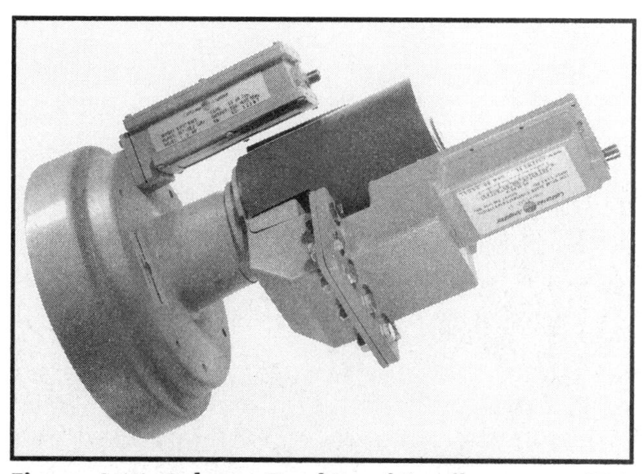

Figure 4-15. CalAmp Dual-Band Feedhorn. *This dual-band feed uses a single servo motor to rotate probes for both C- and Ku-band polarity selection. C- and Ku-band HEMT LNBs are installed on the output flanges. (Courtesy of California Amplifier)*

LNBs

An LNB is the most complex piece of an earth station but most often the most reliable. Its noise temperature should be low enough to yield a good quality picture. C-band LNBs in the 35 to 60°K range are commonly available while Ku-band components have noise temperatures in the 70 to 120°K range (see Figures 4-16 and 4-17).

While a minimum LNB gain of 40 dB is sometimes acceptable, a 50 dB unit should be chosen to guarantee top performance.

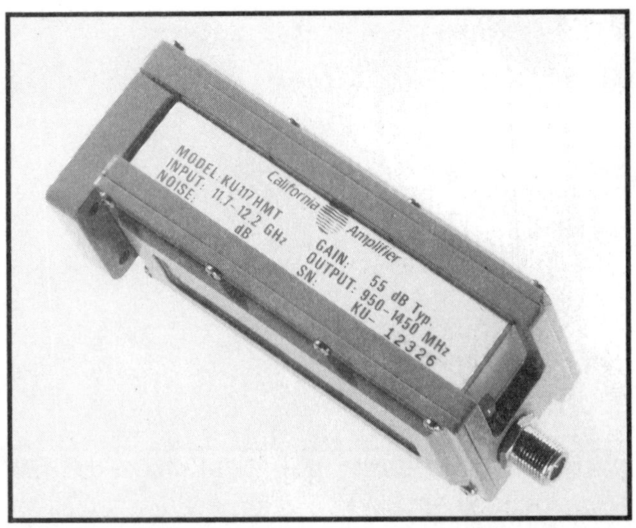

Figure 4-16. Ku-band LNB. *This Ku-band LNB operates in the 11.7 to 12.2 GHz range and has a block output of 950 to 1450 MHz. (Courtesy of California Amplifier)*

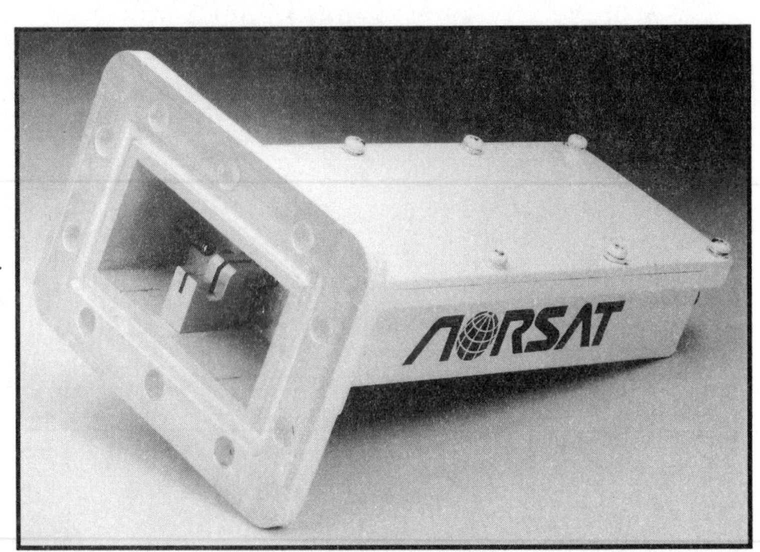

Figure 4-17. C-Band LNB. *LNBs are now available in noise temperatures as low as 25°K. (Courtesy of Norsat International)*

Figure 4-18. Monterrey 90 Satellite Receiver. *This C/Ku-band IRD features picture-in-picture, an audio/video switcher, Dolby™ surround sound, MTS and digital stereo, an adjustable bandwidth, parental lock-out, on-screen menu displays and 100 favorite channel memory. (Courtesy of Chaparral Communications)*

Figure 4-19. Drake ESR 2450 Receiver. *This C/Ku-band IRD features a built-in VideoCipher II module, 36 channel memory, parental lock-out, built-in positioner, remote control, on-screen graphics display and dual C-band LNB inputs. (Courtesy of R.L. Drake Company)*

Receivers and Downconverters

Sometimes it seems that there are as many receivers available as salesmen to sell them. Furthermore, each brand of receivers seems to have a different look and different features (see Figures 4-18 to 4-24).

Video quality can vary substantially between receivers. Some units produce jumpy, flickering or grainy pictures. These characteristics are a reflection of receiver electronic design. If a low threshold has been obtained by the least desirable method, reducing the bandwidth, the result is often fewer sparklies but a picture with less resolution and smeared colors. Phase lock loop (PLL) receivers will normally have lower thresholds than coaxial delay line receivers. However, in more extreme cases when using PLL receivers, bright colors or high contrast scenes can tear or streak. But when using under-sized antennas, PLL receivers can have thresholds 1 to 2 dBs lower than coaxial delay line types and therefore better pictures.

Audio quality also varies between satellite receivers. The audio should be easy to tune over the full 5 to 8.5 MHz range. There should be no buzzing or rasping on weaker transponders with bright colors on the screen. Check both the video and audio quality on transponders which are loaded with audio subcarriers. Each audio subcarrier should be heard clearly and crisply providing the signal strength is adequate.

Receivers also differ in the method used to tune between channels. Continuously variable tuning may be quick but it can be annoyingly inaccurate and determining what channel has been selected may be difficult. Some receivers have an even/odd polarity selection button. This prevents the polarizer from rapidly switching back and forth as all channels are scanned in their natural order, i.e. 1,2,3,4, etc.

If a remote control is desired, it should have the five main functions: channel selection, fine tuning, audio selection, volume control and polarization selection. If any of these are absent a user will have to walk over to the receiver to change the missing

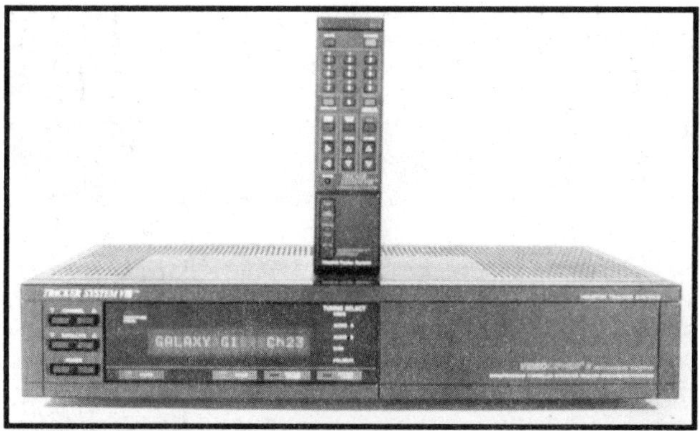

Figure 4-20. Tracker System VIII. *Features of this IRD include an automatic/pre-programmed polarity setting and quartz-locked channel tuning. It can recall 99 favorite video and audio programs, either on C or Ku-band broadcasts. When programming this unit, the receiver seeks the strongest signal, locks onto it on both actuator and polarity settings and then stores these positions in memory. All functions are controlled by a UHF wireless remote control that has a range of up to 65 meters. These functions can be monitored from any television in the home by an on-screen graphic display. (Courtesy of Houston Tracker Systems)*

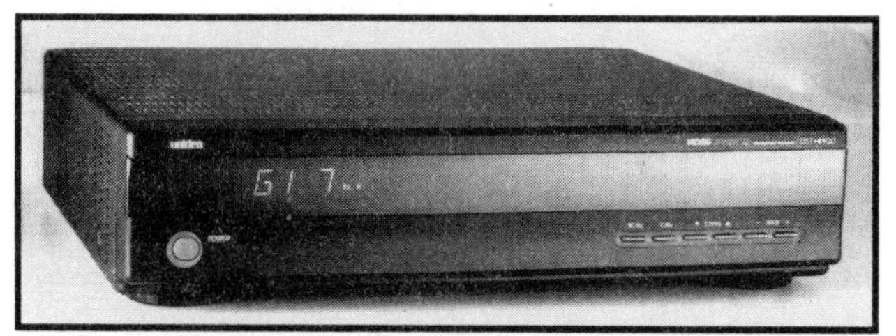

Figure 4-21. Uniden UST-4400 IRD. *This dual-band receiver features high fidelity stereo sound, on-screen graphics and parental lock-out. (Courtesy of Uniden Corporation)*

Figure 4-22. General Instrument 2750R Receiver. *The 2750R IRD is C/Ku- band compatible and accepts a 950 to 1450 MHz input. Its features include digital stereo sound, parental lock-out, 32 satellite and 150 channel memories, favorite channel recall, built-in TI filter, on-screen display and UHF remote. (Courtesy of General Instrument)*

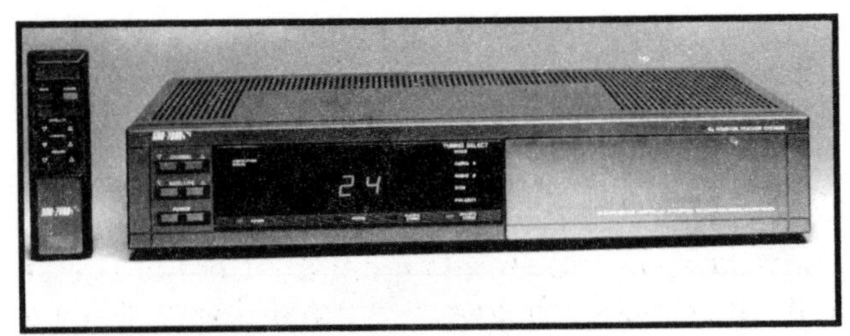

Figure 4-23. Echostar SRD 7000 Receiver. *The Videocipher II equipped receiver features 950 to 1450 MHz block downconversion and programmable video noise reduction circuitry, automatic C to Ku-band antenna positioning which includes an optional C/Ku dish controller accessory, on-screen menu display, phase/locked loop synthesized video and audio with programmable fine tuning, and parental lock-out. All these function are controlled via a UHF wireless remote. (Courtesy of Echosphere Corporation)*

Figure 4-24. STS SR100 MK2 Receiver. *This American market IRD accepts a 950 to 1450 MHz input. Its features include on-screen display, 90 satellite and 40 favorite program recall, parental lockout, full stereo, channel scan and audio muting when changing channels. An optional LCD "smart remote" that control up to six independent audio and video components is available. (Courtesy of Satellite Technology Services)*

function thus eliminating the advantage of the remote control. Some of these functions could be automatically selected by the receiver.

Stereo Processors

Stereo processors, either built into a receiver or stand alone, should be capable of handling at least the two most popular stereo formats: discrete and matrix. The separation between channels should be clear and distinct and the audio quality should be strong and clean with no popping or static audible. There are two ways to tune in audio channels. The first uses a continuous tuning circuit such as a potentiometer. The second and the most accurate method is voltage synthesized tuning. All audio channels are preprogrammed into a memory and selected at the push of a button as displayed by a digital read- out.

Television Monitors

Television monitors accept raw audio and video signals from a satellite receiver and bypass the intermediate modulation step. As a result, picture quality can be better compared to conventional television sets. A satellite dealer may find a good quality monitor useful as a demonstration of how clearly satellite broadcasts can be received and as an installation tool. Table 3-2 below lists some of the monitors available for purchase today.

Warranties

Even the best designed and manufactured equipment can fail. If the factory offers a warranty having too short a period, does not provide a fast turnaround on service or, even worse, is out of business when the equipment fails, a dealer can incur considerable costs.

The typical warranty for satellite TV components offers full coverage for at least 90 days and limited coverage for one year. But there are wide variations. A few manufacturers have extended warranties on electronic equipment lasting up to two years. Some antenna manufacturers warranty their products for five years indicating either faith in their design or an unrealistic hope in the longevity of their product. In any case, responsible manufacturers should stand behind their products with prompt service. Some companies have gone out of business when high equipment failure rates overly taxed their financial resources. In these cases, warranties were only as good as the paper they were written on.

Therefore, a prudent course of action to follow should include careful selection of well-made equipment, a knowledge of the underlying warranties and an evaluation of the manufacturer's stability and reputation. Word-of-mouth information from knowledgeable sources can be invaluable in judging a manufacturer.

Selecting Receivers for International Broadcasts

C-band spacecraft such as Mexico's Morelos, Brazil's Brazilsat and the many North American satellites in orbit can be adequately detected by conventional reception systems. But Intelsat and other satellites which use circularly polarized signals, half transponder formats or unusual frequency allocations place additional requirements on the equipment. In addition, care must be taken to select an appropriate satellite receiver to match the broadcast format of a given country.

Circularly polarized broadcasts can be easily received by retrofitting a conventional feedhorn with a dielectric insert as described in Chapter II. This action may cause some minor losses of signal power which can be compensated for by slightly increasing dish size or by lowering LNB noise temperature.

In order to receive half transponder broadcasts a receiver must be tuned to the 18 MHz bandwidth centered in the lower half of the transponder band. If not, it will detect excess noise from those unused portions of the 28 to 32 MHz band that it normally detects. If necessary, bandwidth reduction can be accomplished by using a 14 to 18 MHz bandpass filter. Remember that as receiver bandwidth is narrowed, C/N rises because the detected noise is reduced. So a smaller dish could be used to view these half transponder broadcasts. But also remember that as receiver bandwidth is reduced, picture fidelity worsens.

Receivers that do not have AFC defeat switches or that are permanently locked onto the standard C-band center frequencies cannot be off-tuned to broadcasts having different center frequencies. Therefore, synthesized tuned receivers which may perform excellently with conventional transmissions might be unusable in certain environments.

Most C-band broadcasts relay their audio information on subcarriers located at either 6.2 or 6.8 MHz. Nevertheless, it is advisable to have a receiver which is continuously tunable over the 5 to 7 MHz audio range for either stereo reception or for detection of audio carriers with uncommon center frequencies. Some international transmissions are relayed with their audio located on separate narrowly tuned carriers, known as single channel per carrier (SCPC) formats. Many receivers today are simply not capable of simultaneously detecting SCPC audio and conventional video. Some Intelsat spacecraft even relay this SCPC information on a transponder different from that which carries the associated video messages.

It is also important to realize that both American and Canadian receivers have usually been constructed solely for processing the 525- line North American broadcast format. Numerous countries in Central and Latin America use the higher resolution 625-line system. The shaping and filtering of a broadcast signal by de-emphasis and low-pass networks can cause some noticeable picture streaking and color smearing when used with 625-line formats. Some receiver manufacturers have designed specially altered receiver circuits which are now available to solve this compatibility problem.

A television or monitor specifically tailored to the particular NTSC, PAL or SECAM format is required for quality reception in any particular country. Excellent multi-standard components are now available if uncertainty exists or if the equipment will travel across international borders. Note that some multi-standard TVs or monitors are not yet capable of resolving problems with the particular variations of PAL used in Brazil and Argentina.

Decoders

In North America two scrambling systems have become standards, the VideoCipher II and II-Plus and the Oak Orion. The latter is used only in Canada and is not available to American consumers. The situation in other regions of the world varies. For example, in Europe a number of systems including Filmnet and EuroCipher are vying to become the accepted standard.

Changes are to be expected. Even in the United States where the VideoCipher technology is now entrenched, the introduction of Ku-band direct broadcast service may bring with it a different, perhaps more secure, scrambling technology.

Most receivers available today in the United States are IRDs (integrated receivers/decoders). While stand-alone VideoCiphers are still available, IRDs overcome any receiver/decoder compatibility problems that may have previously existed.

V. INSTALLING SATELLITE TV SYSTEMS

One of the objectives of this manual is to teach readers how to install home satellite systems that will stand the test of time. Theoretical knowledge, while academically interesting, is of little value to an installer unless the result is a system that works. Furthermore, even the most sophisticated troubleshooting expertise may be overtaxed if the installation was seriously flawed or if unwarranted shortcuts were taken.

In this chapter, the most general installation procedure is presented first to provide a clear overview of the entire process. Subsequently each installation step is examined in detail. Although this text offers numerous examples and details for completeness, a system can be installed with only a rather basic knowledge. This fact is the motivation underlying the sequence in which the information is presented.

For simplicity's sake, some steps in the detailed description are presented in a slightly different order than might occur during an actual installation. For example, the section on aligning the feed precedes the description of mounting actuator arms although it is convenient to hold an antenna in place by installing the arm before the feed assembly. Of course, in some situations, for example, if a fixed dish is being installed, the sequence would be different from a full-blown installation. The exact sequence of steps is often a matter of personal taste: each technician will develop a personal strategy and special methods as he or she gains experience.

A. STEP-BY-STEP INSTALLATION PROCEDURE

A satellite system can be installed by following a series of rather straightforward steps. Although the complexity of each step may vary between installations, most jobs are quite similar. The usual sequence is listed here.

1. Site Survey

The goal of a site survey is to determine where the satellite antenna will be located. Factors that must be considered include (1) finding a position with a clear view of the entire arc of satellites, (2) verifying that terrestrial interference, if present, will not be a serious problem and (3) planning cable runs from the antenna to the indoors electronic equipment.

2. Installing the Antenna Support

Satellite dishes are usually mounted on poles that are set in concrete. However, other types of supporting structures can be successfully used. The critical element of this step is to be sure that the pole is mounted in a perfectly vertical orientation.

Any tilting in the east/west direction can result in poor tracking. While setting a concrete support, install the section of conduit near the antenna base.

3. Trenching and Cable Runs

The route from the antenna to the indoors equipment is then prepared. Conduit is recommended when cable is to be installed underground. Lay the conduit and cable in the trench but do not cover it up. This should be done only when the installation is definitely complete and successful. It is crucial that the conduit be leak proof and it is wise to use conduit of sufficient diameter so extra cable can be pulled in or the old cable can be replaced at some future date. Be sure to use 45° or 90° electrical sweeps on any sections where directions will change.

4. Assemble the Mount

Whenever possible, the mount should be assembled independently of the dish and then lifted onto the pole. Bolting the dish and the heavy mount together and then lifting them onto the pole together can easily warp the reflective surface. This mistake should be avoided. It is also usually easier to set the declination angle on the mount at this stage.

5. Install the Linear Actuator

Installing the linear actuator onto the mount assembly prevents it from flopping about. At this stage, set the elevation angle and actuator position on the mount so it faces either directly upward or as closely to vertical as possible. The dish can then be easily lifted or rolled into place.

6. Assemble the Antenna

After assembly, the antenna can either be lifted onto the mount so it sits horizontally like a bird bath or it can be rolled into alignment with a vertically oriented mount. Secure the reflector to the mount. Then attach the feedhorn support struts or buttonhook.

7. Install the LNB/Feedhorn/Weathercover

After bolting the LNB, feedhorn and weathercover together, lower the antenna by adjusting the elevation angle and possibly by moving the actuator in its clamp so the LNB/feedhorn assembly can be attached to its supports. It is crucial to always install an LNB/feedhorn weathercover to protect the electronics.

8. Electrical Connections

Complete the necessary connection to the LNB, polarizer and actuator at the antenna and to the indoors receiver/polarizer/actuator controller. If the dish will be aligned with a test set-up located outdoors at the dish, temporary connections should be made.

9. Power On and Align Antenna

Check that all wires are connected correctly, then turn on the power. Then carefully set the elevation and declination angles and follow the procedure that aligns the antenna onto the arc of satellites.

10. Fine Tuning

Once a clear signal from each satellite has been detected, the alignment of the antenna and feedhorn can be fine tuned.

11. Waterproofing

All electrical joints should be waterproofed. It is also a wise practice to install an accordion sleeve on a linear actuator arm.

12. Seal all Cable Runs

Finally, bury the length of the conduit run. Be sure that weather covers are installed at the dish and at the entry to the building, if applicable.

13. Program Receiver

Program the receiver by following the steps in the manufacturer's instruction guide.

14. Connect Accessories

Hook accessories such as VCRs or stereos to the satellite receiver. The installation is now complete.

B. THE SITE SURVEY

Every installation must begin with a thorough site survey. This is one of the most critical yet often most neglected steps in a satellite TV installation. The survey is the equivalent of a doctor's examination and diagnosis prior to surgery. No surgeon would ever open up a patient without a complete understanding of what he wishes to accomplish. A competent installer would not cut corners in conducting a site survey.

Three important tasks are completed during a site survey:

- First, a location with a clear view of the entire arc of satellites is found.
- Second, a check for terrestrial interference must be performed.
- Third, the entire installation is planned.

Numerous facts must be established before beginning an installation. It is important to know which satellites and transponders will be accessed (see Table 5-1); what type of equipment should be used; where to place the antenna, how much concrete might be necessary; whether or not to use a long pole, roof, tower or ground mount; what length, route and type of cable is required; the distance from the antenna to the satellite receiver and a host of other questions. The planning process, carried out with the cooperation of the customer, must deal with seemingly minor concerns as well as issues with potential legal ramifications. For example, cable paths must always be clearly marked because the installation crew might be unfamiliar with the site. Furthermore, permits for installing the system and receiving the programming may be requ;ired.

Above all, the customer's wishes must always be considered so that all decisions are mutually acceptable. For example, if the customer does not want the antenna next to her home but 100 meters away, a line amplifier and 12 gauge actuator wire may be required. This may result in a substantial variance from the original bid. The customer's home and yard should be treated as if they were the installer's home. It is a wise practice to respect the saying that "the customer is always right."

TABLE 5-1. SATELLITES in the ARC

Satellite Name	Orbital Location (°W)
Aurora 1	143
Satcom F1R	139
Galaxy 1	134
Satcom F3R	131
Telstar 303	125
Westar V	122
Spacenet 1	120
Morelos F2	113.5
Anik C2	110
Anik D1	104.5
Westar IV	99
Telstar 301	97
Galaxy 3	93
Spacenet III	87
Telstar 302	85
Satcom F4	82
Galaxy II	74
Satcom F2R	72
Brazilsat A2	70
Spacenet II	69
Brazilsat A1	65

Ensuring a Clear View of the Arc

An antenna must have a clear view of each satellite because any obstruction will absorb or reflect microwaves and subsequently lower the detected signal-to-noise ratio. Water is a particularly strong absorber of microwaves, especially in the higher Ku-band frequency range. A reflector installed in winter might have perfect reception until spring when foliage returned to trees that obstructs its view: then pictures would either disappear or deteriorate in quality.

Two basic instruments are required to aim an antenna at any geosynchronous satellite: an inclinometer and a compass (see Figures 5-1 and 5-2). When a polar mount is installed, these instruments are used to aim the mount towards true south and then to set the polar axis and declination angles. When installing a fixed antenna or an az-el mount, each satellite can be targeted by adjusting the antenna's azimuth and elevation angle. The azimuth is measured in degrees of rotation from true north and the elevation in degrees up from the horizon.

A number of aids are available to allow satellites to be targeted. Either the computer program listed in the reference material or the equations in Appendix B can be used to determine azimuth and elevation headings for any point on the earth. (The azimuth headings for a polar mounted antenna can also be useful during installation.) Or the step-by-step procedure listed below can be used.

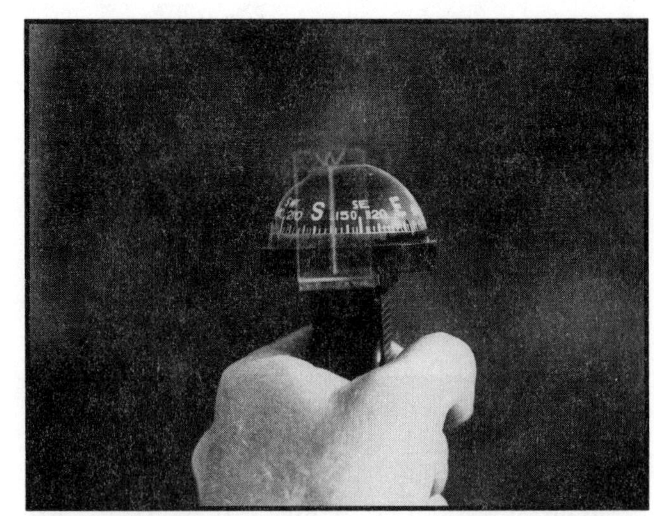

Figure 5-1. A Compass. *This hand-held compass allows the user to aim towards a particular heading while maintaining a view of the satellite dish. (Courtesy of Steve Berkoff)*

Determining a True North/South Bearing

A compass will point along the line between magnetic north and south. In most locations, there is a difference between the bearing to the north and south poles and the needle reading on a compass (see Figure 5-3). Magnetic north, the strongest magnetic center near the north pole, is located near Bathurst Island off the northern coast of Canada. As a result, magnetic variation can exceed 30° in locations such as the American state of Alaska. However, magnetic variation is zero along the "agonic line" that runs just off the west coast of Florida, through Lake Michigan, over the Great Lakes, and to the magnetic north pole. East of this line, the north pole is east of the compass reading; west of this line it is west. Similar agonic lines are also found in other areas of the globe as the accompanying maps illustrate.

A true north/south bearing can therefore be found by using a compass and correcting for magnetic variation (see Figure 5-4). For example, in the American city of Denver, Colorado, magnetic variation is minus 12.5°. So the arc of satellites centered on true south is found by rotating 12.5° east of the south reading on a compass. This is equivalent to a compass reading of a 167.5° (180° less 12.5°) as measured from true north. A similar procedure can be followed at any other point on the globe.

Figure 5-2. Inclinometers. *An assortment of inclinometers are pictured here. The rounded instruments display the reading on an external scale. The smaller device is sighted through to the target. The instrument is the rear is a level/inclinometer with a digital read-out.*

Magnetic variation at any location in the world can be found from the charts shown here (see Figures 5-5 through 5-10) and, in many countries, by requesting the value from a local airport information service. Airline pilots, like satellite TV installers, must also take their bearings from true north and south.

When using a compass, it is important to realize that it can yield inaccurate readings when used in the vicinity of iron-containing objects. Therefore, a bearing taken next to a large truck or any other large steel object could be many degrees off the local average magnetic north/south line.

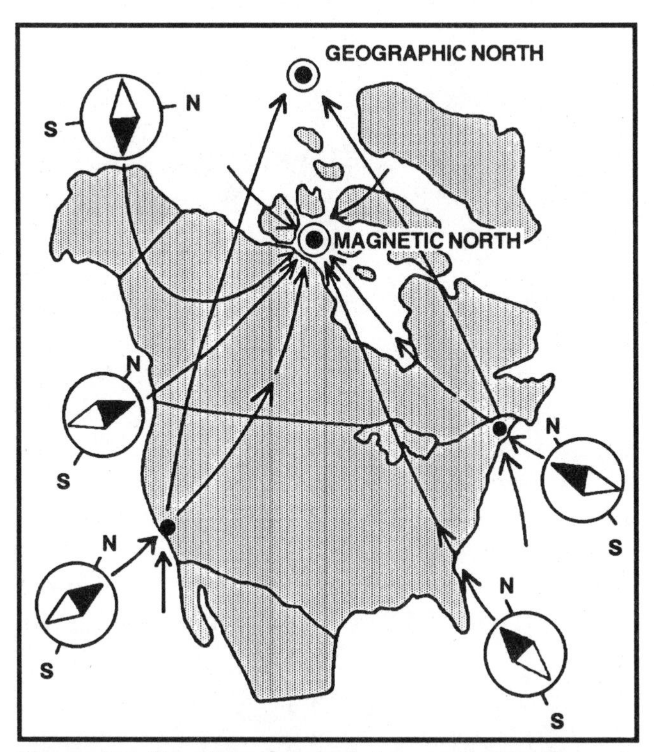

Figure 5-3. True North versus Magnetic North. *In both hemispheres magnetic north is quite far removed from the geographic north pole. In North America it is located just off the coast of Hudson Bay in Canada. This illustration shows the difference between the bearing to true and magnetic north.*

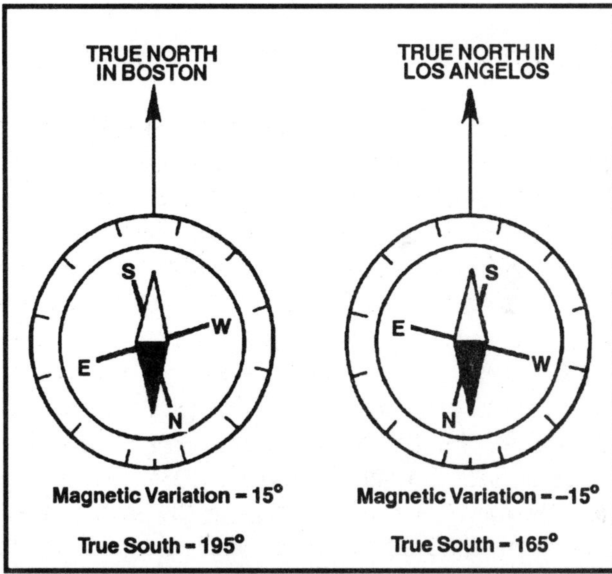

Figure 5-4. Adjusting for Magnetic Variation. *True south can be found by aiming a compass east of magnetic south when the variation is negative and vice versa. Magnetic variation is negative when west of the zero-variation, agonic line at a site north of the equator. South of the equator the settings are reversed.*

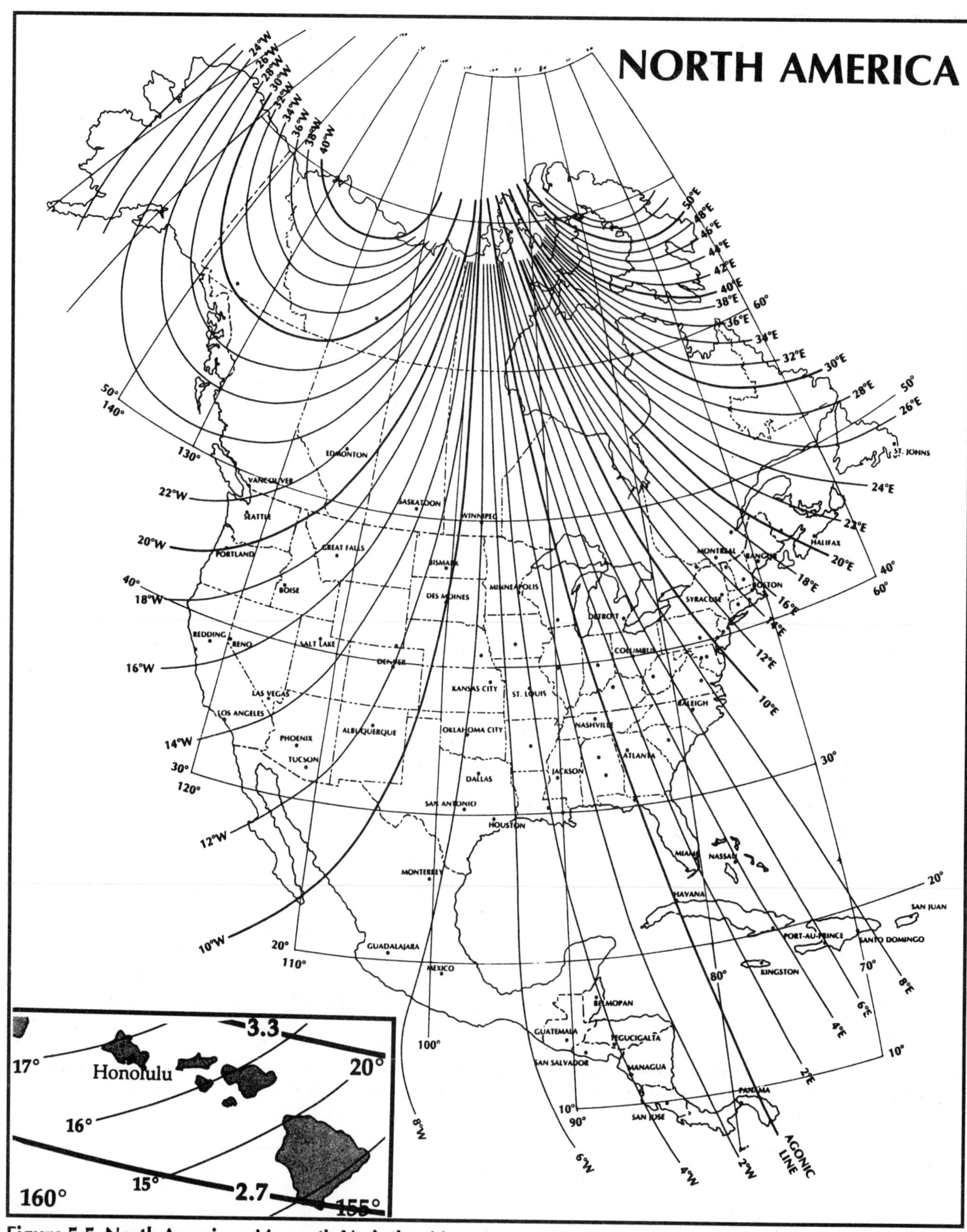

Figure 5-5. North American Magnetic Variation Map. *A compass points to magnetic north and not to true north except along the "agonic line" because of the difference in locations between the magnetic and geographic north poles. North of the equator, a reading taken with a compass west of this line must be corrected by rotating east to find true south. South of the equator the opposite is true.*

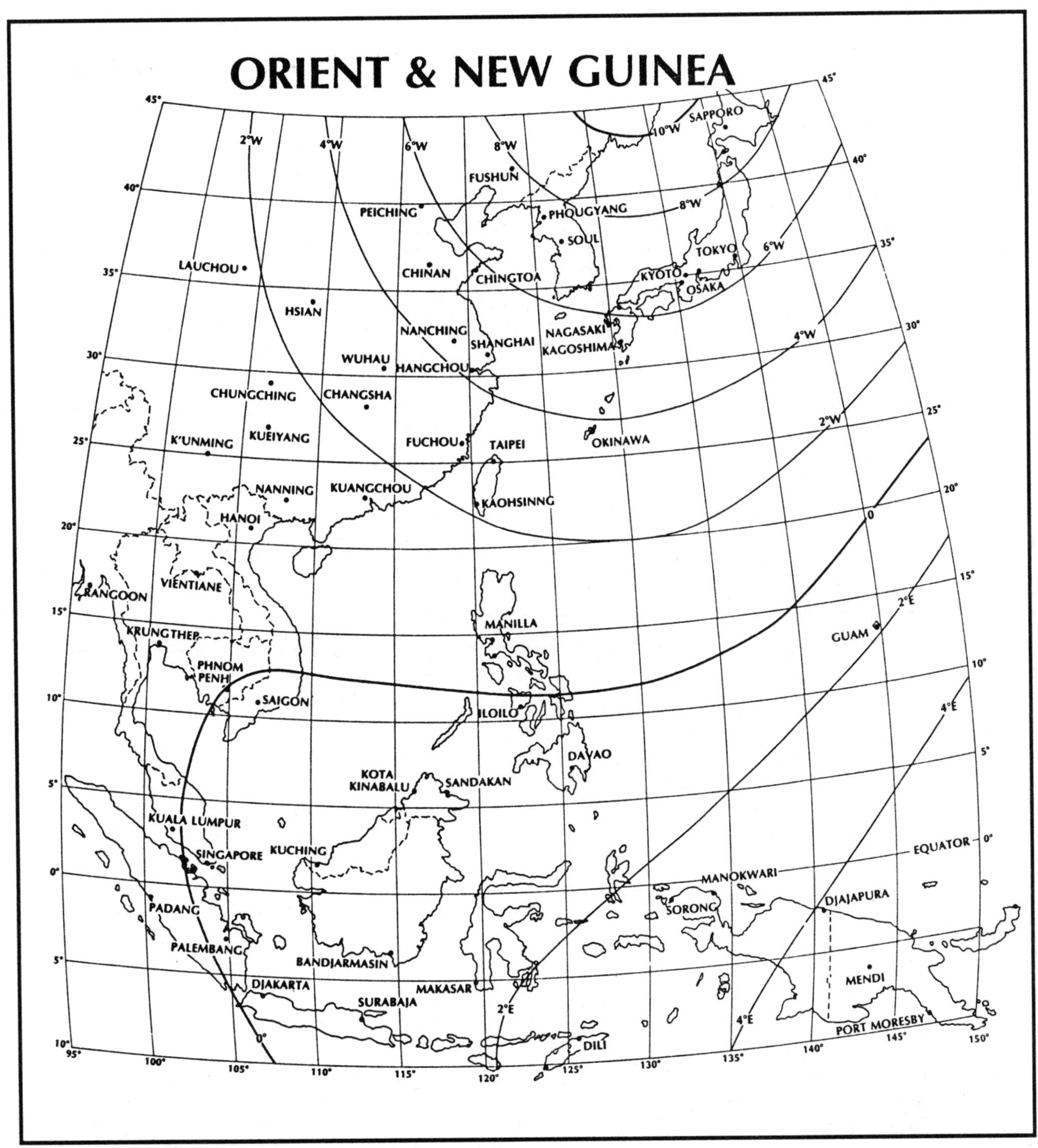

Figure 5-6. Orient and New Guinea Magnetic Variation Map.

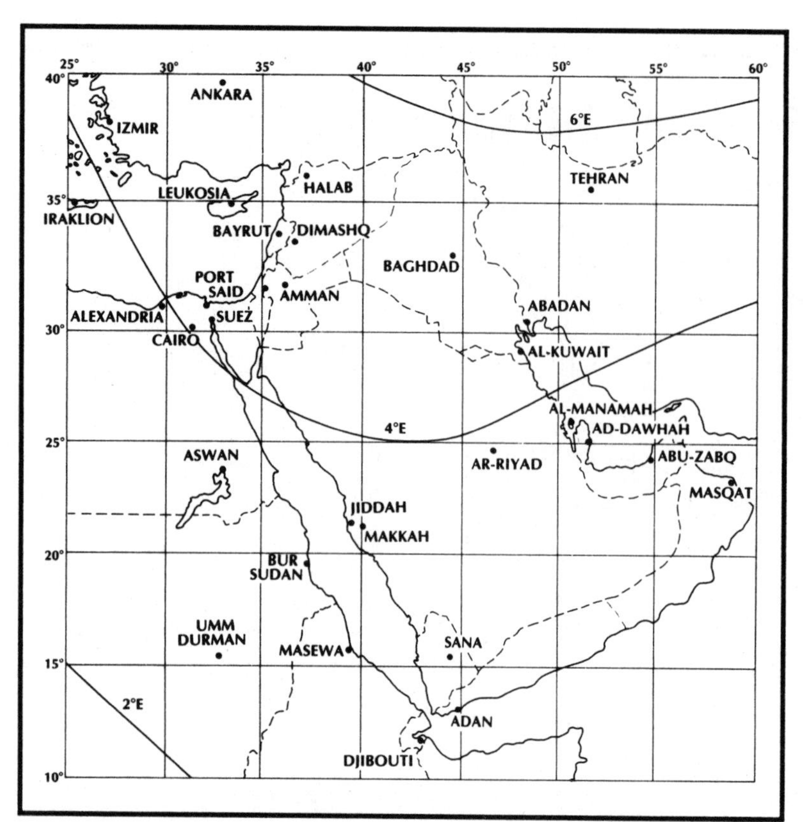

Figure 5-7. Middle East Magnetic Variation Map.

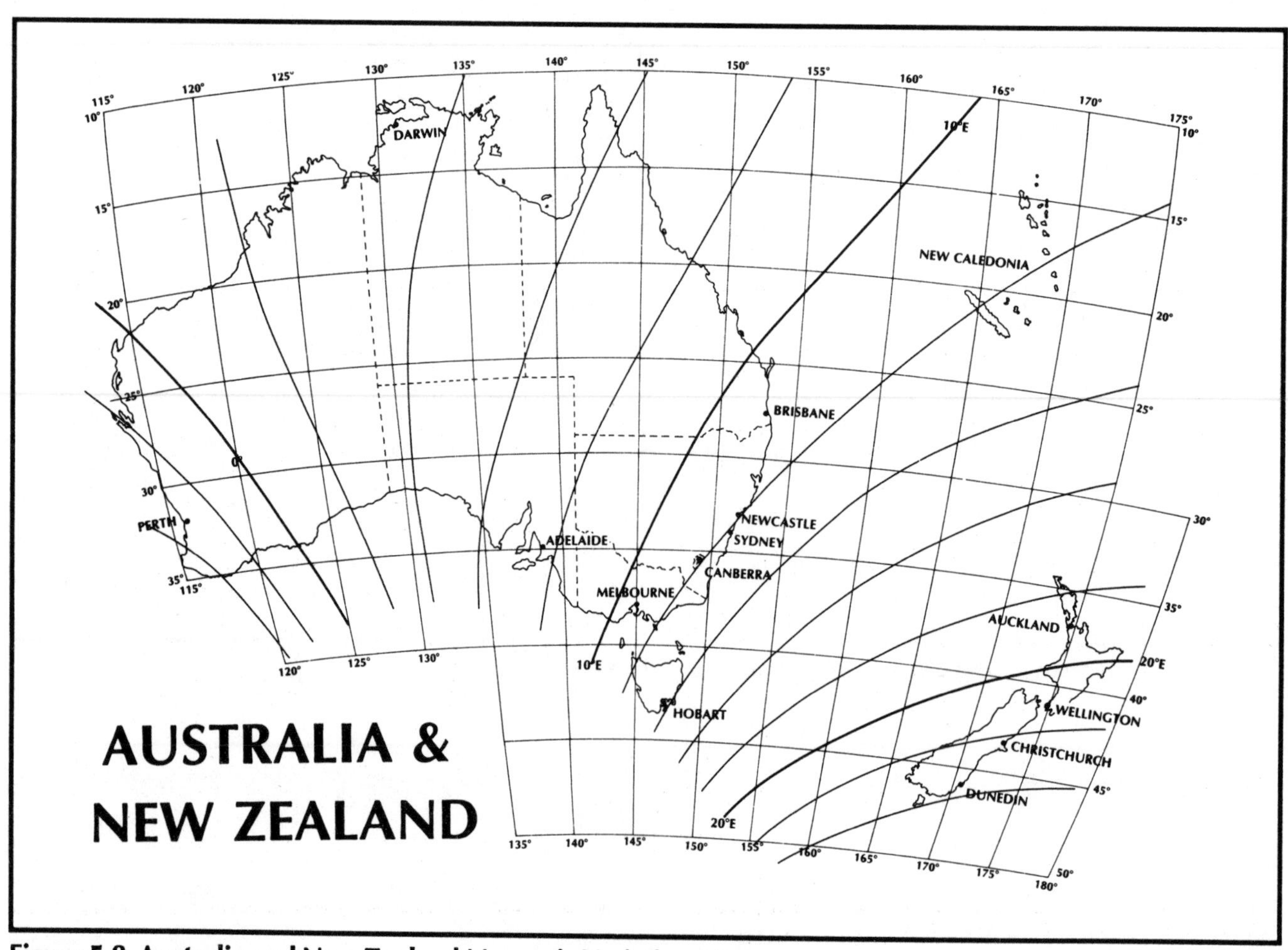

Figure 5-8. Australia and New Zealand Magnetic Variation Map.

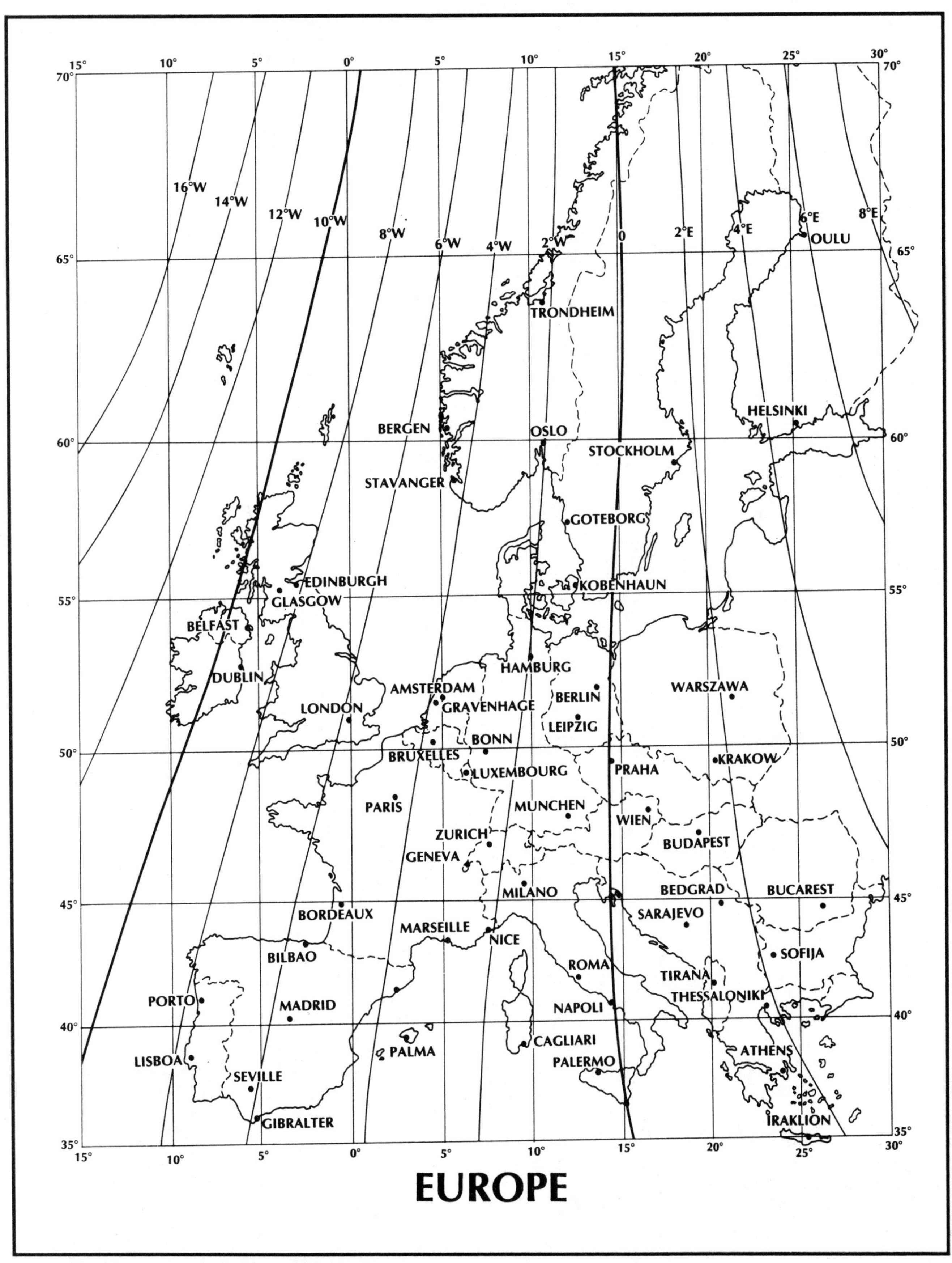

Figure 5-9. European Magnetic Variation Map.

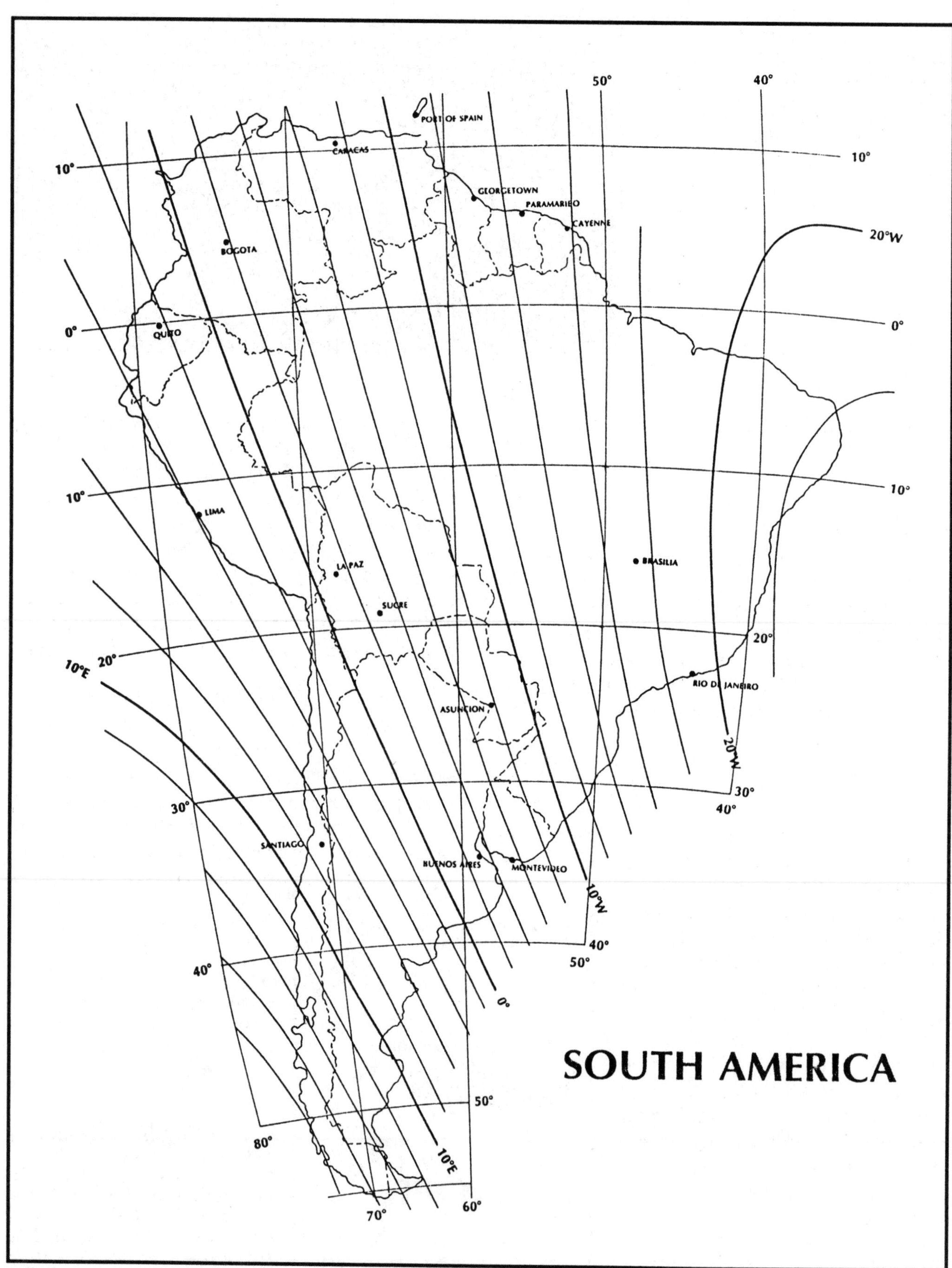

Figure 5-10. South American Magnetic Variation Map.

Figure 5-11. Measuring Elevation Angle. *Elevation angle can be measured by placing an inclinometer on a straight edge and sighting along it in the direction of the target satellite. (Courtesy of Brent Gale)*

Determining the Azimuth and Elevation Look Angles

The azimuth heading towards any chosen satellite can be found from a table or by calculation (see Appendix C for a description of these equations). Although such calculations are based on simple geometry, they might look complex on paper. Using computer programs can simplify this work. The satellite whose angular location corresponds to the site longitude can be found by aiming due south in the northern hemisphere and due north in the southern hemisphere. For example, Anik D1 is located at 104.5°W which is nearly due south of Denver at 105°W longitude.

Similarly, the elevation to any satellite can also be found from computer programs and charts or by direct calculation. For example, Spacenet II has an elevation angle of 30.8° and an azimuth of 138.0°W in Denver, Colorado. During the site survey it would therefore be found by rotating 29.5° east of true south and then by aiming up to 30.8°. A compass would read 125.5° when pointing at Spacenet II. This equals 138.0° less the 12.5° correction for magnetic variation. If a tree or any other obstruction were blocking the view, the proposed installation site would have to be changed.

Elevation angles are easily measured with an inclinometer. This instrument can be placed on a long ruler or any other straight edge and used in a fashion similar to sighting a gun. The ruler is raised until the desired elevation is reached (see Figure 5-11). While sighting along its length, see if any objects are blocking a clear view to each satellite in the applicable portion of the belt. When sighting with some types of inclinometers, a second person reading the scale makes the process a little easier.

Various companies market instruments designed to make sighting the arc even simpler. Some of these products are see-through charts having satellite locations marked. These are mounted on a tripod and pointed due south for sighting the satellites. Others have telescopic view finders which are preset for any geographic location so a user can visually scan across the entire arc (see Figure 5-12).

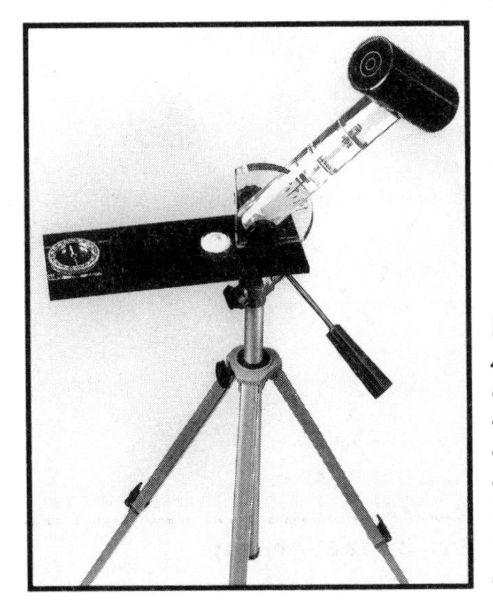

Figure 5-12. A Sighting Tool. *An installer can first set the azimuth and elevation angles to any satellite on the Viewfinder and then check for obstructions along the line to this satellite through the eyepiece. (Courtesy of Focii Manufacturing Company)*

Step-by-Step Procedure to Ensure a Clear View of the Satellite Arc

A rather simple procedure to ensure a clear view of the arc of satellites is outlined here and an example of this method is presented in Figure 5-13.

1. Use a map (Figures 5-5 to 5-10) to find the site longitude, namely its east/west location.

2. Use the map to determine the site latitude, its north/south location.

3. Find the azimuth/elevation chart that most closely matches the latitude and longitude of the site location (Table 5-2). This chart lists compass reading, the azimuth, and the elevation to the satellite closest to due south as well as the most easterly and westerly satellites.

4. Determine the correction to true south by finding the necessary correction on the magnetic deviation chart (Figures 5-5 to 5-10). Subtract values west of the agonic line from the south compass reading (also see Figure 5-4).

5. Correct the azimuth values in the azimuth/elevation chart for the magnetic variation at the site.

6. Use a compass and the magnetic deviation correction to find true south from the site. Aim up towards the most southerly satellite. Note its position.

7. In the same way, aim towards the most westerly and easterly satellites. In your mind's eye, connect an arc between the three satellites and note any obstructions to a clear view.

8. If obstructions to a view of the satellites exist, try the same procedure at another potential installation site.

Choosing the Best Antenna Site

In those cases where an antenna that will track the arc is to be installed, at least three satellites should be sighted, the most easterly, the most westerly and the one highest in the belt. Then a line connecting these three points should be visualized in the sky. The entire arc should be free of obstructions to a clear view of the desired satellites. If there is a question about any other satellite in the arc being partially or fully blocked, it should also be targeted. If a clear view is not possible from one location, then another site should be tested. If no appropriate location which is acceptable to the customer can be found, raising the antenna higher on a roof mount or long pole mount may be the solution. There is, however, a trade off to elevating an antenna above the ground because the higher it is installed, the more susceptible it is to wind loading and interfering signals.

User participation should be encouraged at this point in the installation. The owner should understand that perfect pictures will be obtained only if the view to each satellite is unobstructed. It may be decided, if necessary, that the antenna should be installed in a location where the signal from one or more satellites will be weakened or lost. Or, if there are minor obstructions to a clear view at a preferred installation site, the user may agree to purchase a larger antenna at a higher cost to overcome this partial blockage of signal. In some cases, the extra gain of a relatively large reflector can compensate for signal power lost due to partial reflection and absorption by obstructions.

The installation plan should include informing the customer of any necessary building permits and local ordinances. Distances from fences are usually set by zoning laws and easement restrictions. For example, in Canada the minimum distance to the antenna edge is 5 meters although 4 or 6 meter clearances from the property line are not unusual. Changing an approved site may involve filing for a new permit. Customers generally have very strong opinions concerning the aesthetics of antenna location. In some European countries, local ordinances forbid installing dishes on roofs so eave mounts are the only solution.

Underground utilities such as water mains, telephone lines or electrical cables may be buried at these easement boundaries to conform to local building codes. Planting the pole a few more meters inside the property line than is mandated may therefore be an intelligent strategy.

The site should also be chosen with cable runs in mind. Unexpected difficulties may be encountered. For example, tunneling under a 3 meter wide driveway may prove nearly impossible. If, as a result, the antenna must be located at a less desirable site where it is not hidden from view, it may prompt the customer to purchase a smaller, perhaps lower performance but more attractive brand.

One additional important factor must not be ignored. An antenna needs adequate freedom of motion to track the entire arc. If, for example, it is set too close to a fence, the reflector may either be bent or may demolish the fence before the end of arc limits are set. Nothing can replace careful planning.

SITE: Denver, Colorado

COORDINATES: Latitude - 40°North
Longitude - 104.5°West

OBJECTIVE: Determine look angles to middle, most easterly and most westerly satellites

PROCEDURE: Locate latitude/longitude grid from Table 5-2.

Latitude 40.00
Longitude 105.00

	Spacenet II	Anik D1 104.5°	Satcom F1R
Az	131.5	179.2	226.4
El	30.8	43.7	32.1

Find magnetic variation for Denver equals 12°West from Figure 5-5.

Adjust the grid by subtracting 12° from the azimuth values.

Latitude 40.00
Longitude 105.00

	Spacenet II	Anik D1 104.5°	Satcom F1R
Az	119.5	167.2	214.4
El	30.8	43.7	32.1

RESULT: Use a compass and inclinometer to sight to these three satellites and visualize the arc.

Figure 5-13. Example of Step-by-Step Procedure to Locate Approximate Satellite Positions.

TABLE 5-2. LATITUDE VERSUS LONGITUDE GRID - 30° Latitude

Latitude 30.00
Longitude 80.00

	Spacenet II	Satcom F4 82.0°	Satcom F1R
Az	158.8	184.0	253.3
El	53.0	55.0	18.2

Latitude 30.00
Longitude 85.00

	Spacenet II	Telstar 302 85.0°	Satcom F1R
Az	150.2	180.02	250.0
El	50.9	55.0	22.6

Latitude 30.00
Longitude 90.00

	Spacenet II	Spacenet 3 87.0°	Satcom F1R
Az	142.5	174.0	246.5
El	48.2	54.9	26.9

Latitude 30.00
Longitude 95.00

	Spacenet II	Telstar 301 96.0°	Satcom F1R
Az	135.7	82.0	42.6
El	45.0	55.0	31.1

Latitude 30.00
Longitude 100.00

	Spacenet II	Westar IV 99.0°	Satcom F1R
Az	130.0	178.0	238.3
El	41.4	55.0	35.2

Latitude 30.00
Longitude 105.00

	Spacenet II	Anik D1 104.5°	Satcom F1R
Az	124.5	179.0	233.5
El	37.6	55.0	39.2

Latitude 30.00
Longitude 110.00

	Spacenet II	Anik D2 110.5°	Satcom F1R
Az	120.0	181.0	228.0
El	33.6	55.0	42.9

Latitude 30.00
Longitude 115.00

	Spacenet II	Morelos 1 113.5°	Satcom F1R
Az	115.8	183.0	221.7
El	29.4	55.0	46.3

TABLE 5-2 (continued). LATITUDE VERSUS LONGITUDE GRID - 35° Latitude

Latitude 35.00
Longitude 75.00

	Spacenet II	Galaxy 2 74.0°	Satcom F1R
Az	169.6	178.3	254.4
El	48.8	49.3	9.3

Latitude 35.00
Longitude 80.00

	Spacenet II	Satcom F4 82.0°	Satcom F1R
Az	161.3	183.5	251.0
El	47.7	49.3	16.6

Latitude 35.00
Longitude 85.00

	Spacenet II	Telstar 302 85.0°	Satcom F1R
Az	153.4	180.0	247.4
El	45.9	49.3	20.7

Latitude 35.00
Longitude 90.00

	Spacenet II	Spacenet 3 87.0°	Satcom F1R
Az	146.2	174.8	243.5
El	43.6	49.2	24.6

Latitude 35.00
Longitude 95.00

	Spacenet II	Telstar 301 96.0°	Satcom F1R
Az	139.6	181.7	239.3
El	40.8	49.3	28.5

Latitude 35.00
Longitude 100.00

	Spacenet II	Westar IV 99.0°	Satcom F1R
Az	133.7	178.3	234.7
El	37.7	49.3	32.2

Latitude 35.00
Longitude 105.00

	Spacenet II	Anik D1 104.5°	Satcom F1R
Az	128.3	179.1	229.6
El	34.3	49.3	35.7

Latitude 35.00
Longitude 110.00

	Spacenet II	Anik D2 110.5°	Satcom F1R
Az	123.4	180.9	224.0
El	30.7	49.3	39.0

Latitude 35.00
Longitude 115.00

	Spacenet II	Morelos 1 113.5°	Satcom F1R
Az	119.0	177.4	217.8
El	26.9	49.3	42.0

Latitude 35.00
Longitude 120.00

	Spacenet II	Spacenet 1 120.0°	Satcom F1R
Az	115.0	180.0	211.0
El	23.0	49.3	44.6

TABLE 5-2 (continued). LATITUDE VERSUS LONGITUDE GRID - 40° Latitude

Latitude 40.00
Longitude 75.00

	Spacenet II	Galaxy 2 74.0°	Satcom F1R
Az	170.7	178.4	252.6
El	43.3	43.7	11.1

Latitude 40.00
Longitude 80.00

	Spacenet II	Satcom F4 82.0°	Satcom F1R
Az	163.2	183.1	248.9
El	42.3	43.7	14.8

Latitude 40.00
Longitude 85.00

	Spacenet II	Telstar 302 85.0°	Satcom F1R
Az	156.0	180.0	245.0
El	40.9	43.7	18.5

Latitude 40.00
Longitude 90.00

	Spacenet II	Spacenet 3 87.0°	Satcom F1R
Az	149.2	175.3	240.8
El	38.9	43.6	22.1

Latitude 40.00
Longitude 95.00

	Spacenet II	Telstar 301 96.0°	Satcom F1R
Az	142.8	181.62	326.4
El	36.5	43.7	25.6

Latitude 40.00
Longitude 100.00

	Spacenet II	Westar IV 99.0°	Satcom F1R
Az	136.9	178.4	231.6
El	33.8	43.7	28.9

Latitude 40.00
Longitude 105.00

	Spacenet II	Anik D1 104.5°	Satcom F1R
Az	131.5	179.2	226.4
El	30.8	43.7	32.1

Latitude 40.00
Longitude 110.00

	Spacenet II	Anik D2 110.5°	Satcom F1R
Az	126.5	180.8	220.8
El	27.6	43.7	34.9

Latitude 40.00
Longitude 115.00

	Spacenet II	Morelos F1 113.5°	Satcom F1R
Az	121.8	177.7	214.7
El	24.2	43.7	37.5

Latitude 40.00
Longitude 120.00

	Spacenet II	Spacenet I 120.0°	Satcom F1R
Az	117.5	180.0	208.2
El	20.7	43.7	39.7

Latitude 40.00
Longitude 125.00

	Spacenet II	Telstar 303 125°	Satcom F1R
Az	113.4	180.0	201.2
El	17.1	43.7	41.5

TABLE 5-2 (continued). LATITUDE VERSUS LONGITUDE GRID - 45° Latitude

Latitude 45.00
Longitude 65.00

	Spacenet II	Spacenet II 69.0°	Galaxy 1
Az	185.7	185.7	254.8
El	38.0	38.0	6.0

Latitude 45.00
Longitude 70.00

	Spacenet II	Spacenet II 69.0°	Satcom F1R
Az	178.6	178.6	254.8
El	38.2	38.2	6.0

Latitude 45.00
Longitude 75.00

	Spacenet II	Galaxy 2 74.0°	Satcom F1R
Az	171.6	178.6	251.0
El	37.8	38.2	9.5

Latitude 45.00
Longitude 80.00

	Spacenet II	Satcom F4 82.0°	Satcom F1R
Az	164.6	182.8	247.0
El	37.0	38.1	12.9

Latitude 45.00
Longitude 85.00

	Spacenet II	Telstar 302 85.0°	Satcom F1R
Az	157.9	180.0	242.8
El	35.8	38.2	16.2

Latitude 45.00
Longitude 90.0

	Spacenet II	Spacenet 3 87.0°	Satcom F1R
Az	151.5	175.8	238.4
El	34.1	38.1	19.4

Latitude 45.00
Longitude 95.00

	Spacenet II	Telstar 301 96.0°	Satcom F1R
Az	145.4	181.1	233.8
El	32.1	38.2	22.5

Latitude 45.00
Longitude 100.00

	Spacenet II	Westar IV 99.0°	Satcom F1R
Az	139.6	176.9	228.9
El	29.8	38.2	25.5

Latitude 45.00
Longitude 105.00

	Spacenet II	Anik D1 104.5°	Satcom F1R
Az	134.2	179.3	223.7
El	27.2	38.2	28.2

Latitude 45.00
Longitude 110.00

	Spacenet II	Anik D2 110.5°	Satcom F1R
Az	129.1	180.7	218.1
El	24.3	38.2	30.7

Latitude 45.00
Longitude 115.00

	Spacenet II	Morelos 1 113.5°	Satcom F1R
Az	124.3	177.9	212.2
El	21.3	38.2	33.0

Latitude 45.00
Longitude 120.00

	Spacenet II	Spacenet I 120.0°	Satcom F1R
Az	120.0	180.0	206.0
El	18.2	38.2	34.8

Latitude 45.00
Longitude 125.00

	Spacenet II	Telstar 303 125.0°	Satcom F1R
Az	115.5	180.0	199.4
El	14.9	38.2	36.3

TABLE 5-2 (continued). LATITUDE VERSUS LONGITUDE GRID - 50° Latitude

Latitude 50.00
Longitude 65.00

	Spacenet II	Spacenet 2 69.0°	Satcom F3R
Az	185.2	185.2	251.2
El	32.6	32.6	6.5

Latitude 50.00
Longitude 70.00

	Spacenet II	Spacenet 2 69.0°	Galaxy 1
Az	178.7	178.7	250.0
El	32.7	32.7	7.8

Latitude 50.00
Longitude 75.00

	Spacenet II	Galaxy 2 74.0°	Satcom F1R
Az	172.2	178.7	250.0
El	32.4	32.7	7.8

Latitude 50.00
Longitude 80.00

	Spacenet II	Satcom F4 82.0°	Satcom F1R
Az	165.8	182.6	245.3
El	31.7	32.7	10.8

Latitude 50.00
Longitude 85.00

	Spacenet II	Telstar 302 85.0°	Satcom F1R
Az	159.5	180.0	241.0
El	30.7	32.7	13.8

Latitude 50.00
Longitude 90.00

	Spacenet II	Spacenet 3 87.0°	Satcom F1R
Az	135.4	176.1	236.3
El	29.3	32.6	16.6

Latitude 50.00
Longitude 95.00

	Spacenet II	Telstar 301 96.0°	Satcom F1R
Az	147.5	181.3	231.6
El	27.6	32.7	19.3

Latitude 50.00
Longitude 100.00

	Spacenet II	Westar IV 99.0°	Satcom F1R
Az	141.9	178.7	226.6
El	25.6	32.7	21.9

Latitude 50.00
Longitude 105.00

	Spacenet II	Anik D1 104.5°	Satcom F1R
Az	136.5	179.4	221.4
El	23.4	32.7	24.3

Latitude 50.00
Longitude 110.0

	Spacenet II	Anik D2 110.5°	Satcom F1R
Az	131.4	180.2	15.9
El	20.9	32.7	26.4

Latitude 50.00
Longitude 115.00

	Spacenet II	Morelos 1 113.5°	Satcom F1R
Az	126.5	178.0	210.2
El	18.3	32.3	28.3

Latitude 50.00
Longitude 120.00

	Spacenet II	Spacenet I 120.0°	Satcom F1R
Az	121.8	180.0	204.2
El	15.5	32.7	30.0

Latitude 50.00
Longitude 125.00

	Spacenet II	Telstar 303 125.0°	Satcom F1R
Az	117.3	180.0	198.0
El	12.6	32.7	31.2

Checking for Terrestrial Interference

Many dealers often learn about terrestrial interference the hard way. Typically, once the concrete has set and the installation has been completed they discover that many of the channels are plagued with severe interference. Their profits have been flushed down the drain. This is a perfectly avoidable outcome. Also, just because an installation "down the street" had no terrestrial interference (TI), this does not ensure that the new one will have the same happy fate. Probably the best advice about TI is "know your enemy."

Checking for TI before doing an installation is a protection from hidden costs. It is better to detect TI before than after an installation when filters might have to be purchased. In the worst cases, a system may need to be pulled out at high expense to the dealer. It may be better to turn down a job than to have a dissatisfied customer. Bad news travels faster than good news. A few botched installations can certainly ruin the word of mouth recommendations necessary for a healthy business.

An intelligent strategy in dealing with terrestrial interference is summed up by the title of the *ASTI Manual - The Avoidance/Suppression Approach to Eliminating TI*. The first step to be taken is to avoid the interference by correct antenna siting. TI is very directional and can be reflected from metal objects such as a tin roof directly into a dish. But TI, like satellite signals is also absorbed by many materials. So installing an antenna even in locations where TI levels are high behind natural shields such as trees or a building may be sufficient protection.

If an interference-free site cannot be found, the next alternative is to use filters to "notch out" and suppress the interference. However, notching out the interference also reduces the satellite bandwidth and can make obtaining "studio quality" pictures difficult. A second alternative for suppressing TI is to construct the appropriate artificial shields. This strategy is required in severe cases of TI and it is usually an expensive option. But remember, there is always a solution to a problem.

Figure 5-14. Portable Antenna for Site Checks. *A small portable antenna should be used if there is any uncertainly about obstructions blocking the field of vision of a dish or if TI has been detected by other methods. A 1.5 meter dish is pictured here. (Courtesy of DH Antenna)*

Be aware that some receivers used with LNBs do not have a provision for insertion of TI notch filters. The selected receiver should have an external IF loop. Also, filters for some of the less common IF loop frequencies in use may not be readily available.

Methods to Check for TI

By far the most effective method to check for TI is by testing a portable satellite system at the proposed site. A small dish, low cost system can be easily transported to and quickly set up at the site (see Figure 5-14).

At least three or four satellites should be targeted since TI can be very directional. While interference may not show up, for example, on Galaxy I it may be strongly disruptive on Satcom IV at the other end of the arc. Sometimes rotating an antenna just a few degrees can cause a side lobe to point at a source of interference whereas before none was detected. Try to set the test dish at the same height the final antenna will be placed; if there is interference moving it down only a few feet may suddenly recruit a fence as a good artificial screen.

All transponders on each satellite should be checked since some forms of interference often affect only a limited number of channels. If TI is present, it may be cured by simply relocating the dish. Placing an antenna closer to the ground or, if possible, in a small depression in the ground, reduces its susceptibility to TI. Interference does not travel underground! If a location free of TI cannot be found, the time has come to insert selected bandpass or notch filters and see what happens. Take care to not confuse a weak signal caused by the antenna being blocked by an obstruction with interference. (In those cases where picture quality has deteriorated in an existing installation, also do not mistake obstructions caused by the growth of new leaves on trees for TI. This can be diagnosed easily because all channels would be affected equally.)

An alternative to testing a complete system is to use a hand-held LNB/feedhorn assembly connected to a satellite receiver, TV set or a monitor (see Figure 5-15). It is preferable to have a sensitive signal strength meter in addition to the one often built into the receiver. Turn on all electronics and then scan the test unit in all directions across the sky, not only at the satellites, since TI coming from off-axis directions can be detected by antenna side lobes. The feedhorn must be oriented in both the horizontal and vertical planes so any interfering carriers of both polarities can be found.

Watch both the signal strength meter and the TV screen. If the random white noise or snow on the TV remains unchanged and if there is little or no reading on the signal strength meter then there is no problem. If the screen changes and goes blank, black, or if white or black bars or lines appear and if the meter gives a positive indication then some form of TI is present. If the feedhorn is covered up

Figure 5-15. Testing for TI. *TI can be diagnosed by using an alternative to either a portable dish or more expensive instrument such as a spectrum analyzer. An LNB and feed hooked to a satellite receiver and TV can be scanned across the sky to search for interfering signals (see text). (Courtesy of Steven Berkoff)*

all these symptoms should disappear. If they do not, the test equipment is malfunctioning. Note that if a TV set or monitor is not used, the signal strength meter can be used to detect moderate to heavy TI as indicated by a continuous fluctuation of its read-out. However, this method is not very sensitive to low or moderate levels of interference.

If interference is detected, the feedhorn/LNB should be scanned back and forth to find the direction where the meter reading is highest. This process indicates where the interference is coming from. Orient the feed in all planes to check the polarity of this TI.

If TI is detected, move to another possible site, perhaps one better shielded by natural barriers, and do the same test over again. A logical first choice would be a location where there is an obstruction between the new site and the source of TI. If the interference is coming from behind the antenna, it will not have as great an effect because the dish will block out most of the microwave signal. If no location with a clear view of the arc of satellites and free of interference can be found, it is time to bring in a portable dish to do a more sensitive test. Then microwave and IF filters can be inserted, if necessary, to find a solution to the problem.

The most accurate results can be obtained by using a spectrum analyzer. This device shows the frequency and power of microwaves coming from any direction at any chosen test site. A spectrum analyzer is a very effective tool in checking for both in-band and out-of-band TI. A low cost spectrum analyzer will rapidly pay for itself in saved time for an active satellite dealer.

Another option for a dealer is to chart TI problem sites as described in Chapter III. If a site falls on the line between two microwave relay towers, the dealer will be forewarned when conducting a site survey.

Note that all these methods of testing for interference have their limitations. The information gathered applies only to TI sources which were operating during the site survey. The 5 to 7 P.M. time period is usually when microwave traffic is heaviest. Therefore, an even better test for TI, which can also be an effective sales tool, is to leave the portable satellite system with the customer for a few days. He or she may notice changes in reception quality at different times during the day or night.

None of these methods will protect a dealer or consumer against common carrier relays which will be turned on at some point in the future. The only protection against some of these future sources of TI is to obtain the necessary information from the FCC. The technique of charting the location and paths of all present and future microwave sources in the vicinity of any installation is discussed below.

Customers should be informed that even though TI may not be present today it is always a possibility in the future. Given this possibility, all contracts should have a clause that protects the dealer by either allowing cancellation of the sale if TI is discovered at the time of the site check or by passing along the cost of the necessary filters and installation time to the customer.

Planning the Installation

Clearly, a satellite TV installation should be planned as completely as possible during the site survey. The antenna location must be chosen and coordinated with the cable runs to the receiver and all televisions. How cable will enter the the building must be established. The type of cable and number of splitters and/or taps needed, as well as the necessary connectors must also be determined.

Satellite receivers can be placed anywhere, not necessarily on top of TV sets. If radio controlled actuators and receivers are used, a customer might decide to leave all the indoors electronics in a well-ventilated closet.

It is essential that the customer be consulted and be heard during a planning process. Too many dealers and technicians have preconceived notions and may not pay close attention to the desires of their customers. Every satellite system should be installed as if it were for personal use. That is the route to success. When an installation has been completed, always have the customer sign the site survey worksheet release (see Figures 5-16 and 5-17).

SITE SURVEY WORKSHEET

Job Number______________
Date____________________

Customer Name____
Address____
City____________________________State______________Zip__________
Telephone (Home)__________________________(Work)______________________
Lead Source__________________________**Salesman**_____________________

SITE MAP

Size and Make of Antenna and Mount______________________________
LNB Noise Temperature____
Style and Manufacturer of Feed___________________________________
Receiver Manufacturer and Model__________________________________
Length of Cable Installed____
Length of Conduit Installed_______________________________________
Number of Wall Plates____

Figure 5-16. Site Survey Worksheet. *By documenting everything about an installation and obtaining a customer acceptance signature, the job will be properly competed. In addition, information will be available for any future troubleshooting work that may become necessary.*

SITE SURVEY WORKSHEET (continued)

Make and Model of Actuator____________________

Number of Extra TV Sets Installed____________________

Description of Cable Routes____________________

In Home____________________

Out-Doors____________________

Obstructions to Cable Runs____________________

VCR or Stereo Hook-Up____________________

Type of Mount (pole, roof, etc...)____________________

Types of City Ordinances in Effect and Permit Requirement____________________

____

Type of Warranty____________________

Status of Cable TV Disconnect____________________

Type of Program Guide Requested____________________

Type of Off-Air Antenna and Connection____________________

Method of Waterproofing Used____________________

Number of A/B Switches or Combiners____________________

Date Installation Requested____________________

TI Carriers Noted____________________

Further Comments____________________

____

____

____

____

____

____

____

____

Customer Signature____________________

Figure 5-16. Site Survey Worksheet (continued).

PRE-INSTALLATION WORKSHEET

Customer Name______________________________Customer Number__________

Date__________________

Equipment List:

Number of Items	Item	Manufacturer	Model Number	Serial Number
	Antenna			
	Mount			
	Feed			
	LNB/Feed Cover			
	LNB			
	Coax			
	Receiver			
	Actuator			
	TI Filter			
	Combiner			
	A/B Switch			
	Line Amplifier			
	Splitter			
	Wall Plate			

	Initials
All electronics have been checked before leaving office	______
Back-up electronics are on hand	______
Enough wire has been taken for installation of all TV sets	______
Proper tools are in installation vehicle	______
Customer contacted before leaving office	______
All lights and signals on vehicle and trailer have been checked	______
Correct customer invoice has been taken	______
Pole mount has set and has been checked	______
Connector kit has been checked	______

Time left office______________

Figure 5-17. Pre-Installation Worksheet. *This chart can be used as a guide to ensure that all necessary equipment has been taken to the work site.*

C. ANTENNA SUPPORT STRUCTURES

Satellite TV antennas and their mount assemblies are usually supported on a pole, especially when more than one satellite will be targeted. However, there are now numerous installations in Europe where customers who have purchased systems to view exclusively one satellite, such as Astra, have fixed antennas that are mounted directly on az-el frames. These supports are often bolted onto an eave, a roof or secured to the ground by weights or concrete.

When antennas are being installed on pole supports, securing the pole is critical because an antenna and its mount can often be very heavy and may be subjected to severe winds and other environmental stresses. This is especially the case in North and South America where large antennas are required to receive transmissions. If the pole is solidly set in a true vertical position, all the remaining system adjustments can proceed smoothly. The three basic rules to follow when installing a system, especially a Ku-band TVRO, are stability, stability and stability. Even minor instabilities in the support structure can degrade performance.

Preparing for the Job

Before work begins on the antenna support structure, some preliminary steps should be taken.

Underground Utilities

Gas, telephone, electric and cable television companies often have underground lines in unmarked areas. In many countries these utility companies will send personnel at no charge to locate and mark the route of their cables or pipes. If a utility line is severed during the digging and trenching portions of an installation, the costs in customer dissatisfaction and time lost are high. In addition, the dealer will probably have to pay an hourly rate to the utility company to have the damage repaired.

Pre-Trenching for Conduit Placement

In those instances where a ground mount installation is chosen, once the cable runs have been mapped, a few meters of the trench should be excavated starting at the antenna mount. This will allow the metal or PVC conduit leading to a supporting pole to be set at the same time as other concrete work such as a pad is completed. If this step is overlooked, the cables would lie exposed near the base of the pole and could be susceptible to damage from gardening utensils such as lawnmowers, weed-eaters or shovels.

Pole and Pad Mounts at Ground Level

A variety of construction methods can be used to secure a pole to a ground-mounted antenna site. These include pole supports, pads, pier foundations or combinations of these types. Most require the use of concrete or another strong binding material that will withstand the test of time.

The Fundamentals of Concrete

Concrete is a building material which was invented centuries ago. Its properties are well known and new varieties are constantly being developed for a host of applications. When prepared and hardened properly, it has tremendous strength and is very durable. However, incorrect mixing or setting procedures can result in a weak base which can easily crack or disintegrate. If this were to happen, the antenna base could move and it would be extremely difficult to realign the heavy block of concrete.

Concrete is like a synthetic rock and derives its strength from its ingredients – gravel, sand and cement. When water is added to a cement mixture, a chemical reaction begins. Setting of cement is not simply water drying out. The cement acts like a strong glue which powerfully binds the gravel and sand with other materials within the mold.

During the chemical reaction, heat is generated. If this heat is lost too quickly during cold weather installations, the concrete will not cure properly. Under such conditions it can lose strength and crack. If the surface of the concrete is covered with straw, blankets or other insulating materials during curing, heat will be retained and such problems can be avoided. Calcium chloride can also be added to speed up the setting process; generally 2% by weight of salt is recommended. But be aware that too much calcium chloride can rapidly corrode metal structures such as the supporting pole and also weaken concrete to the point that it degrades to concrete dust within a few years.

Post type foundations can be made from ordinary Portland cement or standard premix. Slab foundations having larger areas of concrete should be made from air-entrained Portland cement and wire screen. This mixture contains additives that allow microscopic air bubbles to be entrained or trapped. These bubbles are like an internal lubricant which makes pouring and spreading easier. They also act like tiny shock absorbers which allow the larger pieces of concrete to expand and contract without cracking.

Both of these types of cement can be purchased in pre-mixed bags or they can be mixed on site. Proportionate ingredients for plain and air entrained concrete are listed in Table 5-3.

Other mixes of concrete or replacements for concrete having specifically designed properties are available. For example, in the United States, QR Inc. sold a special mix sold under the trade name QUIK-ROK which sets up in 15 minutes at temperatures as low as -7°C. Its manufacturers claim that it is stronger than concrete. It expands as it sets, thus forming a tight seal; ordinary concrete tends to contract during setting. Similar characteristics are advertised by the makers of another substitute called Sat-Based Cement. Dish Set™, an alternative to concrete, is a closed-cell, expanding polyurethane foam which sets in about 15 minutes depending upon ambient temperatures. It can be used for pole mounts since it generates force against the pole and the sides of the hole during expansion. It cannot be used in extremely sandy soils which do not support this action. When it is used, the hole must be narrow, 4 to 8 cm on either side of the pole, and deeper than normal for this expansive force to have full effect. All of these quick setting mixes do not allow much time for mistakes; be totally prepared before pouring these materials by using stakes to level the pole.

TABLE 5-3. INGREDIENTS FOR VARIOUS TYPES OF CONCRETE

Type	Proportion	Ingredient
Plain Concrete	2.5	Portland cement
	3	Sand
	5	2 cm aggregate
	1.25	Water
Air Entrained Concrete	2	1A portland cement
	2.25	Dry sand
	5.66	Coarse aggregate
	1.25	Water

Freezing, Underground Water and Stability

Underground water is the chief enemy of any foundation. During winter, water at the base of sidewalks, patios or fence supports may freeze and generate tremendous forces which cause the surrounding rocks and soil to shift. When a thaw occurs, the water percolates away, soil and rocks collapse inwards and the concrete can twist or settle. This heaving effect is usually not too great a bother for fence posts or other such structures. However, if a pole supporting an antenna is moved even slightly away from a vertical orientation, tracking of the geosynchronous arc can be ruined.

The frost line is the depth below which no freezing occurs. In North America, it varies from approximately 2.5 cm (one inch) in southern Texas to over 3 meters (10 feet) in northern portions of the United States. In even more northerly locations such as Alaska or the Canadian Northwest Territories, special construction techniques are always required to manage the difficulties associated with the great depth of the frost line.

Pole or Post Supports

Pole or post type antenna supports are the most common variety in use today, especially for installing larger, C-band antennas (see Figures 5-18 to 5-26). They provide strong, stable platforms. But pole mounts cannot be used in locations where the ground is too rocky to dig the necessary hole or where the water table is too high. Because water is a good conductor of heat, a high water table may mean the frost line extends much deeper than normal in a given region. A pole support should also not be used for antennas larger than 4 meters in diameter since wind generated forces are simply too great. A three-legged tower which distributes the force over its whole base should be installed.

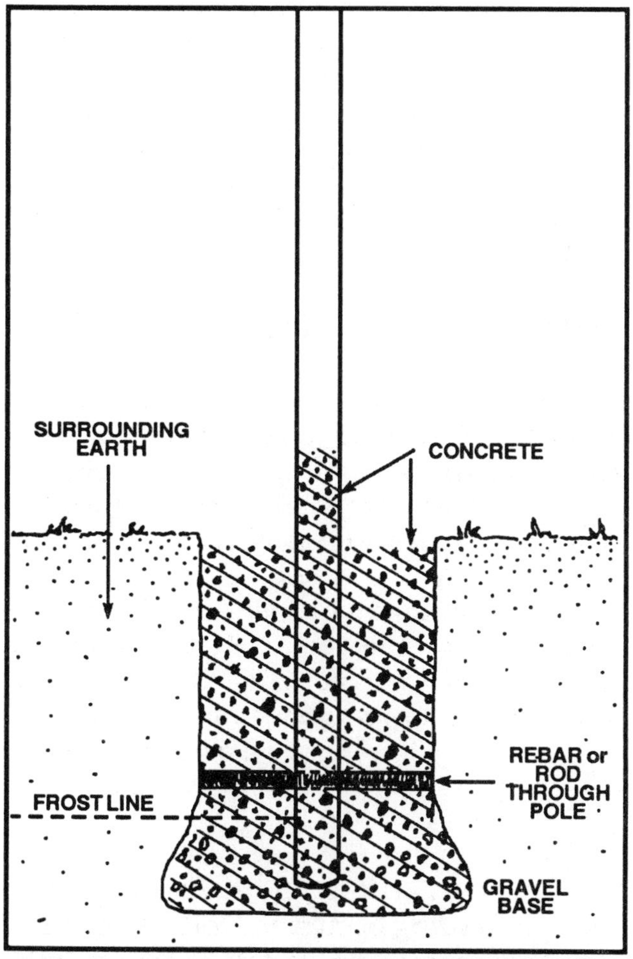

The hole for a pole support should be two to four times the diameter of the pipe. It should extend at least 15 cm (5.9 inches) below the frost line and 2 or 3 cm (1 to 2 inches) of gravel should fill the bottom to allow for proper drainage. Usually between 100 and 300 kilograms (200 to 700 pounds) of concrete are needed for an average pole support.

Pole supports typically extend approximately one meter into the ground and 1.5 meters (5 feet) above. A general rule of thumb is to add 10 cm of extra length in the ground portion for every additional 50 cm of height (or 1 inch for each additional 5 inches of height). For example, a pole extending 5 meters above ground should be planted an additional 70 cm, or 1.7 meters deep because there are 3.5 meters of extra length compared to a standard pole. Poles set in very soft or sandy soil should either be planted deeper or have a wider concrete base.

SURROUNDING EARTH
CONCRETE
REBAR or ROD THROUGH POLE
FROST LINE
GRAVEL BASE

Figure 5–18. Schematic of a Ground-Mounted Pole Support. *A ground-mounted pole is supported in a hole which has a diameter that is typically 2 to 4 times the pipe diameter. The bottom of the hole is filled with about 6 inches of gravel to allow for drainage and is below the local frost line. A half inch metal rod, either welded to or drilled through, extends out from the bottom of the pole to prevent twisting in its concrete base when high winds buffet the dish above.*

Figure 5–19. Preparing Ground for Pole Mount. *After the site survey has been completed and all cable routes have been mapped, the hole for a pole mount can be dug. In general, a few simple tools including a post hole digger, a digging bar, a shovel and a garden hoe as well as a wheel barrel are all that are necessary. The excavated soil should be loaded into the wheel barrel and not thrown on a sod lawn. This simple step makes after installation clean-up much easier. (Courtesy of Brent Gale)*

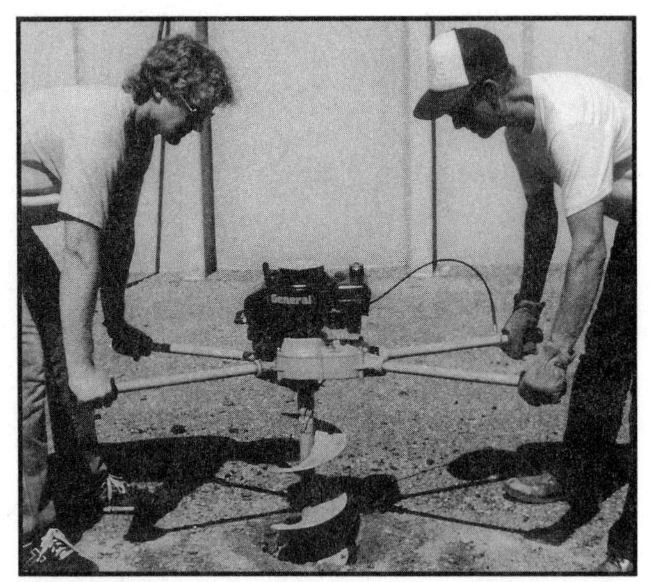

Figure 5–20. Using a Power Auger. *Digging holes with a power auger can save many hours of time and in the process save hands from uncomfortable blisters. Safety must be stressed when using this type of equipment in the vicinity of buried power and utility lines. (Courtesy of Brent Gale)*

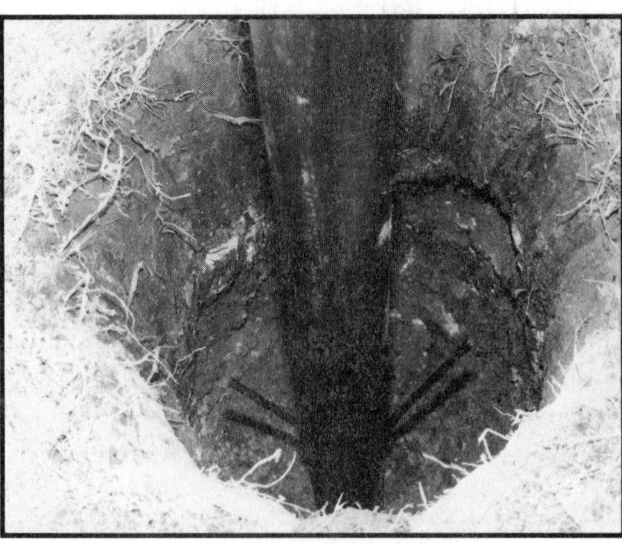

Figure 5–21. Pole-to-Ground Attachment. *It is recommended that construction rebar or a 15 to 20 cm (6 to 8 inch) bolt be welded to the bottom of the pole to keep it from twisting loose when high winds exert powerful forces on the antenna above. (Courtesy of Brent Gale)*

Once the pole is placed into the ground, it can be temporarily supported by a few rocks and by wooden stakes or guy wires. The pole must be perfectly vertical so that the antenna will be capable of properly tracking the entire arc of satellites. A carpenter's level of at least one meter in length or a plumb bob can be used to check this orientation from three or four vertical positions around the pole. These readings should be taken periodically as the concrete is setting. Some believe that concrete should also be added to partially fill the center of the pole for even further stability in high wind areas.

The pole must have sufficient strength to withstand both bending and twisting forces. At least schedule 40 steel poles should be used. Using schedule 80 or 120 steel is even better. (Schedule 120 is like pipe drillers stem used in the oil industry.) Most antennas typically require poles having 7 to 10 cm (3 to 4 inch) outer diameters.

In order to prevent the supporting pole from breaking free from its concrete base and rotating in heavy winds, two half inch (1.3 cm) or larger metal rods at right angles to each other should either be

Figure 5–22. Recommended Pole Depth. *The hole for a standard pole mount should extend at least one meter deep with a diameter of four times the pole diameter. In any case, the hole should be at least 15 cm below the local frost line. (Courtesy of Brent Gale)*

Figure 5–23. Pre-Trenching for Conduit Burial. *Before concrete is poured at least 2 meters of trench should be excavated and the conduit should be aligned within the trench and hole. When the concrete is poured the conduit will be completely buried and will extend parallel to the pole. If it were even partially exposed, cable near the base of the pole could be damaged by pets as well as by garden tools such as weed-eaters or lawn mowers. (Courtesy of Brent Gale)*

Figure 5–25. Stabilizing the Pole with Concrete. *Concrete can be poured down inside the pole to provide additional support. After mixing, concrete should have a thick consistency. Too much water makes it "soupy" and delays the curing process because all the excess water must either evaporate or drain off. Note that the caustic ingredients in concrete can kill grass so care should be taken to avoid spills. (Courtesy of Brent Gale)*

Figure 5–24. The Use of Leveling Stakes. *Leveling stakes serve to keep a pole in a vertical position as the concrete cures. (Courtesy of Brent Gale)*

welded onto or inserted through the bottom part of the pole. These should extends out from 7 to 15 cm (3 to 6 inches) on all four sides. Or a piece of re-bar with similar dimensions should be welded onto it. An alternative to using one continuous pole is to implant a sleeve into which the long pole can be bolted. If designed properly, this type of assembly can allow a pole to be re-leveled in case the base eventually shifts.

Most pole supports have no mechanism for making a leveling adjustment once the concrete is set. This can be a concern because the effects of wind and water can, in time, cause the concrete base and thus the pole mount to shift. This effect is encouraged because a large antenna angled towards the equatorial arc of satellites does not have its weight evenly distributed on the pole. This situation places quite a large bending force on the

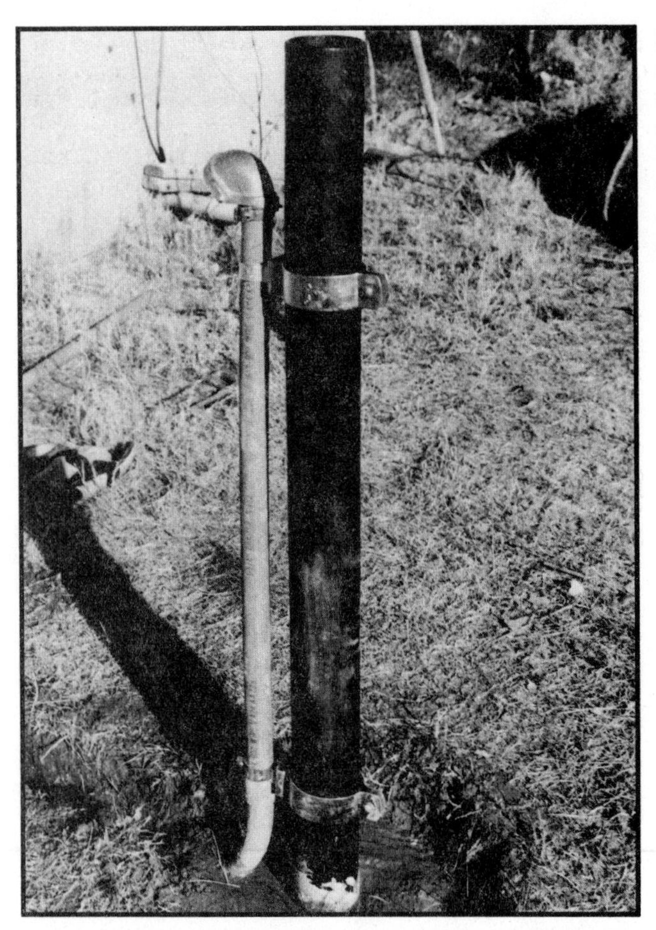

Figure 5–26. Conduit and Pole in Place. *This photo shows the correct installation procedure with conduit extending into the concrete and secured in place with clamps. A weatherhead at the top of the conduit prevents water from entering the run. Conduit near the pole should be set in place when the concrete is poured. The concrete is poured to about 9 cm below the top of the lawn so that grass can be replanted adjacent to the pole. (Courtesy of Brent Gale)*

Figure 5–27. Pad Support on Concrete Patio. *Approximately 2.3 cubic meters (3.3 cubic yards) of concrete were poured onto this concrete patio. The pad was secured by self-anchoring to prevent shifting. The 5 m Scientific Atlanta aluminum antenna rest on an az-el mount. Notice how the cable runs are well protected by conduit. (Courtesy of Brent Gale)*

support. Without a leveling adjustment, the alternatives are to dig up the pole and replant it using additional concrete, install a new pole and cut out the original one, or attempt to level it with a piece of heavy machinery such as a truck or jack. None of these options are attractive.

For the dealer or installer who regularly digs holes for planting poles, a gas-powered auger similar to those used to dig holes for fence posts is probably a good investment. This tool can save many hours of installation time.

Concrete Pads

In cases where the ground is too rocky or hard to allow digging of a narrow, deep hole, where the ground water table is too high, or where an especially large reflector will be installed, a concrete pad should be poured on site (see Figures 5-27 and 5-28).

A pad is constructed by digging a shallow trench, building a wooden form to hold the concrete, adding gravel for drainage, embedding wire mesh for strength and pouring the concrete with the pole and the conduit in place (see Figure 5-29). It is best to recess the pad into the ground so that a lawn can be easily mowed and so that the customer will not stub his or her toe during the night on a concrete obstruction. When installing a pad in a non-level location it is important to excavate the high ground and never to fill dirt into low terrain (see Figure 5-30). Otherwise, the pad is likely to settle and cause the pole to tilt.

Figure 5–28. Pad Installed on Granite Rock. *This pad support was poured on a granite rock outcropping in the mountains of Colorado. A jack hammer was used to drill holes to insert rebar into the granite base. Over 3 cubic meters of concrete were used to pour a flat pad and four 3/4 inch (19 mm) self-anchoring bolts secured the steel mount to the pad. (Courtesy of Brent Gale)*

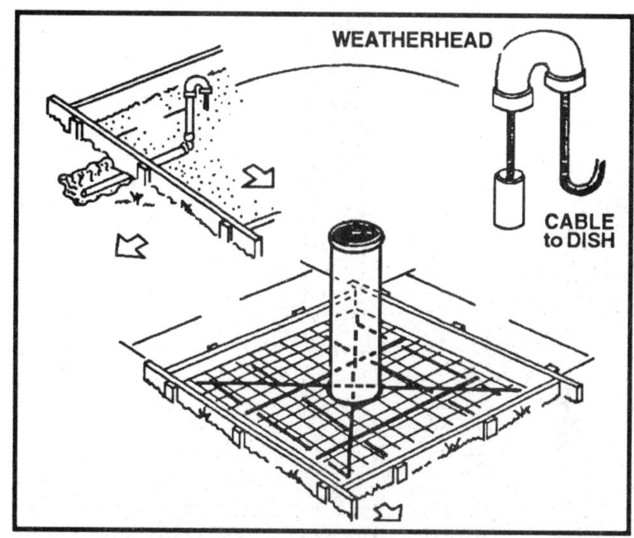

Figure 5–29. Schematic of a Pad Support. *A concrete pad support has steel rebar and wire mesh for strength and a gravel base for drainage. The pole has its weight distributed by extending rebar or channel iron from its base to near the edges of the pad.*

When the concrete has set, the wooden form can be removed. These forms should be made from 5 by 25 cm (2 x 6 inch) green lumber because dry wood will absorb water from the concrete mix and hinder the setting process. Alternatively, the wooden form can be wetted or sealed with a water repellent seal before pouring the concrete. Water should also be periodically sprayed onto the surface of newly poured concrete to improve hardening. This keeps the outside as wet as the inside and thus prevents cracking during setting.

A slab foundation supports an antenna by virtue of its weight and size. Rules of thumb have been developed to size pads. A pad reinforced with mesh should have concrete at least as thick as the pole diameter. Without reinforcement, increase this thickness by 50 percent. The length and width of a pad should be at least half the antenna diameter or the measurement from the pad surface to the base of the reflector. For example, a 2 meter antenna supported by a 9 cm pole should be supported by a pad measuring 1 by 1 meters and from 9 to 15 cm thick.

Wire mesh used for reinforcement should be at least 8 gauge and have holes spaced so that openings are no more than 40 cm on a side. It

Figure 5-30. Pad Support on a Hill. *This pad support is recessed into the ground on a sloping hill. Notice that it is dug into the higher ground, not built onto the lower slope. The pole is secured in about 1.5 cubic yards of concrete by rebar which spans the structure. The concrete is reinforced by wire mesh. Gravel at the base of the form allows for drainage. This pad was installed in the dead of winter when the temperature was -10°F. (Courtesy of Brent Gale)*

should be located at a depth equal to about 1/3 the thickness of the concrete as measured from the pad surface.

A pad mount requires adequate drainage. Sandy soils require less drainage than rocky or clay soils which are both more impervious to water. The gravel bed under the concrete should range from a minimum of 50% to 150% of the concrete thickness. For example, a reinforced pad set on solid rock using a 9 cm (3.5 inch) pole should have about 12 cm (5 inches) of gravel and 16 cm (6.5 inches) of concrete.

The mounting pipe must have two pieces of re-bar inserted through its base at right angles. When these are set with the pole in concrete, they spread the load over the whole base of the pad for stability and prevent rotation of the pole under high wind loads. This re-bar can be tied to the mesh reinforcement with baling wire for further strength.

Three Point Pads and Pier Foundations

Three point pads and pier foundations both bolt onto a tripod assembly which supports the antenna. A three point pad is a concrete pad having either three or more anchor J-bolts preset into the concrete or self-anchoring bolts, also known as redheads, installed when the concrete is still soft by cutting the holes with an electric hammer drill (see Figure 5-31). This is a convenient method because leveling adjustments can be built into the mount (see Figures 5-32 and 5-33). A pier foundation uses

Figure 5-31. Redhead Anchor Bolts. *These specially designed bolts are used after holes are drilled in concrete walls or foundations. The can support thousands of pounds of force.(Courtesy of Brent Gale)*

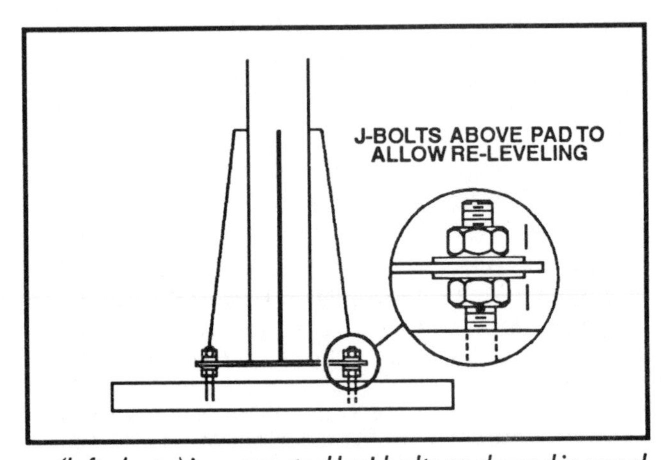

Figure 5-32. J-Bolt Leveling Adjustment. *This 1.2 m antenna (left photo) is supported by J-bolts anchored in a pad. Additional thread extending above the surface of the pad permits re-leveling at any time in the future. (Photo courtesy of Northsat)*

three smaller pads called piers, each one having a small concrete foundation attached to deeply set reinforcement bars and anchor bolts (see Figures 5-34, 5-35 and 5-36). Yet another option is to secure a tower to a pad with three or more bolts (see Figure 5-37).

Anchor bolts should be made of galvanized steel, 2 cm (3/4 inch) thick by 60 to 90 cm (25 to 35 inches) long. After installation they should extend from 7 to 10 cm (2.5 to 4 inches) above the surface of the concrete. This design allows re-leveling of the pole when necessary. If the visible portion of the bolt is greased before pouring the cement, clogging of its threads can be avoided.

A template conforming to the layout of the base of the pole should be used to place the anchor bolts, generally J-bolts, before concrete is poured or while it is still "green" or soft. The bolts should be

Figure 5-34. Three Pier Support. *This 20-foot ADM antenna with horizon-to-horizon mount is supported by a three pier pad foundation. Whenever such a tripod supported polar mount is installed it is important that the back two legs be set closely onto an east-west orientation. Fine adjustments to the mount are then made by sliding the telescoping legs to the final position (Courtesy of Antenna Development Corporation, Inc.)*

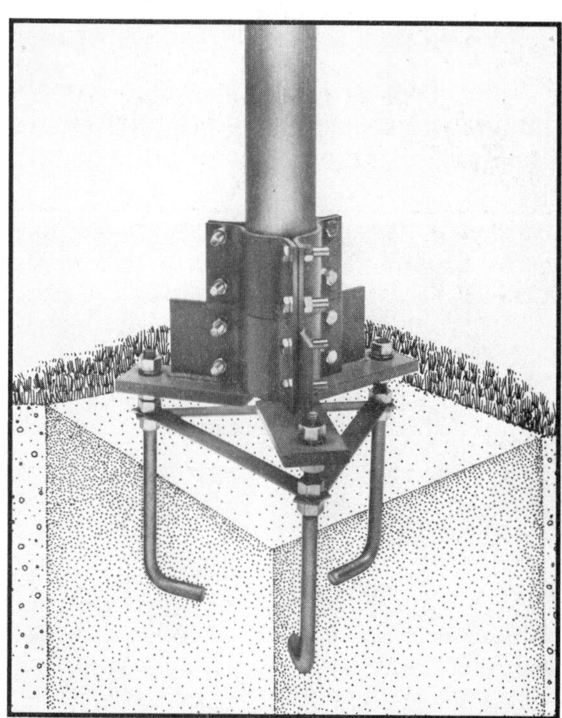

Figure 5-33. Level-Adjusting Pedestal Support. *This support is set onto a concrete base or even an existing solid concrete patio using either the J-bolts provided or redheads. It can be quite useful in cases where the pole in a pad has become loosened. The pole would then be cut off at its base and this support would be installed in the pad near where the pole had previously been located. (Courtesy of Earthbound, Inc.)*

Figure 5-35. Three Point Pad Mount. *A triangular az-el base can be installed with minimal concrete in view. Here, the excavated hole was 1.6 m deep and 1 m in diameter. (Courtesy of Brent Gale)*

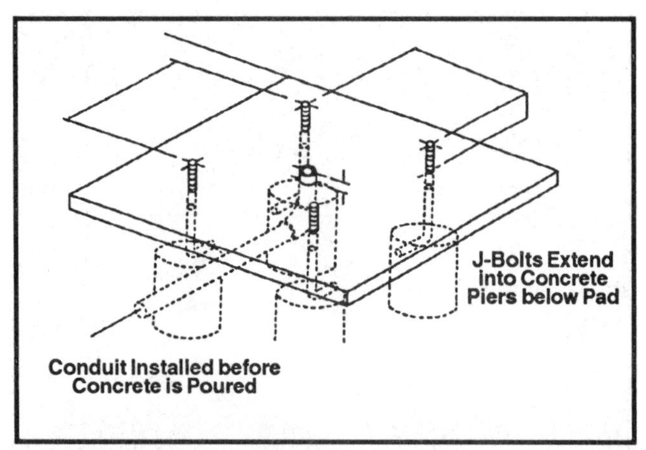

Figure 5-36. Pad Support Using J-Bolts. *This drawing shows the placement of J-bolts that extend into concrete piers below a pad mount. The conduit is also set in place to protect cable runs.*

Figure 5-37. A Three-Leg Tower Mounted on a Pad. *The weight of this 16-foot antenna would be adequately supported by a single pole but requires this more stable, 3-leg tower as a base. (Courtesy of Paraclipse, Inc.)*

carefully kept away from the reinforcing mesh or re-bar since any shifting of this supporting structure during pouring can cause them to move out of place. The bolts should never be welded to the rebar since too much heat will cause them to become brittle. When the concrete is completely cured, 15 cm (6 inch) "redheads" having at least a 16 mm (1/2 inch) diameter can be inserted in a hole made by a hammer drill. Redheads, like some drywall screws or moly bolts, have a collar which folds up when the bolt is tightened, they can support at least 20,000 kilos (more than 40,000 pounds) of force.

When the antenna supporting tripod is mounted, two nuts with lock washers placed both above and below the attachment points allow a fine leveling adjustment during installation and at any time thereafter. When using either a pad or a pier to set the tripod, care should be taken to have the three supporting points as level as possible to facilitate making this vertical adjustment. It is also important to orient the rear legs of this tripod along the east/west axis during this procedure.

Some installers pour concrete at their shops and then truck the completed structure to a site. This technique has the advantage of lowering costs

Figure 5-38. Combination Pole/Pad Mount. *This illustrates a concrete pad poured with J-bolts inserted while the concrete was "green," namely before it had cured. A template of the antenna base should be used to make certain that the bolts are in the correct pattern. If not, it may be necessary to remove them and re-insert redheads at the cost of additional labor and time.*

Figure 5–39. Setting of Concrete Pad. *The 5/5" outer diameter pole could not be installed as deeply in the ground as necessary without conducting a major excavation job. Therefore, a pad mount measuring 10 feet in width was also poured around the top of the pole. Notice the conduit extending up against the pole to provide a protected route for the wiring. (Courtesy of Brent Gale)*

and making winter installations easier. However, slabs can settle in the spring, require special equipment to move them to a site and can be difficult to place once at the site. A 1500 kilo (over 3000 pound) trailer will crush sprinkler pipes in the ground and cannot often be moved into a customer's backyard. Nevertheless, such an approach can be very effective in maintaining business activity during spells of cold weather.

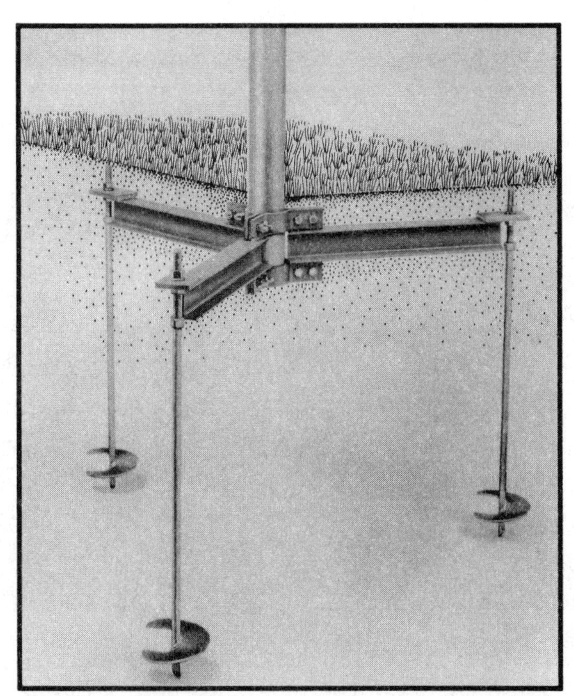

Figure 5–40. Dish Support Not Requiring Concrete. *This system uses a tri-leg steel platform with 4 foot (1.2 meter) long anchors screwed into the earth with a manual driver. Leveling is possible at any time following installation. It can be removed and installed elsewhere by screwing the legs out of the ground. (Courtesy of Earthbound, Inc.)*

Combinations of Pole and Pad Supports

A creative installer can construct any variation of pole and pad supports that a given situation warrants (see Figures 5-38 and 5-39). For example, if an immovable rock is encountered half a meter below the surface when digging a hole for a pole mount, a combination pole/pad support can be created. The pole would still be set into the undersized hole. But a small pad would also be formed around the top of the pole to provide further support. Thus, less concrete could be used and the labor spent in digging the preliminary hole would not be wasted. In this case, the pole would extend further out of the ground than would normally occur and lifting the antenna onto the pole might prove to be a little more difficult.

Ground Supports Not Requiring Concrete

Foundations which use a tripod support design but do not require concrete are also viable structures. Three or more rods can be driven deeply into the ground with a power auger. Then a tripod or even a four-legged assembly can be attached with weather resistant, adjustable bolts. For example, the Earthbound, Inc. system (see Figure 5-40) uses a tri-leg steel platform secured with 1.2 meter long anchors driven into the earth with a power auger. It is interesting that this type of support is not considered a permanent improvement by some building codes because no concrete is used. Thus, additional taxes can sometimes be avoided.

Above-Ground Mounts

As transponder power levels from satellites increase and LNB noise temperatures fall, antennas with ever smaller diameters can receive clear signals. As a result, a variety of alternative supports including wall, roof or eave mounts, towers or long poles are becoming more feasible and common. This is especially the case in Europe where most antennas being installed today are less than three meters in diameter so that eave and roof mounts are the prefered alternative.

In some cases, installing an antenna at ground level is simply not possible because obstructions block a clear view to the satellite arc. In addition, in some countries and locales, real estate is a premium commodity and backyards are small or nonexistent. This presents an installer with two options: choose a different site or mount the antenna at or above roof level. There have even been demanding situations where antennas have been mounted on free-standing towers like some commercial broadcast antennas.

Figure 5-41. Wall Mounted Antenna. *This 1 meter stamped galvanized steel antenna is attached to a wall for reception of European Ku-band broadcasts. (Courtesy of Maspro)*

Roof, Wall and Eave Mounts

Some smaller diameter C-band and many Ku-band antennas are often supported from walls, eaves or mounted on roofs in regions such as Europe where satellite power levels are relatively high and dishes of one meter or smaller diameter are frequently installed (see Figures 5-41and 5-42). Historically, roof mounts have been regarded with caution by many in the satellite industry, especially where larger C and Ku-band antennas have been installed. Roofs are often not designed to withstand weights in excess of normal snow loading or the tremendous upward forces caused by winds acting on a large aperture antenna. For example, an 80 kph (50 mph) wind blowing directly into the face of a 3.5 meter antenna can generate a force of more than 500 kilos (over 1200 pounds). This tremendous force can rip a roof off its rafters and cause even more damage when it finally lands!

Figure 5-42. Roof Mounted Antenna. *This 14 foot (1.2 meter) offset fed dish is designed to receive Ku-band broadcasts. It has 20% of the surface area of a 9 foot antenna and is therefore much less susceptible to wind loading. (Courtesy of Northern Satellite Corporation)*

Mounting an antenna directly onto a roof has some potential disadvantages compared to ground mounting. Any play in the actuator or other components of a roof mount could create annoy-

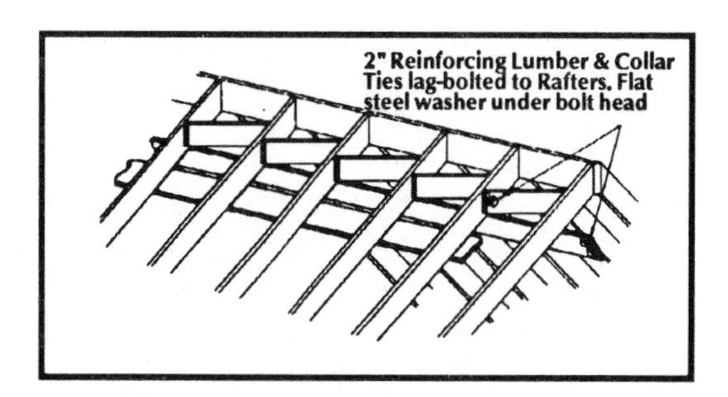

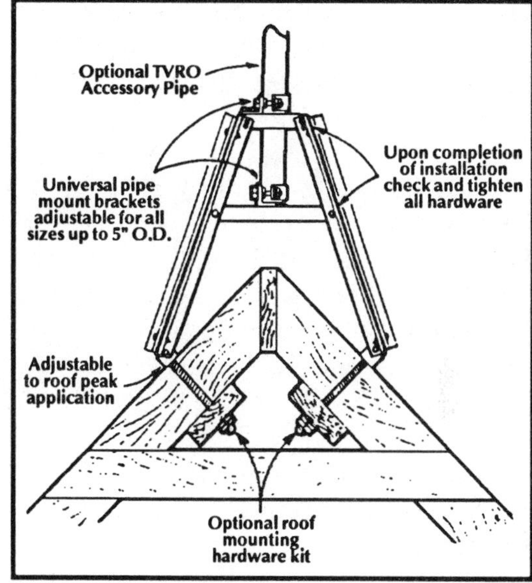

Figure 5-43. Peak Roof Mount. *An alternative to supporting antennas as large as 8 feet in diameter on a roof is a peak mount. Before installation a professional engineer and zoning authorities should be contacted and the appropriate license should be obtained. It is critical that as many as eight beams under the pitched roof be interconnected with additional steel or lumber members. Great care should also be taken to make certain that no water can leak through the holes causing damage to the inside structure and electrical wiring. (Courtesy of Rohn Tower)*

ing sounds that are amplified by their direct connection to a roof, especially in high winds. Some antennas may even resonate loudly in some winds. A typical roof on a house that has some give built-in can act much like the membrane of a loud speaker. This problem would probably not arise where an antenna is solidly attached to a concrete or I-beam roof of a commercial building. An additional hazard is often not considered with dishes mounted far above ground: if the reflector is mounted on a long pole near an area where people walk, snow or ice resting on the antenna may loosen, fall and cause injury to those below. This outcome is less likely when using offset fed antennas which are inclined more steeply towards the horizon than are the prime focus variety. In any case, if damage or injury does occur, both the owner and satellite dealer can be held responsible.

Anyone considering roof mounting a relatively large dish would be wise to hire a qualified structural engineer to analyze the situation. If a roof mount is judged to be feasible, the engineer takes responsibility for making this recommendation. Also, in many countries, a permit is required to change the structural integrity of a building.

The fundamental principle to be followed in designing and installing a roof mount is to tie the support structure in to as many rafters as possible in order to spread the load. A number of companies now manufacture welded steel assemblies with built-in leveling adjustments designed to support poles on most types of roofs (see Figures 5-43, 5-44 and 5-45). The support pads must be attached to a length of lumber or metal inside the

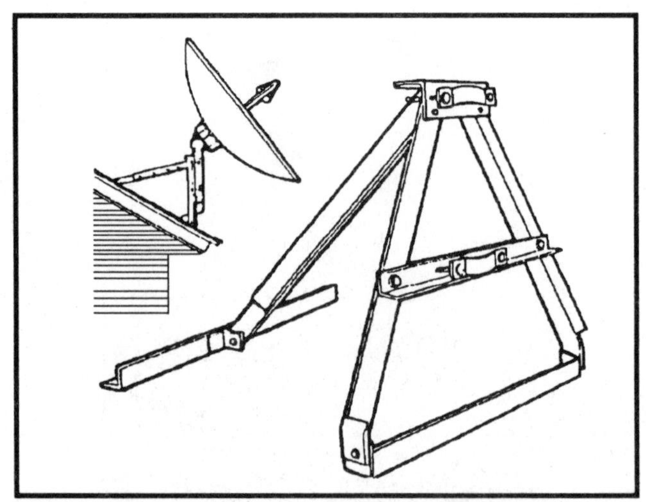

Figure 5-44. Pitched Roof Mount. *When using a side roof mount, the same care should be taken as a with a peak roof mount. As many roof beams as possible should be used and additional support lumber should distribute the weight of the antenna/mount. Roof mounts, as in all satellite TV installations, should be well grounded. (Courtesy of Rohn Towers)*

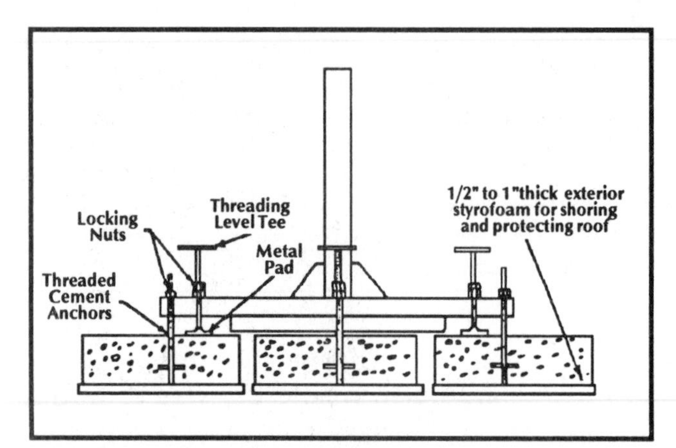

Figure 5-45. Concrete Flat Roof Mount. *This mount is bolted onto four concrete slabs which are spread over a 10 square foot area. "Tea handle" screws are used to allow leveling of a polar mount. No leveling is necessary for az-el type mounts. (Courtesy of Upper Midwest Satellite Supply)*

attic which itself is secured to as many rafters as possible. If the roof has steel beams, it is highly recommended that the mount be tied in to these members. Two or three guy wires, preferably made from a material as strong as aircraft control cable, can be used to secure the antenna for further stability. These will also prevent the reflector or mount from landing a hundred meters away if the worst should happen and it does break away from the roof.

Great care must be taken in breaking roof seals to prevent leaking (see Figures 5-46, 5-47 and 5-48). It is wise to plan this step with the help of a reputable roofing contractor. When seals are broken, the original roofer's warranty is voided and a dealer can be held liable for water damage to the

Figure 5-46. Under-side and Outside View of Spun Aluminum Antenna on a Roof Mount. *This 8.5 foot (2.6 meter) antenna was attached by spanning five rafters with two pieces of 8 inch (20 cm) channel iron both above and below the roof. The resultant "sandwich effect" helped in distributing the antenna and mount weight over a larger surface area. Holes were caulked and then tar was poured on to the roof. If a pitch pan had been installed under the channel iron, water sealing would have been even more effective. (Courtesy of Brent Gale)*

structure or contents of a building. Even small unsealed holes can allow water to penetrate a roof and ruin a ceiling. A lesson in this matter can be learned from local, competent solar installers who have faced and solved related problems of wind loading and water leakage for years.

Roof mounts do have the advantage of being easier and less expensive to assemble than most ground foundations. There are no problems associated with mixing and pouring heavy concrete, with waiting for concrete to harden in harsh winter conditions and with trenching through frozen ground to lay cable runs. In addition, there is a much lower chance of theft when the satellite equipment is mounted out of reach. However, winds are much stronger above a building than on the ground below and installing on a high or steep roof can be dangerous.

Figure 5-47. Flat Roof Pitch Pan. *Before beginning this installation, a roofing contractor was hired to lay down a 4 by 8 foot aluminum pitch pan which was filled with tar after the installation was completed. This sealed the roof and prevented leaks.(Courtesy of Brent Gale)*

Figure 5-48. Flat Roof Mounts. *Both of these antennas were mounted on gravel-top roofs. After the gravel was cleared away a 1/4 inch (6 mm) steel diamond plate sheet was attached to the underlying steel roof members with four 3/4 inch (19 mm) double nutted all-thread rods. Tar was poured onto the roof surface beneath the steel plate and on its top surface after all necessary tracking adjustments had been completed. This installation detail thoroughly sealed the roof and prevented water leaks. A professional structural engineer was consulted before the work was started. (Courtesy of Brent Gale)*

Towers and Long Pole Supports

One alternative to a direct roof mount is a long pole support (see Figures 5-49, 5-50, 5-51, 5-52 and 5-53). In some cases, a long pole can also be somewhat easier to install because the pole is secured in the ground adjacent to a building and is more easily accessible. In addition, building permits are generally not required because no structural alterations other than attaching the pole to the adjacent wall are required. There must be at least one attachment point just where the pole meets the top of the wall, as this point will experience the greatest torque. It is better to weld tabs to the pole instead of using collars which may slip when high winds twist the reflector. The structure of the wall must also be considered in attaching these tabs. There is always the chance that weaker parts of the wall could be damaged when the antenna and pole twist or bend under forces of wind or snow.

In general, for every 10 cm (4 inches) in addition to the normal 1.5 meters standard ground pole length, an extra 50 cm (20 inches) should be added to the normal 1 meter below ground segment. For example, if an 8 meter length were required above ground level, 130 cm (10 cm for each 50 cm of the extra 6.5 meters) should be added to the 1 meter normally below ground. Thus 2.3 meters would be underground and 8 meters would be above ground. Of course, this is a rule of thumb which does not have to be adhered to rigorously.

Figure 5-49. Free-Standing, Long-Pole Installation. *This dish is mounted on a free-standing pole. In such cases, the pole diameter should be roughly double the normal size and a pole reducer must be attached to the top of the pole where the mount attaches. (Courtesy of Brent Gale)*

Figure 5-50. Long Pole Supported by Wall. *This 10 foot antenna is support by a long pole that is securely attached to the adjacent brick wall. The pole must either be welded or bolted to the points of attachment so that it will not twist in high winds. (Courtesy of Brent Gale)*

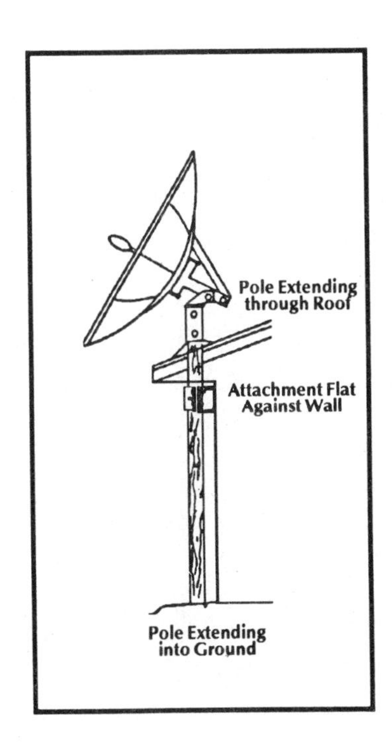

Figure 5-51. Exterior Long Pole Wall Mount. *Two methods to attach a long pole to an exterior wall, bolting the pole directly against the building and extending the pole through a hole in a pitched roof, are shown here. Both poles were secured at their bases in concrete filled holes. Extreme care should be taken when attaching brackets because high wind forces can damage or detach some of the weaker types of wall materials. In order to determine how closely a support may approach a wall, a small hole should be dug at its base because some building footings may extend as far as two feet past the edge of the building. In these cases, a plumb bob can be extended from the roof to the center of the hole in order to determine the required length of brackets to secure the pipe. (Courtesy of Upper Midwest Satellite Supply)*

In addition to securely attaching the pole to one or two points on the adjacent building, it should be oversized in diameter by 5 to 15 cm (2 to 6 inches) for extra stability against twisting and bending. In general, this will require use of a reduction coupling at the top of the pole. Guy wires should be attached to the base of the mount and to nearby support points. Note that a lesson can be learned from the designs of billboard installations which also use long poles. These supports begin with larger diameter poles at the bottom and step down to smaller diameter poles near the top.

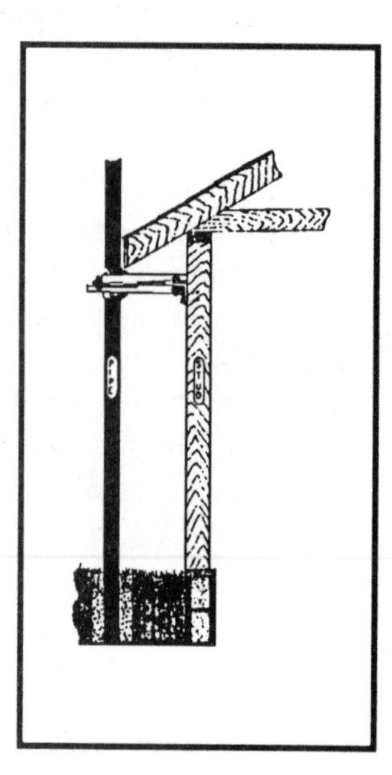

Figure 5-52. Adjustable Exterior Wall Mount. *This mount is designed so that its position can be adjusted beyond the eave of a building. It should be secured to at least two places as well as at its base in a hole filled with concrete. (Courtesy of Upper Midwest Satellite Supply)*

Figure 5-53. Interior Garage Support for Long Pole Mount. *This type of support is fastened to the inside wall studs with two headers located near the top and bottom of the pole. The headers connect to five 2 by 6 inch (5.by 15 cm) wall studs and pass the pole through a hole in the pitched roof. Holes must be drilled through the pipe and multiple bolts should be used to secure the pipe to the V-clamps to prevent it from twisting in winds. The manufacturer of this product recommends that antennas in excess of 1.8 meters in diameter should not be used in this type of installation. (Courtesy of Upper Midwest Satellite Supply)*

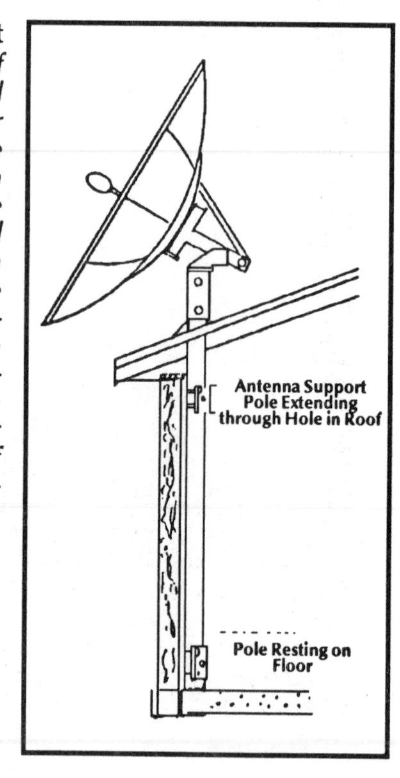

D. CABLE RUNS AND TRENCHING

Cable runs between the antenna and indoor equipment should be safely protected and neatly installed. A variety of techniques can be used depending upon the type of cable required, the length of the run, antenna location, and, when applicable, ground conditions. For example, trenching is not necessary in conjunction with a roof or some types of long pole supports.

Digging the Trench for Ground Mounts

When a trench is necessary, before digging begins the first step is to determine where all the water, electrical, gas and telephone lines are located. In many countries, utility companies will be more than glad to send a crew out to find these lines and even to help dig to ensure that they are not severed. Customer owned underground systems such as lawn sprinklers can be more difficult to pin-point. Often a homeowner will not be sure where the pipes are located. It makes good sense to proceed with caution. In those regions where sprinklers are common, having parts on hand for repairing pipe can be a great time-saver.

The depth of the trench depends upon many factors. If rigid conduit is used there is less risk that an avid gardener will slice a shovel through the cable. If the cable is running through an unused portion of the yard, it should be buried at least 15 to 20 cm (6 to 9 inches) and preferably 30 to 45 cm (12 to 18 inches) under the surface. Electrical codes in many countries specify required burial depths (see Figures 5-54 and 5-55).

Figure 5-54. Installation of Conduit. *This 2 inch conduit was placed in an 18 inch trench that had been excavated by a power trencher. Notice the weatherproof cap and the drip loop at the point of entry into the building. (Courtesy of Brent Gale)*

Figure 5-55. Trenching Machine. *This machine can dig a 1, 2 or 3 inch wide trench with a depth of 8 to 12 inches at a rate of 20 to 25 feet per minute. It can be a great time saver in certain situations but can damage sod on a well-kept lawn. (Courtesy of T.H. Riley Manufacturing)*

It makes good sense to keep a record of the length, route and gauge of buried cables. This information on a map of the site can be invaluable if an update or repair is required at some future date. The customer will also find a copy of this record useful if landscaping changes are made.

Leaving unburied or unsecured cables lying around is an invitation to an animal or someone with a lawnmower or a destructive curiosity to cause damage. As discussed in the following section, it is not advisable to splice damaged cables.

Whenever a trench is being cut across a lawn, care should be taken with the sod. It should be cut with a sharp knife or a flat shovel to a depth of at least 7 or 10 cm (3 to 4 inches) on both sides of the trench. The sod should then be carefully rolled in sections and laid next to the trench. All loose earth should be placed in a wheelbarrow since it is often difficult to clean loose fill from the surrounding sod with a rake after the trench is refilled. Any sod left above ground for periods longer than a day must be watered to avoid damage because it can easily dry out and wither.

A trench must not be refilled until the installation is completed and it is ascertained that there are no defects in the cable. This precaution will prevent unnecessary re-excavation of a deep trench. Refilling the trench is one of the last steps in a successful installation.

The Use of Conduit

Many direct burial, multi-conductor cables do not necessarily require encasement in a rigid, protective conduit. However, conduit does provide extra insurance against damage (see Figures 5-56 and 5-57). In most situations using conduit is strongly recommended. For example, it provides the necessary protection for runs under driveways or roads. If rodents are common, conduit can prevent them from chewing through cables. In areas with excessive ground water, conduits can protect against water damage. Be aware that there have been cases where improperly installed conduit has done just the opposite and has become a water trap. As a result, the only solution was to dig up the conduit to drain the excess water.

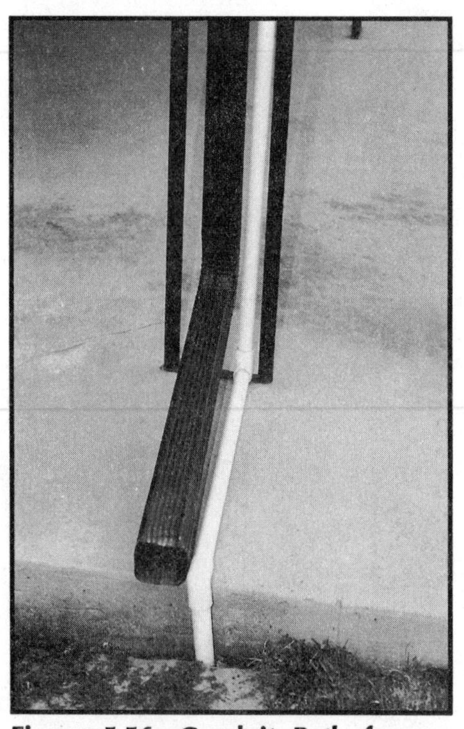

Figure 5-56. Conduit Path from a Long Pole Mount. *This conduit was routed from the roof along a gutter and then into a buried path. Its path along the drain was chosen so that passers-by would not trip over the conduit and perhaps damage the cable. (Courtesy of Brent Gale)*

Figure 5-57. The Use of Conduit. *Conduit protects cables from any damage that could be caused by rodents as well as from moisture ingress at points where the cables enter and exit the ground. A weatherhead must be installed to prevent moisture from entering the conduit at both the dish, as illustrated here, and at the entry to the building. (Courtesy of Brent Gale)*

Local building codes in some countries may require the use of conduits for cables that carry power above a predetermined level. In these cases, white PVC water pipe or black rolled flexible tubing is not permitted. Only rigid varieties like gray PVC or aluminum are permissible. The reason for this regulation is that flexible conduit rusts and breaks through in a relatively short time. It also should never be used in satellite installations.

Conduit should be a least twice the diameter of the enclosed cables. And installing sweeps instead of right angles at corners makes pulling cables into and out from conduits easier. Since sweeps are not available in standard white PVC used for plumbing lines, either gray PVC or standard metal electrician's conduit should be used. Some types of conduit are designed to be bent into shape by applying heat when they are wrapped with a material similar to heat tape. It is important to realize that if more than four sweeps forming a full 360° turn are used throughout the cable path, it would be impossible to push or pull cables into or out of the conduit. If this happens, one major advantage of using conduit is eliminated. The proper installation of conduit allows removal and replacement of cables years after burial without the need for excavation.

Precautions should be taken to prevent water ingress, especially at the antenna. Sections of conduit can be tightly glued together. Weatherheads or 180° sweeps attached to vertical conduit should have removable screw fittings and should be used at all exposed ends. If the cable ever has to be removed from the conduit and replaced, the screw fittings can be undone to facilitate this task. The use of weatherheads as well as correct bonding at joints is very important since any water build-up can ruin cables and corrode connectors. In those cases where the conduit runs downhill from the antenna to a building, improper installation could cause water to be channeled directly into its interior.

When using PVC type conduit, care must be taken to avoid dripping the solvent or cement onto cable insulation. Some types of insulation can be dissolved by these solvents and damage to cables can result.

When pulling cable through conduit, do not lay the cable on the ground, as it can pick up debris and make the job more difficult. A "fish tape," which is a piece of rigid wire, can also be used to pull cables through conduits. It is inserted at one end and the cable is attached to it at the other end. Cables can also be lubricated with a liquid soap to allow them to slide through the conduit more easily. Of course, this soap must be cleaned off where the cable enters a building. If the cable is too short to span the entire length between the antenna and receiver, never splice two pieces together. Start again. Run a complete and sufficiently long cable. Splicing would not only increase resistance to current flow and potentially cause impedance mismatches, but would also present a perfect opportunity for cable breaks to occur in the future or for water to enter and short out the system. The splice is also likely to pull apart when cable is being fed through a conduit. In those rare cases when a splice is absolutely necessary, the splice should be in an above-ground weatherproof box and weatherproofed with silicon grease.

All the cables can be attached to the supporting pole with wire ties or hose clamps. Drip loops can also be formed where the cable terminates to channel water away from the actuator motor. A drip loop is simply a loop in the cable which has its lowest point below connectors, a hole in the wall or an opening in the bottom of a waterproof box. Any water from rain or melting snow and ice tends to collect at this low point and drip away (see Figure 5-58).

All of the cable used for both pole and roof mounts running from the antenna to the building entry should be placed in electrical conduit. This will prevent the cable from being damaged by ultraviolet light and by other enemies such as birds and squirrels. Many building codes also mandate the use of conduit. Using well-installed conduit also results in a neater and more professional installation.

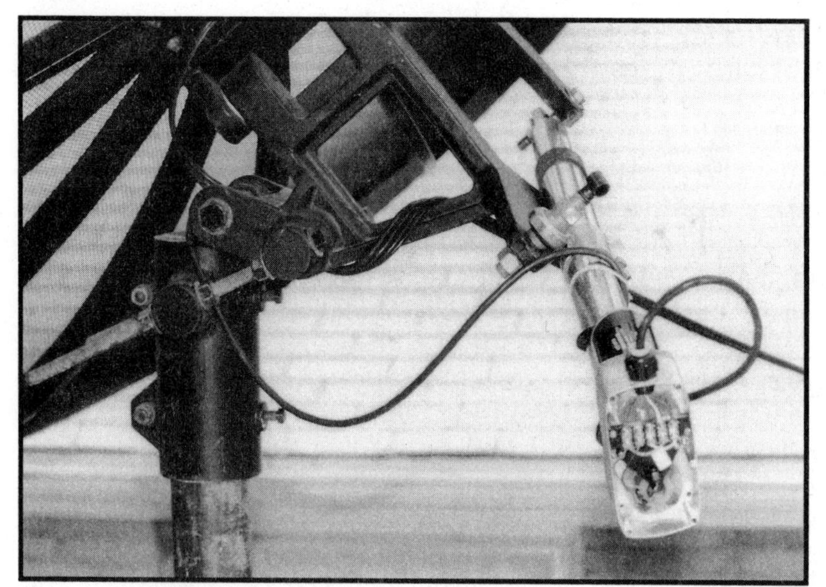

Figure 5-58. A Drip Loop. *A drip loop allows water build-up caused by melting snow, rain or condensation to drain away from components such as actuators. Notice how the cable here has its lowest point below the actuator motor.*

The Entry into the Home

Most customers prefer to have all cables hidden from sight. Holes drilled in both exterior or interior walls should be as small as possible and well caulked after the installation is completed. If the cable entry is through a crawl space, holes may be drilled below the ground level only if they are well sealed. In general, above-ground cable entries are recommended. Always have sets of masonry and wood bits on hand. Many older homes have very thick walls so that extra long wood bits (up to 45 cm or 18 inches) should be available during an installation. Holes should be drilled at least 30 cm (12 inches) above or below and away from any electrical outlets to avoid hitting power lines. Exercise caution when drilling near bathrooms, kitchens or other places where plumbing may be close and where substantial damage can be done.

Drill from the inside to the outside of a wall, not the other way around, and only after all measurements have been taken and the path of the bit is known. When possible, use wallplates similar to those on electrical outlets. A wide variety of brands are available at most electrical parts distributors. Cables fed either into an attic or crawl space can usually be fished up or down into a wall to make an unobtrusive entry. Difficulties can be encountered if the walls have cross-supports between studs.

When carpet must be disturbed, cut a small "X" with a razor blade or an "exacto" knife through both the carpet and padding so later if the system or cable is moved the rug can be easily repaired. A drill can often catch on the rug pile and tear a wide area open or ball up padding. Never run cables under a rug. The tacks used to attach the rug to the pad and floor can also easily short cable out. Cables under rugs cause bumps and make tripping all too easy.

When cables are routed through either crawl spaces or attics, they should be attached to rafters or floor joists. Special staple guns are made to fit onto the cable surface to avoid damaging the conductor. For example, the Arrow T-25 staple gun used with 14 millimeter (9/16th inch) round-top staples, which are made especially for RG-59 cables, makes this job easy. The T-15 staple gun can be used with 18 or 20 gauge wires and cable. Small plastic clips which are designed to be nailed onto a wall are another cosmetically attractive option.

One final note: never run cables parallel to electrical lines. Doing so can result in interference such as 60-cycle hum, especially when grounding is inadequate.

E. ASSEMBLING THE ANTENNA, FEEDHORN & LNB

The antenna, feedhorn and LNB are the most critical components of a satellite system. If they are not selected and installed properly, using the best and most expensive satellite receiver and modulator in the world will not improve picture quality.

Assembling the Antenna

Antennas generally are accompanied with a set of assembly instructions. These instructions should be carefully read at least twice before beginning an installation to determine if any special tools or manpower will be required or if there are any unanswered questions. Usually, most of these questions can be settled by simply beginning the assembly, but occasionally a call to the distributor or manufacturer is necessary. Be sure that all parts are accounted for before proceeding to the job site.

Figure 5-59. Lip-Sighting a Dish. *Sighting along the edge of a dish to see if the line from the edge closest to the eye is parallel to the one furthest away can be an effective method to check that the dish is not warped. (Courtesy of Brent Gale)*

Each variety and type of antenna has some unique aspects to its assembly. Spun or stamped reflectors are most often delivered in one piece ready to be bolted to their mounts. In cases where holes have to be drilled to attach a mount, it is critical that they be centered. Do so by crossing two strings each tied to two opposite corners of the mount. Lining up their intersection with the antenna center will properly align the mount. Spun reflectors have been formed in a lathe and therefore always have a hole located precisely at their center. Stamped antennas often have a hole for the buttonhook or an identifiable center point remaining from the manufacturing process. It is good practice to sequentially tighten all bolts or screws on any antenna only after all are in place to avoid warping the reflector. It is just as important not to over-tighten these bolts. This could cause a dimpling or indentation and a substantial loss of gain. Rubber spacer washers or grommets are sometimes included to act as shock absorbers and should be used when provided. If the antenna is aluminum and the mount steel, it is even more important to insert these rubber grommets to prevent electrolytic action between these dissimilar metals. In time, this type of corrosion would weaken or even destroy the points of attachment.

There are two accurate and very simple techniques used to check that antennas are not warped either during manufacturing, shipping or assembly. Lip-sighting an antenna can be accomplished by sighting along one rim so that the other edge appears in the same plane. Both rims should line up as two straight lines one on top of the other. If not, the antenna is warped (see Figure 5-59). The crossed string method involves using two or three strings taped or otherwise fastened over the outer edge of an antenna. If these strings either mesh together or are separated, the reflector surface is warped. They should just lightly touch one another (see Figure 5-60). Using three crossing strings makes it easier to locate the center point. If there is some warping, loosening and re-tightening the bolts which hold the antenna structure can often be an effective method to realign the surface. A warped antenna will not perform up to par.

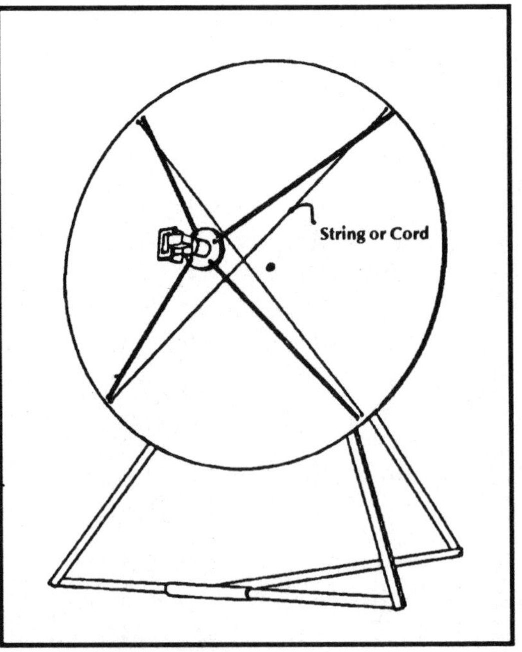

Figure 5-60. Stringing a Dish. *Running two or three strings attached to the edges of a dish is, like sighting a dish, an effective method to determine if the reflector is warped. If these strings are either meshed tightly together or separated, the antenna is warped.*

Fiberglass antennas should be assembled on a flat surface so that all sections fit together with the rim lying perfectly flat. Again, do not tighten bolts until the assembly is completed. This procedure allows each section to be lined up to smoothly interface with the adjacent panels. Some fiberglass reflectors can be assembled on the mount. In this and all other cases, the bolts should be tightened progressively from the center to the outside to make the antenna expand outwards from the center and maintain a true parabolic shape.

Mesh reflectors can be assembled on any surface but completing the work on grass or a soft material such as the cardboard boxes used for shipping avoids scratching the paint. It may also be convenient to rest the central hub on a stand such as a small table or a garbage can to make attaching all the ribs, panels and bolts or screws easier. If wind clips are provided they should be used, especially in areas having potentially strong and gusty winds. There have been cases of panels popping out partially or completely in a wind storm. Some installers run a small wire through each section near the rim to provide further protection against this type of damage.

Mounting an Antenna on the Pole

Except for bulky fiberglass antennas, most can be easily lifted by two people. Large fiberglass antennas in excess of 12 feet (3.7 meters) in diameter usually require four or more people to share the lifting. In some cases, a crane may even be required to complete the job. This extra weight has prompted some manufacturers of fiberglass antennas to design their products so they can be assembled directly on the mount after it has been set on a pole.

Most antennas can be installed by first setting the mount onto the pole. This is recommended to avoid warping the reflector surface which could occur when it is lifted with a heavy mount attached (see Figure 5-61). Then the elevation adjustment is set so that the mount either points directly upward or as close to the horizon as possible. The antenna can then either be lifted like a bird-bath onto the mount or rolled into alignment with the mount to the point where securing bolts can be inserted.

In particular, the surfaces of spun or stamped antennas can be easily deformed, dented or scratched so they should always be placed in their final position after the mount has been attached to the pole to avoid bending the rim at the points of lifting. Damage can also occur until the actuator arm is attached since the antenna can often move freely and strike other objects.

Figure 5-61. Preparation for Lifting Antenna on to Mount. *This mount has been prepared for the antenna. In this case, the dish can be rolled into place for attachment of the bolts. One technician can use this method to install even large antennas. Alternatively, the mount could have been adjusted to point directly upwards. Then the dish would be "bird-bathed" by lifting it onto the mount.*

Assembling the Feedhorn/LNB Structure

The feedhorn and LNB should be bolted together before being installed on the buttonhook or tripod support. Two important rules should be observed. First, never touch or bend the probes on either component. They are finely tuned at just the correct location and any tampering will increase the VSWR and lower the signal-to-noise ratio. Second, always use the gaskets provided for insertion between the flanges. These protect against water ingress. But never use any sealant between these gaskets because there must be metal-to-metal contact in these joints. The distance between these joints should not be increased as this would upset the impedance match between the waveguides. The bolts around these flanges should also be evenly spaced. They should first be tightened by hand and then with a spanner.

The throats of most feedhorns have plastic covers, called insect covers, with small holes to allow any condensed water to escape. These covers should always be used. Occasionally, when a feedhorn has been left uncovered, wasps or other creatures have built nests inside. Needless to say, such an obstruction would absorb microwaves and cause complete loss of reception.

Correctly Aligning the Polarity Selection Probe

When using servomotor feeds, align them so that their 270° range of rotational motion is correctly centered. Note that some older feeds had only 180° of motion so that more care must be taken in aligning the feed (see Figures 5-62 and 5-63). For linearly polarized signals, the center must be halfway between the directions of the two polarities. For circularly polarized signals, the feedhorn probe must be able to move at least 45° on either side of the dielectric slab as described below. The range of motion of a servomotor can

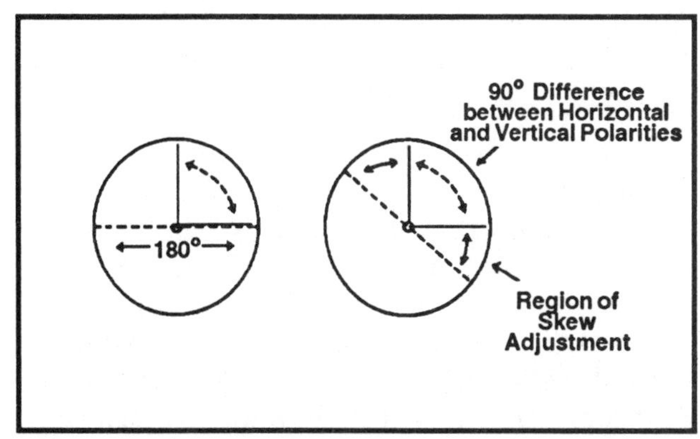

Figure 5-62. Servo Motor Mechanical Limits in Older Feeds. *Most older servo motor feeds have a 180° range of motion. If the mechanical limits do not lie past the orientation of both vertically and horizontally polarized signals, the probe will not be able to be properly aligned and the motor might be burned out in the attempt to move the probe to the correct position. Modern servo motors now have a 270° range of motion and the alignment problem is less critical.*

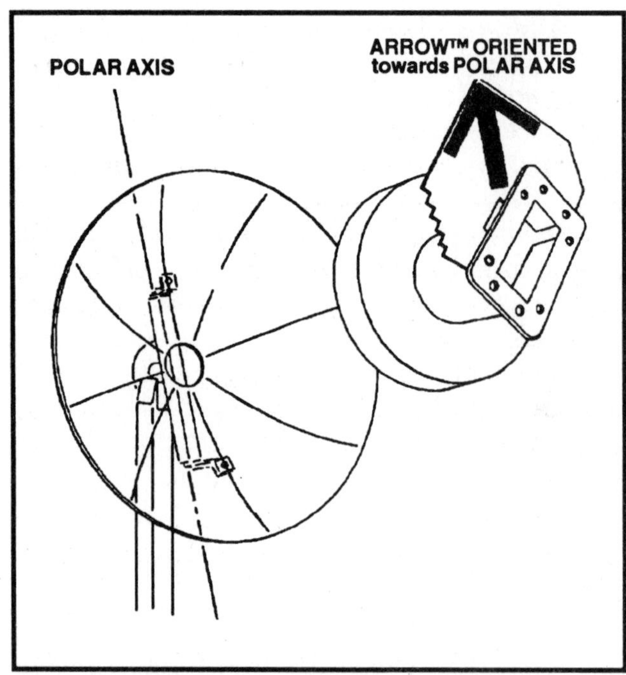

Figure 5-63. Orienting the Feedhorn/Polarizer. *Chaparral Communications makes a useful device called the Arrow™ which aids in properly aligning a ferrite or servo motor feed. When the feed has been properly installed, the point of the arrow is lined up with the polar axis on the mount. This guarantees that the motor will not be driven up against a mechanical limit and be damaged in the attempt to line up the probe with the direction of polarization. The one diagrammed here is used for orienting the Twister™ and Polarotors™. (Courtesy of Chaparral Communications)*

be observed by hooking it to a test receiver when limits are set. When properly installed, the motor should have the freedom to rotate the full 90° between positions of horizontally and vertically polarized signals and still have extra room to move past either end of this range. This will permit the necessary skew adjustment when tracking between satellites transmitting different orientations of linear polarization. Most importantly, it will prevent the motor from running up against a limit and burning out either the motor or the timer in the receiver. This installation procedure also guarantees that the probe motor is functioning properly. Signs that a feed does not have an adequate range of motion are snowy pictures or floating lines (hum bars) appearing on the television screen (see Chapter VIII for more details).

Most manufacturers include a template to be used in correctly aligning the probe. One of two basic methods is employed. Either the template fits onto the WR-229 waveguide or onto the servomotor. The arrow on the template is lined up so it points upwards along the polar axis and the motor is in approximately an 11 o'clock position. An exception is the Calamp/ADL feed whose waveguide is aligned with the polar axis.

Retrofitting for Detection of Circularly Polarized Signals

Some types of scalar feedhorns which select between linear polarities by probe rotation can easily be retrofitted to discriminate between left and right-handed circularly polarized (LHCP and RHCP) signals. This is done by correctly inserting a dielectric slab (see Figure 5-64).

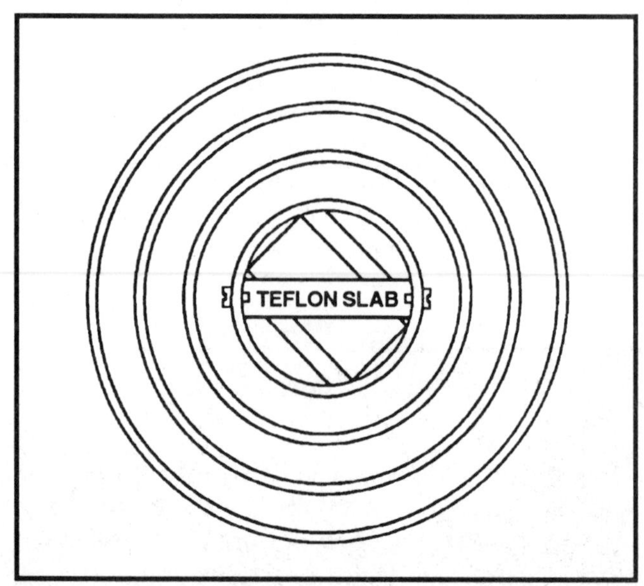

Figure 5-64. Feedhorn Modification for Detection of Circularly Polarized Signals. *A linear feedhorn can detect right and left hand circularly polarized signals if a slight modification is made in the circular waveguide in front of the transition to the rectangular waveguide. A teflon slab inserted at a 45° angle in the circular waveguide transforms a circularly polarized signal to a linear format.*

In order to maximize reception of both linear and circular polarizations by the same feedhorn, the probe position must first be set for best reception of either linear polarization from a low power transponder. The slab is then inserted parallel to the direction of the feedhorn probe to guarantee that linearly polarized signals are attenuated as little as possible. LHCP and RHCP satellite transmissions can then be received when the probe is rotated to a position 45° from either side of the dielectric slab. It is important that the feedhorn be oriented to allow the probe movement to 45° from both sides of the slab as well as to both horizontal and vertical polarity positions. This requires aligning the feedhorn so that the dielectric slab is set about 15° or 20° closer to what normally would be the center of the probe's range of motion.

The teflon insert converts circularly polarized signals to linearly polarized ones which are then detected as usual by the feedhorn. If properly installed, losses can be kept to an acceptable minimum on detection of the linearly polarized signals. Dielectric inserts, available from some feedhorn manufacturers, usually come with easy-to-read instructions for their installation and use.

Properly Aligning the Feed

Properly adjusting the feed is critical to TVRO performance. This involves correctly setting the focal length and aligning the feed along the antenna axis. In order to maximize the performance of the entire system, great care should be taken to ensure that this step is properly completed.

Always follow the manufacturer's installation instructions for measuring the correct focal length. Each antenna has a unique f/D ratio and focal length (see Figure 5-65). To illustrate, if a one meter reflector has a 30 cm focal length, the distance from the center surface of the antenna to the front edge of the feedhorn must be precisely 30 cm. If the antenna were warped this could change the position of the focal length and cause a marked impairment in reception. To avoid this outcome, the procedures for assembling and testing the antenna reflective surface discussed earlier should be carefully practiced.

Centering the feedhorn is just as critical as setting the focal distance. At least three methods can be used:

1. Measure from the outer lip of the reflector to the throat of the feedhorn from four different points at 3,6,9 and 12 o'clock positions. The resulting measurements should be very similar in each case. If not, the feed should be adjusted until the measurements are nearly identical (see Figure 5-66).

2. Cross the antenna using two pieces of string at right angles to each other over its surface. This is similar to the method outlined above to check for warping. Spun aluminum, hydroformed or other reflectors having a

Figure 5-65. Setting the Focal Length. *Setting the correct focal length is very important in optimizing reception of satellite broadcasts. This measurement should be taken from the antenna surface to the front of the feedhorn throat. The manufacturer of each type of feedhorn generally specifies the precise point at the throat from which to measure the focal length. (Courtesy of Brent Gale)*

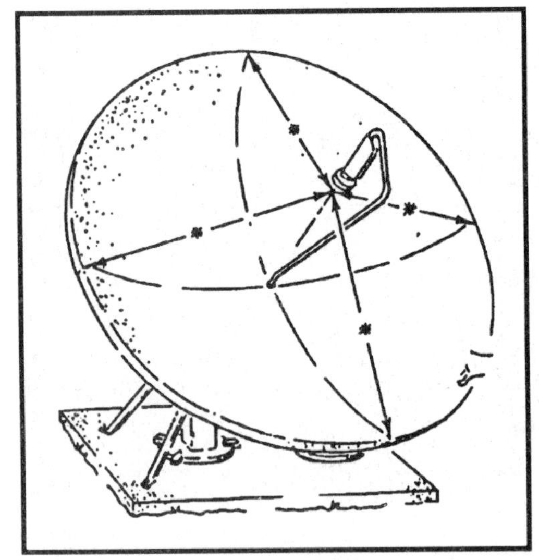

Figure 5-66. Centering the Feed. *The feed can be easily centered by making sure that three or four measurements from points around outer edge or rim of the dish to the feed are of equal distance.*

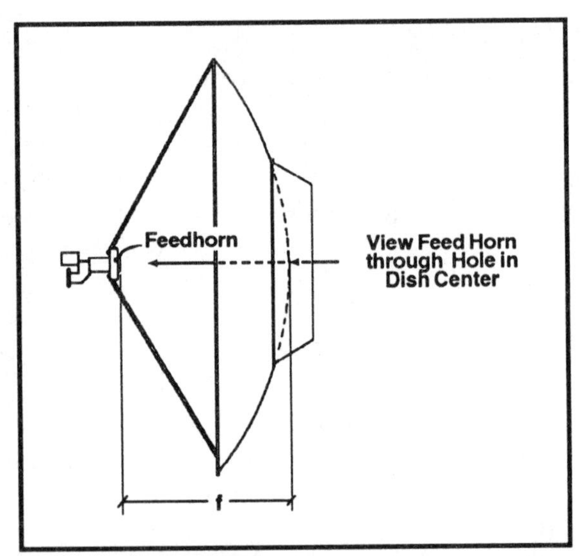

Figure 5-67. Sighting a Stringed Dish. *When looking through the hole in the center of many dishes to the point of crossed strings, the feedhorn should be located directly on center. This is similar to sighting through cross hairs on a gun. It is crucial in this process that the strings span the full diameter of the antenna so they will cross in its center.*

hole in their center can then be tested for feed alignment by stringing two additional pieces of twine behind the antenna. They should cross over this hole. Sighting along both cross- points should target the center of a properly aligned feedhorn (see Figure 5-67).

3. Use a focal finder that inserts into the feedhorn throat. The telescoping portion, similar to an automobile antenna, should touch just at the center of the dish (see Figures 5-68 and 5-69).

When using a buttonhook design feed support, three small cables with S- hooks that attach to the feedhorn at the 12, 4 and 8 o'clock positions must be used to improve stability. The other ends of these cables are attached to the antenna with metal clips that slide onto the outer lip of the reflector. This "wind kit" makes centering the feed by tightening or loosening the different threaded eye hooks easy.

Figure 5-68. Using a Focal Finder. *This photo shows a focal finder inserted into the throat of a feedhorn. It points directly at the center of the antenna as it should when the feed is correctly aligned. If a tripod instead of a button hook feed support had been used, the focal finder telescoping arm would have extended all the way to the center point of the antenna. (Courtesy of Brent Gale)*

Every feedhorn is designed to function optimally at a rated f/D ratio. For proper performance, the feedhorn must be mated to antenna design and construction. Tests have shown that a mismatch between feedhorn and antenna could cause in excess of a 1 dB drop in performance. That extra gain can be the difference between having a few sparklies or no sparklies in a poor footprint area or when using a small aperture antenna.

Figure 5-69. Focal Finders. *This photo shows focal finders fitted into the throats of two feedhorns. The telescoping rods are fully retracted here. (Courtesy of Natropolis International)*

F. INSTALLING ACTUATORS

Step number one in installing actuators is reading the instructions. No one knows better than the manufacturer how many problems a poorly installed actuator can cause. The instructions clearly spell out what mistakes to avoid as well as correct installation methods.

Mechanical Assembly

Linear Actuators

Once the antenna is mounted on its pole, the actuator arm can be attached. The end of the actuator with the ball joint is usually attached to the reflector; the other end with some type of pivot clamp is attached to the mount (see Figure 5-70). This mounting hardware is installed to assure that adequate lateral movement is allowed so binding is avoided. This is crucial because lateral forces on the inner tube can make it bend and seize up. These forces can also cause deterioration and potential failure of the internal O-ring seal which protects against water and dirt entry between the inner and outer tubes. Never weld an actuator directly to the rear of an antenna!

The arm should also be attached so it forms a minimum of at least a 30° angle with the rear surface of the antenna at all points in its sweep. Having a smaller angle causes the motor to work harder because the arm has less leverage. Actuator arms vary in length from 30 cm (12 inches) and less to over 132 cm (52 inches). The larger and heavier the antenna, the longer the arm required to provide adequate leverage to easily move the antenna. This arm should be installed as parallel to the plane of rotation as possible to minimize forces which could cause it to bend.

Figure 5-70. Attachment of Actuator Arm. *The actuator arm attaches to the mount at one end and to the dish frame at the other. In this installation, both joints can rotate and the upper joint has lateral movement so that any potential binding is prevented.*

It is crucial to be aware that the installation of actuators on the west coast of North America differs from systems on the east coast. There are few C-band satellites west of Satcom F1 at 139°W while many satellites that relay video signals are clustered in the east. For example, the longitude of Los Angeles is 120°W, a difference of only 19° from Satcom F1. New York on the opposite side of the continent is located at 74°W, a full 65° from the same satellite. By comparison, a site in New York is only 5° removed from Spacenet II at 69°W.

An actuator installed on the west coast should be attached to the right side of an antenna from the perspective of an observer at the back. During installation, the clamp on a fully retracted actuator should be tightened only when the antenna is swiveled as far east as possible. The antenna should then track from the far eastern satellite to a westerly position about 25° past true south. When installing an actuator on the east coast, the actuator should bolt to the left rear side of the antenna and the end of the fully retracted arm should face the antenna as far west as possible. If the actuator is installed on the wrong side in either the far east or west, the end result could be a bent tube or a system that cannot receive all satellite transmissions.

Install the actuator with the above details in mind. First, make sure that the actuator tube is fully retracted. Next the actuator clamp should be positioned so that the antenna can be aimed just past its most westerly or easterly target, depending upon which side the actuator is mounted. This is accomplished by allowing the actuator tube to move in its clamp while the antenna is rotated into position. Then tighten the collar bolts after the antenna is properly positioned.

There is an exception to the above procedure. A movement called rock-over can occur if an oversized actuator has been installed or if the actuator tube is accidentally fully extended as a result of action by a user or the wind. Rock-over prevents the actuator from retracting and will usually bend the tube as well as a portion of the antenna or its mount. During installation, align the antenna beyond the lowest satellite and make sure that the actuator has mechanical limits.

Most ball jack actuators can exert 700 kilos (1500 pounds) or more of force. Acme arms can generate about 350 kilos (750 pounds) of force. This force applied via the reflector outer rim could easily cause damage to surrounding objects. Be sure that the antenna has a full range of movement without coming into contact with any obstructions such as trees or fences. In addition, the actuator must not bind on any part of the mount or reflector over its full range of movement, even when inadvertently driven past the preset electrically programmed east and west limits.

Sufficient slack cable should be left hanging between the actuator and the mount so a drip loop can be formed. The antenna should be able to move through the entire arc without tugging on this cable and thereby eliminating the drip loop.

Preventing Water Damage

Water is the chief enemy of actuators. If it manages to accumulate in the motor housing or in the gear box it can cause corrosion. More immediate damage results if water freezes so that motors and gears seize.

An actuator should be mounted according to its manufacturer's instructions so that with the motor facing either upwards or downwards the drain holes in the housing are at the bottom. If there are no drain holes, drill one or two 5 mm openings into the housing at the lowest point. This will allow water which enters by condensation to escape before it accumulates.

Another place where water can enter is the opening between the outer and inner sleeve. Most manufacturers provide a rubber fitting, called a shaft wipe, to protect this opening. Further protection can be realized with a neoprene accordion weatherboot. Because neoprene is transparent to sunlight, the solar energy more easily evaporates moisture. There should be vent holes at both ends of this boot to allow water and air passage so that before any water accumulates it can evaporate and escape (see Figure 5-71). If a rubber boot is used to cover the motor housing, make sure that it also has large drain holes. If not, it can do more damage than good.

Remember that the actuator is the sole piece of equipment responsible for a great deal of mechanical work. If not properly installed, failure will probably result. An estimated 75% of all service calls on C-band systems have been for actuator related problems, especially in cold climates! These statistics are expected to be closely matched in Ku-band home satellite systems.

Horizon-to-Horizon Mount/Actuator

A horizon-to-horizon mount has a built-in actuator. This combination of mount and actuator is much simpler to install than the linear type and is also much less susceptible to water damage. The reflector can first be affixed to the mount which is then lifted onto the pole. In this case the weight of the mount must be supported so the dish will not be distorted. Alternatively, the mount can be directly lifted onto the pole and then the dish can be attached to the mount.

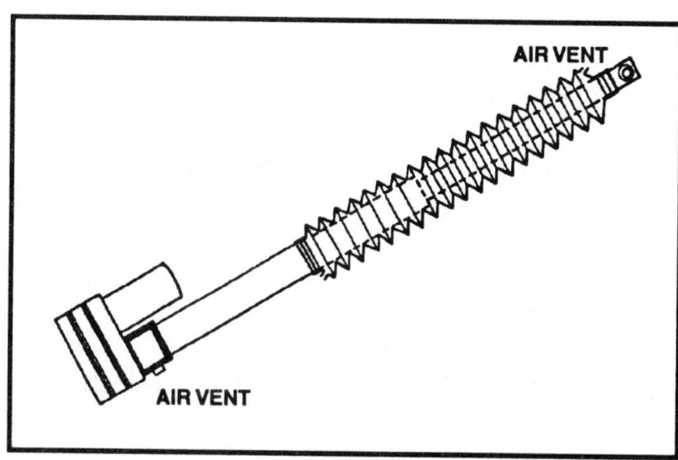

Figure 5-71. An Actuator Boot. *A neoprene actuator boot should have air vents at top and bottom and should attach securely to both the inner and outer actuator arms.*

Wiring Actuators

Most direct burial, multi-run cables provide the necessary five-wire conductor needed to connect most actuators (see Figure 5-72). This includes two #14 gauge wires carrying typically 36 Vdc to power the motor. If the east button is pressed on the internal control box and the antenna moves west, or vice versa, reversing these wires will correct the situation. Three #20 gauge wires in a shield are used to send the counting pulses between the actuator sensor and indoor control box. These wires are shielded and grounded in order to prevent spurious voltage spikes from being mistaken for counting pulses when using controllers that expect similar pulses from Hall effect transistors or reed switches. Note that only two wires are required when using reed switches.

It is important to use the seals provided to protect the points of cable entry against water intrusion. These should be waterproofed with a material such as coax seal or weather strip caulking.

Figure 5-72.Wiring an Actuator. *Five wires connect this actuator motor/sensor to the indoors controller. Two of the heavier gauge wires on the right are for motor power. If they are interchanged, the west/east movement will be reversed. The ground and two lighter gauge wires on the three left terminal strips are used for control voltages or pulses. Notice how the incoming cable is sealed against water entry by a tightly fitting plastic cap. (Courtesy of Brent Gale)*

Safety Considerations

Installing and repairing actuators is a potentially dangerous activity. Actuators operate with the highest current and voltage of any TVRO component, from 1.5 to 6 amps of current and typically either 24 or 36 Vdc. If a live circuit is touched by an installer who, for example, has wet feet and happens to be at ground potential, the shock would certainly be uncomfortable and possibly deadly. It takes only one milliamp of current to override the heart's electrical pulses.

The key to avoiding electrical shocks is to take steps not to be at ground potential. Carrying a rubber mat to work on is a good habit to develop. If the old rule "always keep one hand in your back pocket when working on live circuits" is followed, the chances of receiving a fatal shock will be reduced to near zero. Never work on a TVRO if an electrical storm happens to be in the vicinity.

Using an ac power strip with a built-in circuit breaker known as a ground fault interrupter is an additional safety precaution. The satellite receiver, actuator and TV should all be connected via this outlet. If a ground problem such as a short circuit through a person's body develops, the breaker will open before any harm is done to either the person or the electronic equipment.

Actuators also have the potential of being a fire hazard because they draw a relatively large amount of current. This danger is recognized in some countries like Canada where actuator power supplies must be enclosed in a separate container to meet government safety standards. It is generally a wise decision to install actuators as well as satellite receivers in a non-flammable environment having adequate ventilation. Actuators have been known to overheat and cause melting of carpet fibers.

G. ELECTRICAL CONNECTIONS

Poor wiring techniques or improper connections can ruin an installation. A poorly grounded cable or a leaky connector may be sufficient to ruin reception of a satellite broadcast. Installers should be familiar with the correct techniques for installing all connectors that may be encountered. Fortunately, only a few types of connectors and cables have become standard in the satellite TV industry, so acquiring the necessary tools and knowledge can be rather simple.

Connector Types and Techniques

The most familiar types of connectors encountered in TVROs include F-connectors, N-connectors, RCA or phono jacks, moly plugs, the solderless lug terminal and Scotch Locks. Other varieties often encountered in video accessory systems include K10, K14, BNC and DIN connectors. Each of these has certain uses and may require special tools to be properly installed.

F-Connectors

F-connectors are the industry standard used for attaching coaxial leads to the LNB, satellite receiver and television sets (see Figure 5-73). There are a variety of types used for both RG-6 and RG-59 coax. Some come with the crimp ring as a separate part.

It is imperative that these connectors be correctly installed particularly when working with higher frequency signals which are all too easily attenuated. F-connectors are attached by first strip-

Figure 5-73. F-Connector. *F-connectors are used with coaxial cables such as RG-59, RG-6 or RG-11 which are rated for operation at frequencies below 1,500 MHz.*

ping off about 2 cm of the outer insulating layer. The ground braid is next trimmed to a length of about 3 mm and folded back over the outside insulation past the 2 cm point. Then about 10 mm of the white center insulating jacket is removed from the inner wire. Use a knife and scrape the inner conductor wire clean to guarantee that a good electrical contact will be made. The crimp ring, if a separate part, is twisted on before the body of the connector. The connector should be twisted onto the coax so that the inner white dielectric jacket is flush with the inner shoulder of the F-connector. Finally, a crimp tool is used to bind the crimp ring onto the ground wire and outer sheath (see Figure 5-74).

When installing an F-connector, it is important that a crimp tool and not simply a pliers be used for the final step. This special tool compresses the connector onto the jacket equally on all sides. A pliers would distort the shape of the inner insulating jacket, alter the electrical characteristics of the cable and result in signal losses due to impedance mismatches.

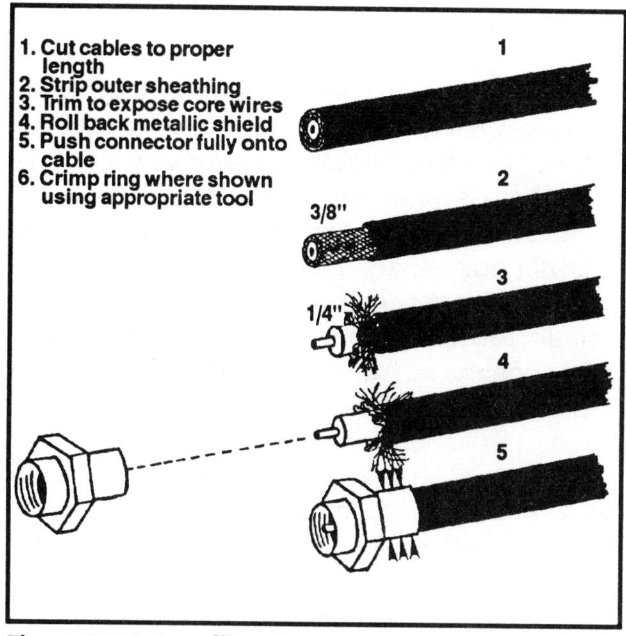

Figure 5-74. Installing F-Connectors. *This diagram illustrates a step-by-step method to correctly install F-connectors. During this process, it is important not to score the central conductor wire because this would affect the current flow at the high frequencies used in satellite television.*

The use of push-on F-fittings should be avoided. They do not provide as tight an RF connection as threaded connectors and can potentially let radiation leak into or out from the attachment point. They are also easier to accidently pull off than the threaded type when, for example, furniture is being moved.

F-connectors can be joined together with couplers called F-81 barrels which have a female lead at either end. This is not a recommended practice except when absolutely necessary as it causes some loss of signal at the junction and creates another possible route for moisture entry.

Tools Required: Scalpel, Wire Strippers, Cutters, Crimping Tool

N-Connectors

N-connectors are rated for higher frequency use and are similar to BNC fittings. These are heavy-duty connectors that are weatherproof when properly installed. The center pins must be soldered to the center wire of the coax (see Figure 5-75).

N-connectors were used in earlier LNA systems and, in Europe, in conjunction with LNBs until about 1987. These are rarely found on modern satellite television systems. However, some brands of equipment sold on the European continent such as the NEC and certain MASPRO receivers as well as some commercial systems still employ this fitting.

Tools Required: Scalpel, Wire Strippers, Cutters, Soldering Iron, Solder

Figure 5-75. N-Connector. *N-connectors are used to join 50 ohm cables such as RG-213 or RG-214 which are rated to carry frequencies in the C-band range of 3.7 to 4.2 GHz. They are more difficult to install correctly than F-connectors.*

Phono Connectors or RCA Jacks

RCA jacks are often required for relaying audio and video outputs from receivers to stereo processors or to external modulators. They should never be used for higher frequency signals because they are not shielded as well as F-connectors. Most phono connectors need to be soldered onto a dual run wire or coax. A simple way to avoid the difficulty associated with soldering is to use phono-to-F-connector adapters. Varying lengths of cables having pre-attached phono connectors can also be purchased at audio or electrical supply stores.

Tools Required: Scalpel, Wire Strippers, Cutters, Soldering Iron, Solder

Nylon Block Plugs

Nylon block plugs (the Moly plug is a familiar variety) are occasionally required in installing satellite receivers or actuators. These have internal pins which are inserted after the wires have been attached. These pins easily fit into the housing but are nearly impossible to remove without an extraction tool specifically designed for this task. When Moly plugs are installed, be careful to seal them against water entry.

Tools Required: Stripper, Cutter, Crimper and Extraction Tool

The Solderless Lug

Solderless lugs are designed to be used with screw-on terminal strips, often found on rear panel outputs on satellite receivers. The wire fits in one end and is either crimped down with a standard tool or soldered in place (see Figure 5-76). Such lugs are recommended for attaching wires to screw-on terminals. Bare wires can fray and, in time, even short out to adjacent terminals. As a minimum, if lugs are not used, wires should be twisted and tinned before being screwed onto a terminal.

Tools Required: Wire stripper/cutters

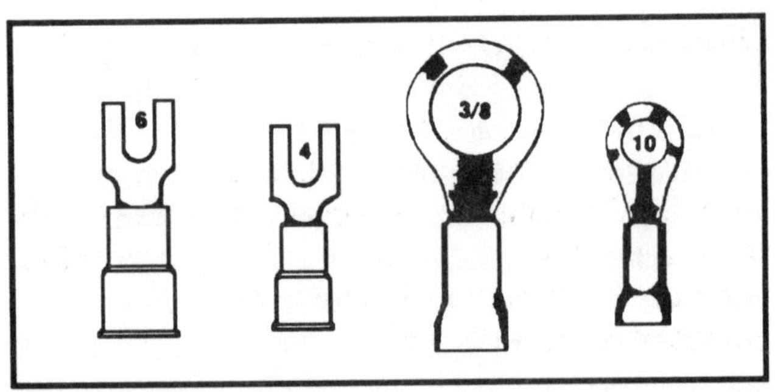

Figure 5-76. Solderless Lugs. *This type of connector can be used on any equipment that has a screw-type barrier strip. Attaching conductors to this connector will prevent stray wires from touching an adjacent terminal and shorting out.*

Scotch Locks

Scotch locks are another similar type of connector for joining two wires together. These crimp-on connectors are filled with a flooding compound which is squeezed out during the crimping procedure. A good mechanical and well-protected electrical connection results. These connectors are ideal for use in attaching polarizer wires (see Figure 5-77).

Tools Required: Stripper and Crimper

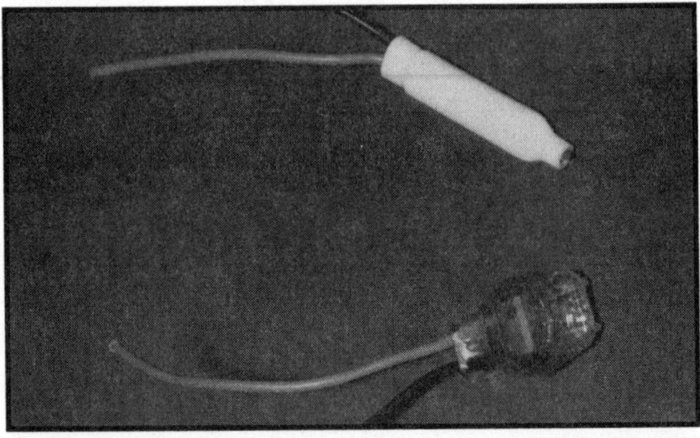

Figure 5-77. Scotch Loks. *These connectors are self-stripping and self-sealing with an internal insulating silicon-filled, moisture-resistant enclosure. Either two or three wires can be joined by simply inserting the wire(s) and crimping with a pliers.*

Other Connectors - DIN, BNC, PL259

DIN connectors, soldered to very thin wires on a multi-wire cable, are standard audio/video fittings. BNC connectors are commonly used with video recorders, generally as video in and video out. PL259 connectors have been used on older brands of video recorders.

Tools Required: Scalpel, Wire Strippers, Cutters, Soldering Iron, Solder

Additional Pointers on Installing Connectors

When stripping insulation or ground braid from cables and wires, be careful not to score the internal conductor. This would weaken it and could cause a break at some later time. Also, be careful, especially when peeling the outer insulation off coaxial cables, that the tiny pieces of wire do not fall into a vent in the actuator controller or satellite receiver. Such wires could easily short an internal component and burn out the unit.

As a general rule, never splice cables. However, some cables can be extended with the appropriate adapters joined to fittings on both pieces of cable. If the point of union is exposed to weather, use both coaxial sealant and a rubber boot or shrink fit tubing.

Adapters are also available to join different types of connectors. For example, an RCA to F-connector adapter has a male F-fitting at one end and a standard phono jack at the other. This particular adapter is often needed when attaching stereo processors or external modulators. It is much easier to make a cable with F-fittings at either end and use an adapter than to install a phono jack at one end and a F- connector at the other.

The Final Wiring

The wiring sequence to be followed depends upon how the final adjustments of antenna position will be made. If a test setup at the antenna is first used, temporary wiring is required. If the antenna will be aligned using walkie-talkies to communicate from the indoors receiver to a crew member outdoors, the final wiring is completed without an intermediate step (see Figure 5-78).

An alternative to using walkie-talkies while still being able to avoid wiring a temporary set-up at the antenna is to use an remote control extender such as an Extra-Link™ or similar product when a two-run coaxial ribbon is installed. Two-run ribbons have an extra coaxial cable that may later be used to hook up a Ku-band LNB for dual-band reception. This extra run can be attached to the TV output on the satellite receiver and to the remote extender and a small TV at the other end.

In those cases where the antenna will be adjusted using only the indoor satellite receiver, the LNB and polarizer should be connected before it is raised into position to track the arc. The electronics at the focus are more easily reached with the antenna angled down to either horizon.

The three polarizer leads should be twisted tightly together with the matching three leads on the cable and joined with Scotch locks. If not, they should be soldered and then covered with wire nuts or other types of crimp-on connectors. These are then covered with "shrink spaghetti" which, like shrink tubing, contracts when heated by a small flame or hair dryer. The F-connector output from the LNB should be waterproofed with a material similar to coax seal. Finally, the five-wire actuator lead should be hooked up and double checked to ensure that a proper connection has been made. All weatherproofing should be done only after the installation is complete.

The satellite receiver has an F-connector input for the IF signal which falls in the range of 950 to 1450 MHz (950 to 1700 MHz in European systems). Note that some European LNBs have an N-connector output which must be connected to an N-to-F adapter and subsequently to a similar connection at the receiver. Power and RF signal are relayed between the LNB and satellite receiver on a signal coaxial cable. The rear panel of receivers also usually has a terminal strip for polarity power and control. This strip has three connections: +5 Vdc, pulse control and ground outputs. Actuator motor wires are attached to either the control box or to those receivers having built-in actuators.

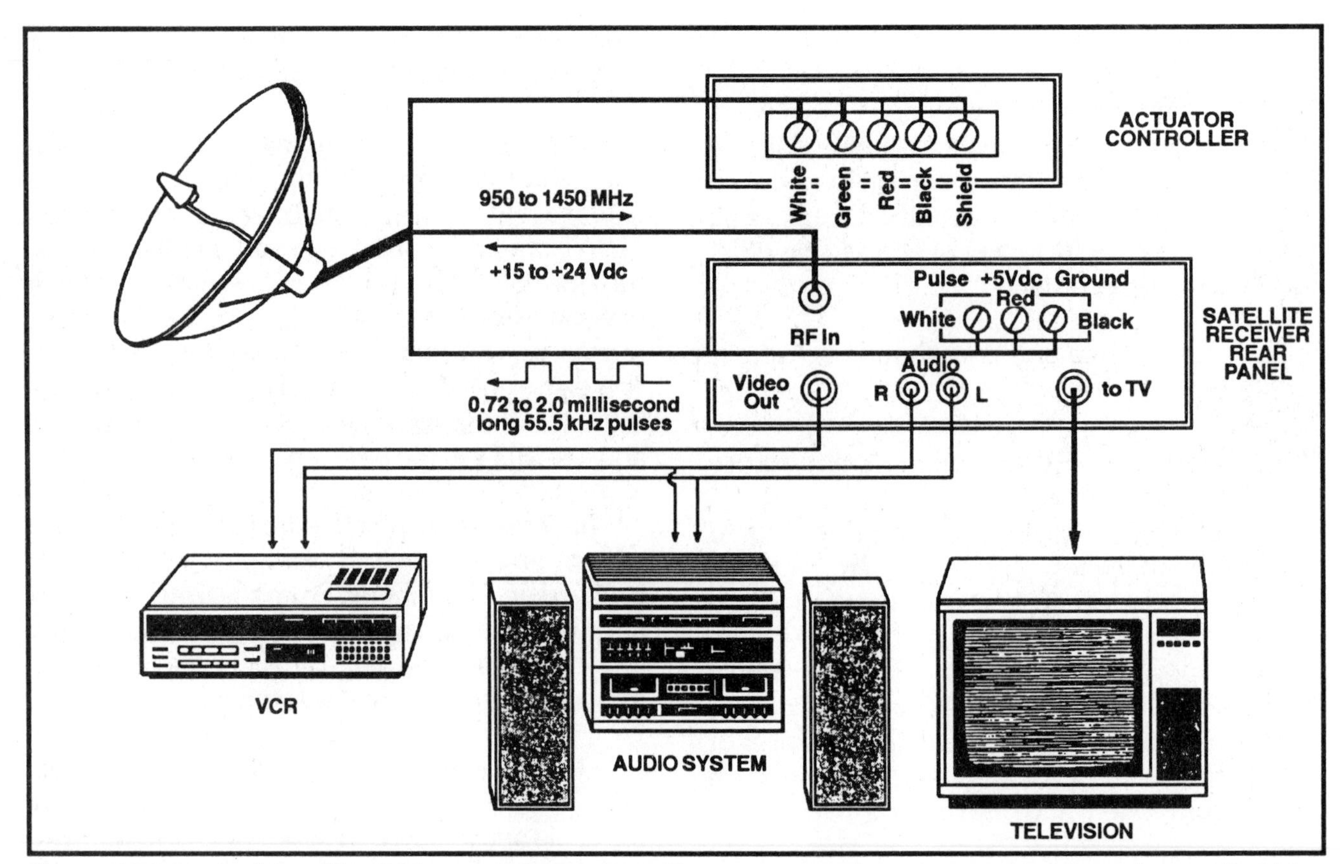

Figure 5-78. Typical Satellite TV System Wiring. *A satellite receiver can be connected to a complete audio/video home entertainment center. This illustrates how a system is wired. The receiver relays voltage to the LNB and receives RF signals on a single coaxial cable. Three wires send 5 Vdc pulses every 18 milliseconds (which equals 55.5 kHz hence the name 555 timer chip) to the polarizer. The position of the servo motor which moves the probe is controlled by the width of these pulses which vary from 0.72 to 2 milliseconds. The actuator controller transmits 6 Vdc to and receives pulses on the three sensor wires. It also sends 36 Vdc via two motor wires to control movement of the actuator. The receiver here has a built-in stereo processor which hooks directly into the home audio receiver via its output audio terminals. Alternatively, the baseband output could be used to drive a stereo processor. The VCR can be connected to either the TV set or video output ports.*

As a final check, every connector should be carefully examined prior to mating to ensure that the center pin has not been broken off or badly scored and that it is centered and properly extended. There will be no electrical contact if the pin does not extend far enough, and damage may occur if the pin is not centered or extends too far. LNBs damaged by off-center pins are typically not covered under warranties. It can also be very difficult to locate a poor connector once an installation is completed.

Note that experience suggests that any spare or extra wires at the feed assembly, actuator motor or satellite receiver should not be cut off. These can be taped to the multi-run jacket and kept as spares for possible use at a later date.

Safety Considerations

Satellite receivers typically draw up to a maximum of one ampere of current to provide a range of regulated power outputs. The polarizer usually requires +5 Vdc while internal circuitry generally requires a regulated +12 Vdc, +18 Vdc and often +23 Vdc. Other voltage outputs are not uncommon.

While it is rare that these voltages are sufficient to cause serious harm, the same precautions taken when installing actuators should be followed. Better be safe than sorry. Many receivers have built-in actuator controllers and, therefore, are potentially dangerous.

Protection Against Power Surges and Voltage Fluctuations

A direct lightning hit is a rare occurrence that can not only destroy the electronics in a satellite system but can also set a house on fire. However, most hits are indirect and their effects are seen as power surges. Standard home power provided to many rural regions is often not very "clean" and produces similar but much smaller voltage transients which can cause a receiver or actuator microprocessor to lose its memory.

Power or voltage fluctuations are common in rural areas where the majority of satellite receivers are used. Variations typically range from 10 to 20 Vac but can be as high as 30 to 40 Vac during storms or times of peak summer heat. They will be more prevalent if a TVRO is located near the end of a power line or if there are high-current loads between the installation site and the power source.

Whatever the source of voltage spikes or fluctuations, it is better to take precautions during installation than to be forced to make a time-consuming service call at some later date. Most satellite receivers and actuators have built-in protection in the form of voltage regulators and filters. These will be examined in more detail in Chapter VIII. Although TVRO warranties do not cover "abnormal conditions of operation" such as being hit by lightning, or other acts of God, a technician can take some very effective preventative measures during installation.

Ground Rods

Ground rods are constructed from long narrow poles of copper-clad steel which are driven into the ground to provide an escape route for lightning. The effectiveness of such "lightning rods" depends upon the ability of the soil to pass current, the soil resistance, and details of their connection to the antenna and mount. An eight-foot (2.5 meter) ground rod will usually be sufficient to handle most lightning strikes.

Most soils have sufficient conductivity so that a well-made ground connection will effectively shunt voltage transients caused by nearby lightning strikes. Soils which have little clay or loam but are mainly rocky may need to be treated with materials like rock salt, copper sulfate or magnesium sulfate to increase their ability to pass current.

A 2 to 2.5 meter (6 to 8 feet) lightning rod should be driven into the ground about 30 cm (12 inches) or so away from the supporting pole foundation. A minimum of a 6 mm (1/4 inch) copper or 9 mm (3/8 inch) diameter aluminum wire is bolted onto the ground pole. A tight clamp, often provided with the rod, is used for making contact at the top end. This wire should follow a smooth curve; any sharp bends will impede current flow. For the same reason, if knots are tied in regular power cords, voltage transients will have a harder time finding their way into electronic components.

An effective method of planting a grounding rod is to use a pipe with a cap on one end as a driving tool. If the soil is difficult to penetrate, water can be added to the hole as the rod is driven. The top of the rod should be planted at least 15 cm (6 inches) below the ground for best results. This will also protect the customer's lawnmower as well as the customer's feet from damage. A driver sleeve is also available for use with a standard jackhammer to plant the rod. This is a required tool in many cases.

Probably the best alternative to using a ground rod is to run a heavy wire, #8 gauge or larger, from the pole or antenna to the location where the building's electrical circuits are grounded. This should effectively shunt any excess currents to ground. Paint must be removed from these points of contact prior to attaching the ground strap so that a good electrical contact is made.

Surge Protectors

Surge protectors can also be used to shield electrical components from voltage spikes (see Figure 5-79). There are many varieties available ranging from simple units which plug into a power line to more expensive ones requiring the installation talent of an experienced electrician. Any microprocessor controlled satellite receiver or actuator is similar to a home computer in that its memory can be erased, interrupted or altered by power surges. A surge might cause channels to change, the antenna to move unpredictably or a carefully programmed memory to fail.

Figure 5-79. A Surge Protector. *This type of surge protector has an internal power break that can be reset in case a strong voltage transient appears on the power line and trips the unit.*

A surge protector consists of a combination of varistors and filters. A varistor allows voltages up to a pre-defined value to pass without attenuation but transfers large spikes directly to ground through the third prong on the wall outlet. Some brands can handle up to 50,000 volts and are able to respond in microseconds. Electrical filters employ capacitors to shunt any rapid changes in voltage to ground but do not affect the passage of direct currents. Good quality surge protectors incorporate varistors and filters rated to respond rapidly to large surges. However, the least expensive brands may lose their ability to operate after only a few large spikes.

All TVRO components should be joined together to one common ground as further protection against voltage spikes. Unless a separate ground line connects all components to a common point at the antenna, the ground on a three-prong power cord must always be used. In those cases where only two-prong outlets are available, a ground relief adapter can be used. Paying good attention to grounding procedures will also minimize the possibility of an annoying type of interference known as hum bars (see Chapter VIII for more details). Further protection is provided in some brands of LNBs that have "grounded probes" that can withstand short duration power surges of up to 1000 amps.

Constant Voltage Transformers

When a receiver often overheats and shuts down, a constantly high or low line voltage may be the cause. It may be necessary to install a constant voltage transformer to counteract this problem. This component can also be useful in improving the performance of TVs or monitors.

Protection Against Overheating

Actuators and satellite receivers generate enough heat to cause problems when they are not properly vented. At the very least, the electronics may not function at peak performance or could shut down or blow a fuse. In the worst situation, a fire may result. These components should always be placed on their feet above a flat surface, and a 7.5 cm (3 inch) air space should be provided above and to their sides. Care should be taken never to cover vent openings or install the receiver or actuator in a non-ventilated wall location or directly on a thick pile carpet.

H. DECODER INTERFACING

Hooking up a descrambler or a decoder to a receiver is not a difficult process. However, the interface between decoders and receivers has sometimes proved difficult. For example, in North America when the outputs of some earlier receivers were fed into the VideoCipher II decoder, signal levels were not compatible. Today most satellite receivers in North America feature an integrated VideoCipher II decoder so this is no longer a concern.

Clamped Or Unclamped Input?

Descramblers typically need an unclamped composite baseband input signal to operate properly. Most receivers have such an output accessible on their back panel. The need for an unclamped video input stems from the fact that the receiver clamp circuit can distort a scrambled signal.

In most receivers, the clamp/unclamp switch is located on the back panel. In others, the clamp can be selected from a front panel switch or via the remote control when programming the channel parameters.

Baseband Or Composite Video?

Most descramblers can decode the unclamped baseband or unclamped composite video. Many of the newer models do not rely on this carrier to descramble the signal and therefore the use of unclamped composite video suffices. The baseband signal is unfiltered and not de-emphasized. If a descrambler did not have any filtering and de- emphasis circuitry, the colors and definition of the descrambled picture signal would be non-linear and therefore the picture would be somewhat too sharp in appearance.

I. ALIGNING THE ANTENNA ONTO THE SATELLITE ARC

Aligning the antenna and actually seeing television from Ku-band satellites is by far the most exciting installation step. A technician can complete the necessary adjustments to prepare for receiving signals from one of two locations: indoors or outdoors. Once the final wiring is finished, a set of walkie-talkies can be used to communicate between one person at the receiver/positioner and another at the antenna. Alternatively, a complete test setup can be wired at the antenna (see Figure 5-80). The latter method is recommended when weather permits. There is less possibility for mis-communication and, if necessary, just one person can complete the installation.

In either case, the initial test setup should include a satellite receiver and a small television set or monitor. An actuator is required when installing a tracking system and a signal strength meter may be quite useful when fine tuning the alignment. Note that the signal strength meter built into many brands of satellite receivers can be adequate during the initial alignment procedures. Some better equipped professionals often have a portable spectrum analyzer or another type of sensitive signal detection instrument that can also be used at this stage and when fine tuning the system. Such sensitive equipment is not at all mandatory.

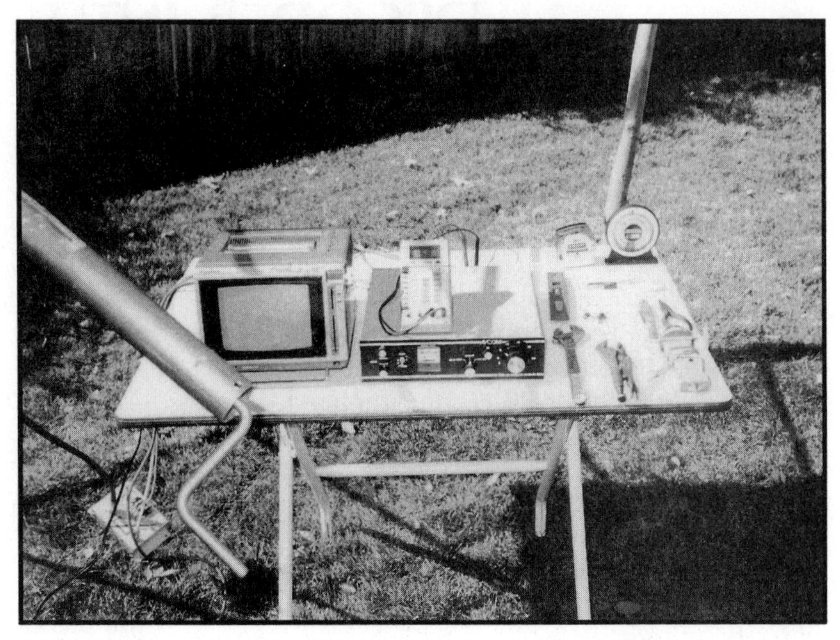

Figure 5-80. An Outdoors Installation Set-Up. *All the equipment required to align a dish onto the arc of satellites is set up on this table next to the dish. It includes a portable television set, a satellite receiver (preferably one with a built-in signal strength meter), a volt/ohm meter, an inclinometer, a compass, a crescent wrench and assorted tools. During alignment, the TV screen can be viewed for direct feedback on antenna movements.(Courtesy of Brent Gale)*

An alternative to hooking up the antenna to an actuator during dish alignment is to remove the housing and motor from the actuator arm and then to use a small hand crank to move the antenna between satellites. It is quite useful to have a small monitor or TV during installation. This eliminates the need to disconnect a customer's set and lug it out to the site. Using a familiar TV known to be working avoids the possibility of a malfunction occurring with unfamiliar equipment. It is also important to have a supply of 75 to 300 ohm transformers on hand. These are necessary if the user's TV set is an older model and does not have F-connector attachments.

The antenna and feed can be aligned using the receiver and actuator that are being installed. But if difficulties with any electronic component arise, it is useful to have a spare, shop-tested receiver and actuator on hand for use as substitutes.

Fixed Antennas

In some installations an antenna is aimed at just one satellite and a positioner is not required. This is particularly the case today in Europe where tuning to just a single powerful Ku-band satellite such as Astra can be sufficient to provide a complete range of television entertainment. In these cases, a satellite can often be easily targeted knowing just its elevation and azimuth bearing and establishing a vertical plumb line. The elevation and azimuth angles to any satellite from any location can be easily calculated from the equations presented in Appendix B.

Nevertheless, there are two reasons why it is important for an installer to be familiar with installing systems that are capable of tracking. First, there may be situations where locating true south may prove to be difficult and some of the techniques outlined in the sections below may prove quite useful. Second, as more satellites are launched and as the curious customer learns of the diversity, those having fixed systems may very well want to install tracking systems at some point in the future. Many of the techniques and pointers outlined in the sections below apply to both fixed and tracking systems.

Testing Components Before Setting Angles

By far the easiest method to test if equipment is working is to check out each electronic component on a working system back at the shop. Taking the necessary half hour of time to do so in a familiar environment can eliminate many installation headaches.

A technician also can use pre-alignment tests to determine if the equipment being installed is working as expected. First be sure that the television set is tuned to the same channel as the modulator output. If necessary, the health of the TV can be checked by viewing a local over-the-air channel. Next, attach a coaxial cable between the TV set input and the RF output on the satellite receiver. Set the audio mode selection to the mono position if there is a choice between this and stereo.

If the cable from the LNB to the receiver input is disconnected and if the receiver and TV are on, the screen should be black or solid white. If not, the receiver has a short or the cable between it and the television is open. It is important always to turn the power off and unplug a receiver before disconnecting or hooking up any connectors since damage can be caused by shorting a center lead to ground.

Finally, hook up the LNB. The screen should display white noise and a loud hiss should be heard. If not, the LNB is either defective or is not being powered by the receiver.

Another option is to use a voltmeter to check for the necessary 15 to 24 Vdc at the receiver output and LNB input. If the proper reading is indicated, the components are probably good. Next check that the voltage at the LNB input is adequate. This will require installing a 2-way, dc-passive power splitter that allows current to pass in both directions. Alternatively, disconnect the cable and measure the power between its ground and center conductor. If power is not getting to the LNB, the cable or one or both connectors are faulty. These and other troubleshooting procedures are described in more detail in Chapter VIII.

Aiming a Tracking Antenna

Three adjustments are required on most polar mounted antennas to ensure proper tracking of the arc of satellites: north/south heading, declination offset angle and polar axis angle. It is usually simpler to set these parameters when the antenna is at its highest position and is aimed towards the center of the satellite arc. This can be accomplished by turning on only the actuator and moving it to this position with the east or west buttons, or by using a hand crank. Once this has been accomplished, check that the base of the antenna mount still has a vertical orientation by placing the inclinometer on its surface. The alignment of the mount may have been altered from the weight of the dish.

North / South Orientation

A polar mount must have its axis aligned with the north/south axis of the earth in order to be able to detect all satellites in the viewable arc. This is easily understood by visualizing an antenna at the equator. If it rotates in an axis aligned with the center of the earth, it will correctly scan the circle of satellites in the sky directly above from horizon to horizon.

Most antennas have a flat plane on the mount which can be used as a sighting reference. A handheld compass with cross-hairs is the most effective type for lining up with the north/south plane. Remember that a correction for magnetic variation is necessary. West of the line of zero variation, the agonic line, rotate the antenna and mount east of magnetic south by an amount equal to the variation. East of this line, rotate it west to correct for variation.

The Polar Axis Angle

The polar axis angle is set equal to the site latitude. This targets the antenna directly out into space along a plane parallel to one passing through the equator as described in Chapter II. Most antennas have one or two long threaded rods which are used to adjust polar axis angle. An inclinometer resting on the axis bar or back part of the mount is used to set this angle (see Figure 5-81).

Declination Offset Angle

The declination offset adjustment lowers the antenna's sight from a plane parallel to the equatorial plane down to the arc of satellites. The declination angle is greater in locations farther away from the equator. Table 5-4 shows how declination

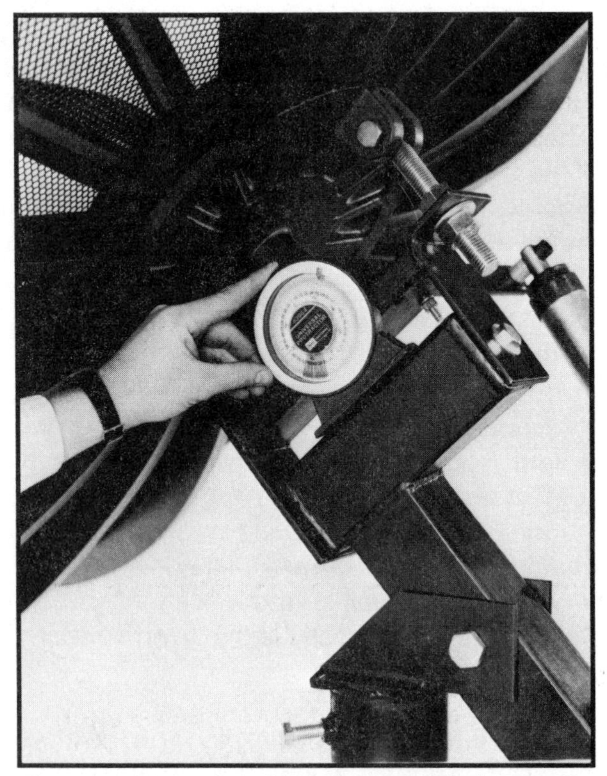

Figure 5-81. Setting the Polar Axis Angle. *The polar axis angle is set by placing an inclinometer on the polar axis bar and then raising or lowering the antenna so the angle equals the site latitude. (Courtesy of Brent Gale)*

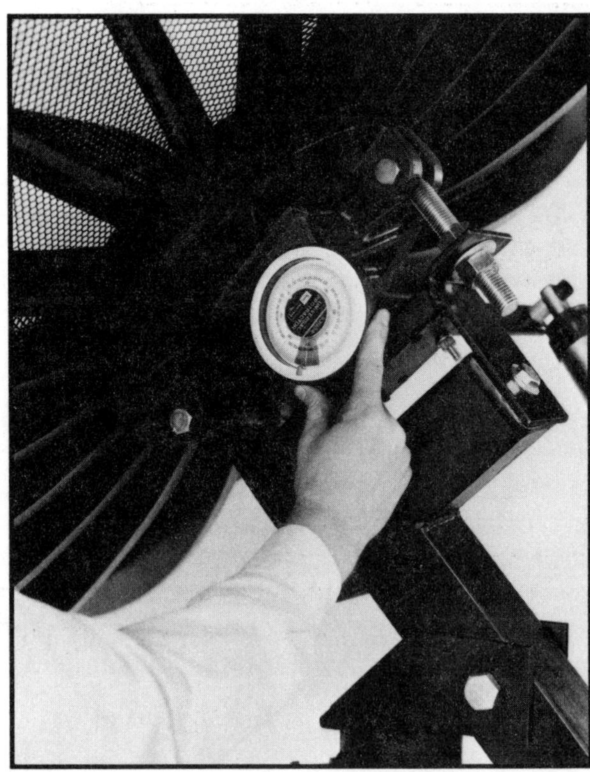

Figure 5-82. Setting the Declination Angle. *The declination offset angle is set by placing an inclinometer on a flat rear surface of the dish that is parallel to the line between the rims and adjusting the declination bolts until the reading equals the site latitude plus declination. (Courtesy of Brent Gale)*

varies with latitude. These values can be used to set declination in either the northern or southern hemisphere.

Declination offset angle is measured with an inclinometer. The difference between two readings, one on the main part of the mount, the axis bar, and one on a flat spot on the back of the antenna, should equal the declination offset angle (see Figures 5-82 and 5-83). The easiest way to set the declination offset angle is with an inclinometer placed on a back surface which is parallel with the face of the antenna or a flat board placed in a vertical direction spanning the antenna rims. This reading should be set equal to the sum of the site latitude plus the offset angle. For example, in the American city of Denver, the polar axis angle is set equal to its 40° latitude. The declination offset angle is 6.26 degrees so the antenna should be set at a 46.26° elevation. Similarly, at a latitude of 35°, the declination offset angle is 5.64°, so the final elevation is 40.64°.

When the modified polar mount geometry is used, as explained in Chapter II, these declination offset angles as well as the elevation angle must be slightly altered (see Table 5-5). This adjustment can be made by directly measuring the required angles or by finely tuning antenna alignment by the methods described below.

Most antennas have continuously adjustable declination offset angles that are changed by moving the assembly on all-thread bolts. However, some antennas manufactured for C-band operation had a preset declination offset angle allowing no adjustment. This type cannot be used when tracking Ku-band satellites which require a finely tunable adjustment.

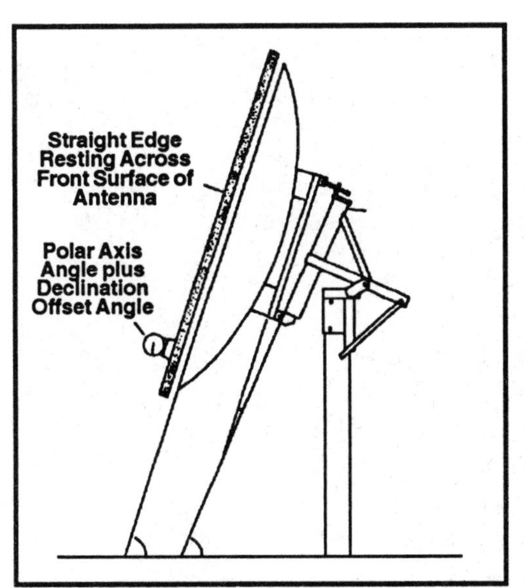

Figure 5-83. Using Antenna Face to Set Declination Angle. *If a flat surface cannot be found on the rear of the dish or on the hardware that supports the declination adjustment, the declination angle can be set by placing an inclinometer on a flat board that rests the face of the antenna.*

TABLE 5-4. POLAR MOUNT ELEVATION AND DECLINATION

Site Latitude	Elevation	Declination	Site Latitude	Elevation	Declination
0	0	0	34	34	5.510
1	1	0.178	35	35	5.641
2	2	0.355	36	36	5.770
3	3	0.533	37	37	5.897
4	4	0.710	38	38	6.021
5	5	0.887	39	39	6.142
6	6	1.063	40	40	6.260
7	7	1.239	41	41	6.376
8	8	1.415	42	42	6.489
9	9	1.589	43	43	6.600
10	10	1.763	44	44	6.708
11	11	1.936	45	45	6.813
12	12	2.108	46	46	6.799
13	13	2.279	47	47	7.015
14	14	2.449	48	48	7.112
15	15	2.618	49	49	7.205
16	16	2.786	50	50	7.296
17	17	2.952	51	51	7.385
18	18	3.117	52	52	7.470
19	19	3.280	53	53	7.552
20	20	3.442	54	54	7.632
21	21	3.603	56	56	7.782
22	22	3.761	58	58	7.792
23	23	3.918	60	60	8.047
24	24	4.073	62	62	8.162
25	25	4.226	64	64	8.265
26	26	4.377	66	66	8.357
27	27	4.526	68	68	8.437
28	28	4.674	70	70	8.505
29	29	4.819	72	72	8.562
30	30	4.961	74	74	8.608
31	31	5.102	76	76	8.643
32	32	5.241	78	78	8.666
33	33	5.377	80	80	8.678

An Alternative Method to Set Angles

There are some alternatives to the method suggested above for setting the tracking angles. For example, the arc set kit consists of three levels that are set on a "mother dish" and then used to set angles on any other dish within a radius of approximately 50 miles (see Figure 5-84).

Powering On and Aligning Onto the Arc

The time has come to watch satellite television. Check all connections and leads one final time before turning the equipment on. In rare cases, a satellite will happen to be targeted by the initial adjustments and clear pictures will immediately appear on the TV. Congratulations! However, while locating a C-band satellite by simply scanning across the arc may be rather easy at this point if all angles have been carefully set, targeting Ku-band spacecraft can be more demanding. This is especially the case when using larger diameter reflectors having very narrow beamwidths, because the antenna must be aimed so very accurately.

TABLE 5-5. MODIFIED POLAR MOUNT TRACKING ANGLES

Latitude	Polar Axis Angle	Declination Angle	Latitude	Polar Axis Angle	Declination Angle
1	1.02	0.15	32	32.60	4.66
2	2.05	0.30	33	33.60	4.79
3	3.07	0.46	34	34.61	4.91
4	4.10	0.61	35	35.62	5.04
5	5.12	0.77	36	36.63	5.16
6	6.15	0.91	37	37.63	5.28
7	7.17	1.06	38	38.64	5.40
8	8.20	1.21	39	39.64	5.51
9	9.22	1.36	40	40.65	5.63
10	10.23	1.54	41	41.65	5.74
11	11.27	1.66	42	42.65	5.85
12	12.29	1.81	43	43.65	5.96
13	13.31	1.96	44	44.66	6.07
14	14.33	2.11	45	45.66	6.18
15	15.33	2.29	46	46.65	6.28
16	16.38	2.40	47	47.65	6.38
17	17.40	2.55	48	48.65	6.48
18	18.42	2.69	49	49.65	6.58
19	19.44	2.84	50	50.64	6.67
20	20.43	3.02	51	51.67	6.70
21	21.47	3.12	52	52.66	6.79
22	22.49	3.26	53	53.65	6.88
23	23.51	3.40	54	54.65	6.97
24	24.52	3.54	55	55.61	7.11
25	25.51	3.73	60	60.56	7.51
26	26.56	3.81	65	65.49	8.84
27	27.57	3.95	70	70.41	8.11
28	28.58	4.08	75	75.32	8.33
29	29.60	4.21	80	80.22	8.48
30	30.57	4.40	85	85.11	8.57
31	31.59	4.53			

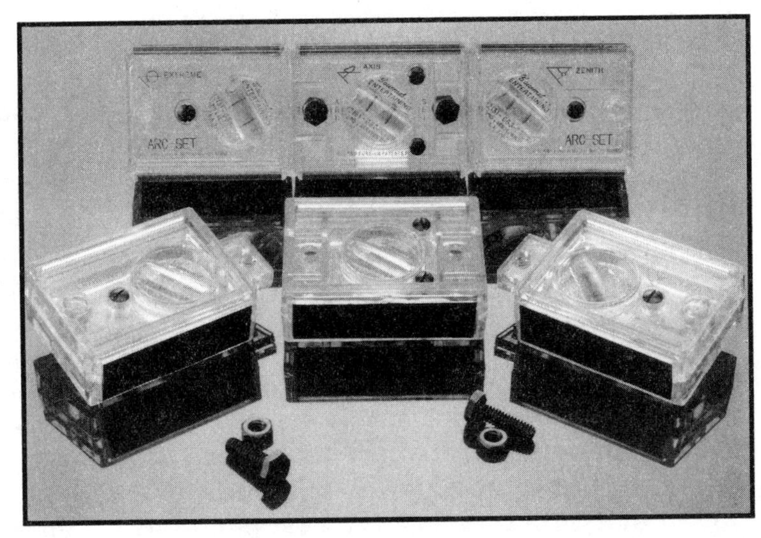

Figure 5-84. Arc-Set Alignment Tool. *This installation tool is set by "capturing" three critical angles which are copied from a properly aligned, local "mother" dish. It can also be used to troubleshoot installed dishes located within a 50 mile radius of the "mother" antenna. It can free an installer from the need to fine tune the north/south position with a compass because one of the plumb adjustments allows for setting north/south orientation. (Courtesy of Gourmet Entertainment)*

For the technically inclined and equipped, the alignment procedure can also be aided with a portable spectrum analyzer that is connected to the LNB output. Its sensitivity should be set to maximum and its range should be set for 950 to 1450 MHz. As the antenna is tracked across the arc, any signal received is seen as a clear blip on the screen. If more than one transponder is active and if a satellite is on target, more than one peak would be visible. The center frequency of these peaks can be determined and they can be compared to those of the broadcast channels expected from a given satellite.

Often a good starting point is to target a satellite which has many active transponders, preferably one near due south. For example, in North America, Galaxy I, Galaxy 3 or Satcom F3 are heavily used. When the alignment process begins with an active satellite located near due south, it is easier to arrive at an accurate polar axis angle setting. For example, in Denver at 105°W, Anik D1 at 104.5°W is nearly due south. During this entire procedure, it is necessary to have either a satellite TV guide or an excellent memory so that those transponders which are active during the installation can be located and identified.

Tuners on many brands of C and dual-band receivers can be set in the scan mode. This causes all channels to appear on the screen in rapid succession; it is a very useful feature for locating active transponders. As a reflector is slowly tracked across the arc, all channels on each satellite targeted for the preset polarity will flash across the screen.

Whether or not the scan mode is used, attempt to find a picture by tuning through all active transponders. As soon as anything other than white noise appears, the process can begin. All that is needed to start the alignment fine adjustments is just a semblance of a picture. If the receiver does not have a scan feature, check the TV guide and pick an active channel on one of the satellites which is broadcasting at the time of the installation. Then aim the antenna as closely as possible to this satellite and begin the hunt.

In those rare cases when a receiver signal strength meter is "maxed out," it can usually be set to a lower sensitivity by either a front panel button or a back panel or internal potentiometer setting. Occasionally, these adjustments may not be sufficient to lower the meter sensitivity when testing outdoors at the antenna since cable runs are very short and signal-to-noise levels are higher than usual. Most meters built into receivers are set to compensate for losses in the typical 30 meter run of coaxial cable. Inserting a coil of 30 meters of cable at the antenna may reduce the signal strength enough so that the meter returns to the readable range.

Once a rough picture has been detected, try to improve it by moving the antenna ever so slightly east and west and by adjusting the polarity and video fine tune. Note, if a side lobe has been targeted by mistake, no adjustments will improve the picture quality. It may be necessary to move the antenna until the picture disappears and then reappears more clearly. During this procedure, be certain that the receiver's automatic frequency control (AFC) is off. Also, make sure that the polarizer probe is free to move so that the skew can be adjusted. The probe must also easily move between even and odd channel positions.

Once the clearest possible picture has been obtained by using the standard mechanical adjustments, gently push the antenna up and down from its front bottom rim. Take care not to bend the reflector. If the picture improves, adjust the elevation in that direction until the best picture is received or until the signal strength meter has the highest possible reading.

At this point, the final feedhorn adjustments should be made. A more detailed discussion of these methods is presented in the following section. It is important that once the feed has been peaked, it remains in this optimal position so that the antenna can be aligned as closely to the arc as possible.

Next, target a second active satellite near one end of the arc. Either make sure that the receiver is set to an active transponder on this spacecraft or use the receiver scan setting. It will probably be necessary to readjust the skew and video fine tune as well as the polar axis angle once a picture is seen. Note the movement needed to improve pictures. If it is necessary to raise or lower the antenna, the north/south axis usually needs some minor correction. For example, if pictures had been peaked on Galaxy 1, a satellite in the far west, but the reflector needed to be lowered after being swept east to Satcom F4, then the antenna and mount would have to be rotated slightly west. The geometry underlying these adjustments is outlined in Figure 5-85.

North/south re-adjustments should be made in very small increments, on the order of 1 or 2 mm (1/16 inch) as measured by the movement of the mount relative to the support post. This fine tuning can most easily be accomplished by partially loosening the mount-to-pole bolts and by pushing on the outer rim of the antenna where leverage is greatest. Pushing from the mount is more difficult and can cause jerky, unpredictable movements. It is also wise to use a piece of chalk to mark the starting point so that if too much movement occurs, the process need not be started over from scratch. An antenna can often be quite heavy and should not be allowed to pull the mount forward on its supporting pole. The bolts should be only slightly loosened. Even so, the fine adjustments are often made with the mount tilted slightly forward so that when the bolts are tightened, the mount will pull back and the antenna will be pulled slightly off target. Be aware that this can happen and that in most cases a slight decrease in elevation will correct the error. Do not make more than one adjustment at a time; it can only cause confusion.

After the antenna is tracking the arc reasonably closely, the initial fine tuning should begin. Accurate readings are now taken from the signal strength meter. It is important during these steps that the scale chosen on the meter be left unchanged so that comparisons are possible between these readings before and after changing any settings. For example, assume that the meter read 6.23 on Galaxy 1, transponder 20. Then the antenna was moved to Satcom F4 and fine adjustments in north/south position and polar axis angle were made. When the antenna was returned to the same channel on Galaxy 1, the new reading should be compared with the old one on the same transponder. After readjusting to obtain the best signal, this reading should be at least as high as the old one. For the same reason, it is also important to keep the skew adjustment knob unchanged when satellites are transmitting linearly polarized broadcasts. This is possible if transponders on satellites having the same polarity format are being compared.

When the antenna is accurately aimed, maximum signal strength readings should be observed on all satellites in the arc. Generally this can be accomplished by adjusting just the north/south position and the polar axis angle. However, the declination offset angle will occasionally need some small readjustment. The accuracy of this iterative fine tuning procedure is better if satellites at far ends of the arc are used. If a stronger signal can be obtained on any satellite by pushing either up or down on the antenna, the arc is still not perfectly targeted.

For further clarity, this procedure is outlined step by step in Table 5-6. If this table is understood, accurate tracking can be easily accomplished.

Six common tracking problems are illustrated in Figure 5-85 for locations north of the equator. South of the equator reverse the sense of these drawings. The best way to visualize these is to picture the arc of satellites as one half circle and the tracking movement of the antenna as another. Both of these circles must be aligned for perfect tracking. It really is that simple. It is important to understand that none of these procedures will work if the pole is not perfectly plumb or if the mount is not sitting vertically on the pole. If problems are encountered, check these mechanical settings. Be perfectly sure that the antenna cannot be driven into a table, toolbox or even an installer's vehicle by the actuator while it is being repeatedly scanned across the arc.

This fine tuning process is completed by repeating the necessary adjustments at both ends of the arc. Satellites at the center of the arc should also be targeted to ascertain that all the fine adjustments are correct. Tighten all bolts and nuts and then recheck to make sure that doing so has not slightly altered the elevation alignment.

During this procedure, familiarity with the programming relayed by each active transponder on every satellite is invaluable. By far the easiest way to learn this information is by having experience with a working satellite TV system. It becomes clear what to expect in a relatively short time. Become familiar with those satellites and transponders having test patterns with the satellite name and transponder number. These are invaluable landmarks in the search.

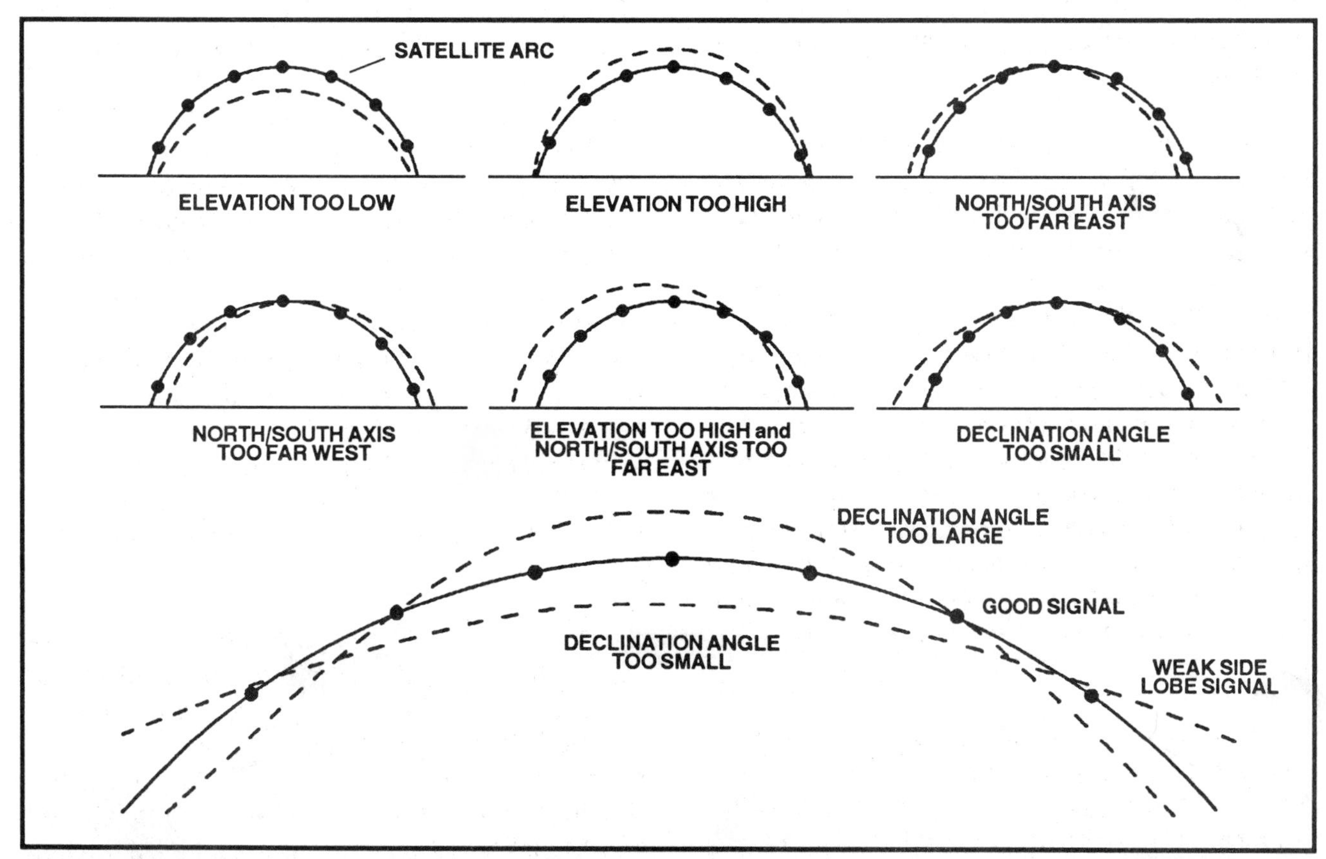

Figure 5-85. Common Antenna Tracking Problems and Solutions. *Most tracking problems are associated with an incorrect north/south orientation. However, if the declination angle has not been correctly set, tracking will also be incorrect. Lining up an antenna with the geosynchronous arc of satellites is simply a matter of lining up two half circles, the one in which the satellites are located and the one scanned by the polar mount.*

TABLE 5-6. THE BASIC ANTENNA ALIGNMENT PROCEDURE.

1. Carefully check that the feed system is centered and located at the correct focal distance. It is important that the feed be peaked before tracking is started.
2. Set the two tracking angles, polar axis and declination offset angles, and align the antenna polar axis in a north/south direction as well as possible. It is easiest to adjust the polar axis and declination offset angles on a satellite high in the arc, i.e. at the center of the arc, and to adjust the north/south orientation during the following step on a satellite lowest or at either end of the arc. The simple rule is elevation adjustment on a high satellite; north/south adjustment on a low satellite.
3. Check all electrical connections and turn the power on.
4. Aim the antenna at a preset position or slowly track across the arc to find even the faintest picture. Attempt to target a satellite located as close to due south as possible so that the polar axis angle can be accurately adjusted.
5. Once even a semblance of a picture has been detected, move the antenna slightly east and west and adjust the polarity and video fine tune controls on the satellite receiver to get the best possible reception. Then make any adjustments necessary in the polar axis angle to further improve reception.
6. Re-peak the feed system to maximize the signal strength reading. Once this has been accomplished, do not change the feed position again.
7. Move to another satellite, preferably at one far end of the arc, and attempt to tune in a second channel. During this and procedure number 3, having a receiver with a scan setting may be very useful. Otherwise, set the receiver tuner, both frequency and polarization, on a channel known to be active on the satellite being targeted. If signals from this transponder cannot be located on this satellite, move to another one closer to the first successful target. If three or four satellites in a cluster but no others further away in the arc can be found, the north/south alignment is probably substantially off position and needs correction.
8. Tune in the second satellite as well as possible with the video and polarity fine tune controls and east/west scanning.
9. Next, determine if the antenna has to be pulled up or pushed down to improve reception. If neither, move to another satellite at the other end of the arc and try again. In those rare cases when all satellites are targeted accurately, jump to step number 12. If adjustment is necessary, consult the common tracking error figure to determine which direction to rotate the north/south axis.
10. Make a small movement of the north/south axis in the correct direction. Remember to mark the original position in case the mount moves too much.
11. Return to the original satellite and repeat this procedure to further zero in on the arc. It makes sense to do this for one satellite at the far eastern and far western end as well as one in the center of the arc for best results.
12. Next, if possible, choose one satellite at the middle and one at each end of the arc, all three of which have the same polarization format and a common active transponder. Adjust the skew and video fine tune controls one last time to maximize the reading on the signal strength meter on one satellite. Do not touch these controls until this procedure is completed. Repeat steps 4 through 9 again to get maximum readings on all three satellites. At this point, some adjustment to the declination offset angle could possibly be necessary. It will probably be necessary to move back and forth between the three transponders on the three satellites until all three readings are consistently at their maximum value. The accuracy of this procedure can be improved by using signal measuring instruments.
13. As a final test, tune to the lowest power transponders in the arc. If excellent pictures can be obtained on these as well as on satellites all across the arc, the adjustments are optimized and the arc is correctly targeted.
14. Tighten all bolts carefully, watching to be sure that the signal strength meter maintains its maximum readings.

Aligning a Polar Mount Without a Compass

After some experience in the field, many installers have developed useful shortcuts to align a polar mounted antenna onto the arc of satellites. One such method that allows a complete installation to be accomplished using only an inclinometer is described below.

First, two satellites that can be used as reference points are located. A satellite east and another west of due south are chosen. Note that due south always corresponds to a satellite located at an orbital slot equal to the longitude of the TVRO. So, for example, in Denver, Colorado, due south would correspond to an orbital slot of 105°W. The closest satellite to this point is Anik D1 at 104.5°W.

The antenna is aimed in a fashion similar to aligning an az-el mount. The elevation angle of the first reference satellite is set. Then the antenna is rotated around the azimuth axis, i.e. the mount is rotated on the supporting pole, until a signal from this satellite is detected. The most common technique is to slowly rotate the dish and watch the television screen for even a flicker of a picture. This method can be preferable to using a spectrum analyzer because it allows visual confirmation of picture content to verify that the satellite observed is the actual target. When the signal is at a maximum, this first point is noted on the pole with a marker or tape.

Next, a second satellite is located by the same procedure. If the two satellites are at equal distances from due south they would have equal elevation angles. This would simplify the aiming procedure but is not crucial. Once the elevation is set and the second satellite is found, its position is also marked on the pole.

The calculation of the heading for south is now simple. The circumference of the base pole is next measured. This distance divided by 360° gives the linear distance per degree. Since the azimuth bearings to the reference satellites are known, the linear distance of due south from each of the reference points can easily be calculated. Once this is done, the mount axis can be rotated to a true south heading as measured from the previously marked bearings from either of the two reference satellites.

This technique is simple to implement. It is also quite well suited to cold weather installations since most of the effort and time spent in tweaking a polar mount is focused on finding a due south heading. In contrast, adjusting the elevation and declination angles can be accurately and rapidly accomplished using an inclinometer.

The azimuth and elevation bearings to the two reference satellites should be calculated before traveling to the installation site. This can be done using the equations in Appendix C or via a computer program such as the one listed in the reference materials.

An Example

The example of this simple technique in Table 5-8 uses the satellites Westar 5 at 122.5°W and Galaxy 3 at 93.5°W. The same procedure would obviously be applicable in any location on the globe. Tables 5-7, 5-9 and 5-10 can be useful guides in determining the information required in this procedure.

TABLE 5-7 . OUTSIDE DIAMETER, CIRCUMFERENCES and DISTANCES

Outside Diameter Inches	(mm)	Circumference (mm)	mm per degree
3.5	88.9	279.3	0.7758
4.0	101.6	319.2	0.8866
4.5	114.3	359.1	0.9975

TABLE 5-8. FINDING TRUE SOUTH WITHOUT A COMPASS – AN EXAMPLE

	SATELLITE 1	SATELLITE 2
	Galaxy 3(93.5°W)	Westar 5(122.5°W)
Azimuth	162.4°	206.1°
Elevation	42.2°	40.3°

Site Location: Latitude – 40.0° N
Longitude – 105° W

Mount base pole outside diameter: 88.9mm (3.5")

The data above can be used to calculate the following:

Angular distance from satellite 1 to true south = 180 – 162.4 = 17.6° from west
Angular distance from satellite 2 to true south = 206.1 – 180 = 26.1° from east
Mount pole outside circumference: 3.1415 x 88.9 = 279.3 mm
Distance per degree: 279.3/360 = 0.78 mm per degree
Distance on pole of true south from satellite 1 = 0.78 x 17.6 = 13.7 mm to west
Distance on pole of true south from satellite 2 = 0.78 x 26.1 = 20.3 mm to east
Elevation of satellite at true south: 30.26°

TABLE 5-9. INFORMATION REQUIRED TO FIND TRUE SOUTH BEARING WITHOUT A COMPASS

	Satellite 1 (East of south)	Satellite 2 (West of south)
Name		
Azimuth		
Elevation		
Channel & Polarity		

A1 = Angular distance of satellite 1 from true south
= 180° – Satellite 1 azimuth

A2 = Angular distance of satellite 2 from true south
= Satellite 2 azimuth – 180°

Cp = Pole outside circumference
= 3.142 x outside diameter

Ddeg = Angular distance per degree
= Pole outside circumference/360

Ddeg * A1 = Distance of true south from satellite 1

Ddeg * A2 = Distance of true south from satellite 2

TABLE 5-10 . A RAPID POLAR MOUNT INSTALLATION TECHNIQUE

1. As part of the pre-installation check list, before traveling to the site the data listed in Table 5-9 should be found and recorded. Azimuth and elevation angles can be calculated from the equations in Appendix B or from the computer software listed in the reference materials.

2. Calculate true south using the simplified true south derivation procedure as outlined in this section. Set the mount facing true south.

3. Set the dish on the mount so that it is parallel with the east-west line. Then adjust the polar axis angle equal to the site latitude angle. Add the modified polar mount adjustment angle, if further accuracy is desired. Finally, set the declination angle of the mount.

   Note that the elevation of a satellite whose longitude equals the site longitude is the highest point on the geostationary arc and due south.

4. Look up or calculate the declination angle for the site location. Some mount manufacturers provide a chart of declination angles versus declination bar length. This data makes an installation simpler because it is often easier to measure a linear distance rather than an angle. Set the declination angle.

5. The resultant elevation of the dish should equal the calculated value of elevation angle for the satellite due south. If the elevation angle of the dish is not equal to this value, recheck the angles. In situations where the declination angle is given on a declination angle versus declination bar length chart, the problem is likely to be the polar axis angle. In situations where the declination angle has to be measured, the declination angle should be first to be checked for accuracy.

   Note that polar mounts having fixed declination angles should be avoided at all costs. No fine tuning is possible.

6. If these steps have been correctly implemented, the mount should be properly adjusted. A quick scan across the geostationary arc should verify this tracking.

   Note that with smaller diameter base poles, the mount may be off true south by a degree or so. Given that the polar axis and declination angles were correctly adjusted the southern orientation of the mount probably still must be fine tuned.

   The procedure is designed to reduce the initial errors to a controllable level. However, the steps outlined above in the basic antenna alignment procedure and the fine tuning procedure with test instruments can be combined to develop a very skilled and effective alignment strategy.

Final Feedhorn Centering Adjustments

Final feedhorn centering adjustments should be made once a clear picture has been observed. If the feedhorn is precisely at the focal point, picture quality will be at its best. Moving this assembly away from its correct position will degrade picture quality by lowering gain and increasing side lobe power. Note that if the reflector is warped the optimal reception point may be centimeters away from the recommended focal point. It is critical that this problem be corrected, preferably during the early stages of the installation.

Moving the feed assembly closer to the antenna will decrease the signal power relative to side lobe power. This shows up as a decrease in signal strength. Moving the feed away from the antenna past the prime focus will also decrease the ratio of signal to side lobe power as the signal strength decreases and as more noise from the surrounding ground is detected. If both the TV picture and signal strength meter are carefully monitored at the same time for these effects, the feedhorn can be located precisely at the focal point. The payoff in improved picture quality will be more than worth the trouble.

If the feedhorn is improperly aligned, there can be substantial degradation in system performance. For example, a loss of 2 dB in signal stregnth is equivalent to reducing the performance of a 3 meter antenna to that of one having a 2.5 meter diameter. In addition to centering the feedhorn and correctly setting the focal length, the body of the feed should not be twisted relative to the face of the reflector. If this does occur, additional gain can be lost from shadowing and mislignment.

Some manufacturers of earlier C-band TVROs set the focal length so that no adjustment was possible. While this design detail simplified and speeded installation, the ability to fine tune the feed system was lost. Fine tuning capability is quite an important feature.

Alignment with Test Instruments

Sensitive measuring tools can be used to monitor signal strength. This allows for more efficient use of the available equipment so that even finer adjustments can be made to the feed position, north/south orientation, polar axis angle and declination offset angle. If the antenna mount has been designed to allow sufficiently fine adjustments in the two angles, the system can be aligned to incorporate the modified polar mount geometry settings and tracking can be extremely accurate.

Using test instruments can be an alternative to aiming at the satellite belt by observing a television screen. The choice between these two strategies is a matter of taste. In general, using both the television and test instruments simultaneously can be a very effective approach.

Signal power can be measured at the LNB output. It is unrealistic to expect to measure power levels at the LNB input since instruments to do so are very expensive. One of the most effective tools that can be used is a portable spectrum analyzer. However, there are other choices. The first device available to measure power levels from C- band downconverters was an adapter called an "RF head" or a "carrier level detector" which plugged into an inexpensive voltmeter and gave this tool the capability of reading 70 MHz signals. (The model "82RF" from John Fluke Manufacturing was the pacesetter.) Neither a voltmeter, which is designed to detect much lower frequency signals, nor a field strength meter, which monitors AM signals on a cable TV system, is capable of detecting 70 MHz signals from C-band downconverters. The RF head cannot measure signals having frequencies much in excess of 70 MHz. This rules it out for use with LNBs relaying intermediate frequencies in excess of 900 MHz.

Two other instruments to monitor signal power levels at the LNB output were originally developed for C-band use in North America. These provided attractive alternatives to signal strength meters built into receivers which are usually relatively crude instruments not really suited to squeezing out the last dB or two. The Pico "Peaker," one of the earlier C-band models, has a sensitivity adjustable from +30 to -80 dBm. Another, the "Squawker" has a variable sensitivity digital scale to handle from -7 to -48 dBmv but also has an audible tone output that increases in frequency as signal strengths rise (see Figure 5-86). The digital scale allows greater accuracy in reading powers than an analog scale. Both these instruments can also be used with signals up to 1500 MHz in frequency.

Such instruments must be protected from the power being relayed from the satellite receiver to the LNB (see Figure 5-87). A directional coupler which passes dc between two ports but isolates the third test port should be used in those lines also carrying dc power. While the Squawker is not protected, the Pico Peaker has a built-in coupler so LNB output can be connected via a short jumper cable to one port of the test device while the cable from the satellite receiver is connected to the other port; dc protection is provided internally.

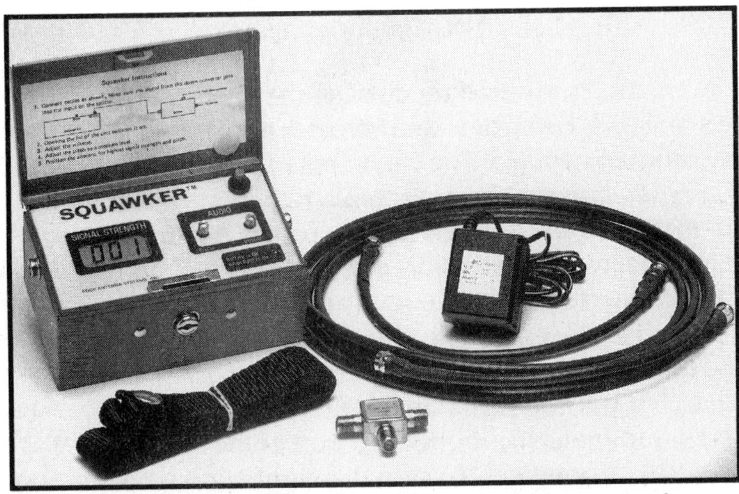

Figure 5-86. The Squawker. *This instrument has a battery powered digital signal strength meter as well as an audible tone which increases its frequency as signal power increases. This can be a very useful feature to aid in peaking an antenna. (Courtesy of Focii Antenna Systems)*

Figure 5-87. The Bultz Meter. *The input to this meter is a coax line from the LNB; the output is to the satellite receiver. It has a range of 10 MHz to 1.9 GHz. (Courtesy of Natropolis Industries)*

Spectrum Analyzers

A spectrum analyzer displays the relative levels of signals being detected across the entire satellite frequency band. The alignment procedure for commercial installations generally requires the use of excellent electronic test gear. Today, relatively high quality but reasonably priced analyzers have fully adjustable frequency ranges and signal sensitivities. For example, the Avcom PSA-35 can be connected via a 2-way power passing splitter to the LNB output. Its internal battery can provide power to the LNB. The second port of the splitter can be connected to an instrument such as a Squawker which offers both an audible tone and digital read-out of signal strength. Another option is to connect the second port directly to a satellite receiver. Then its signal strength meter can be used in conjunction with the analyzer for an accurate determination of satellite position. In this latter case, the receiver, not the Avcom, would power the LNB.

A spectrum analyzer is a very useful tool. It can be set to display the entire frequency band or can be used to zero in on the structure of a message transmitted on a single carrier. Even the most minute microwave signals can be observed. It can be used to determine if an LNB is functioning properly. When the LNB is connected, its noise floor should increase if all is well. It can even be used to troubleshoot hand-held remotes, which also relay RF signals. More details on this instrument are presented in Chapter VIII.

J. FINE TUNING THE SATELLITE RECEIVER

Once the antenna has been accurately aimed onto the arc of satellites, the satellite receiver can be moved indoors to its final location, if this has not already been done, and fine tuning can be completed. First, make sure that each active satellite in the entire arc can be tracked from the actuator controls. Then the receiver IF gain should be adjusted to receive the optimal picture across all transponders. In those installations where linear polarities are being received, some receivers have back panel or internal potentiometers that adjust the skew control so that the knob is near the center of its range of movement. Also make sure that the polarity probe is not sitting at the end of its range of movement by observing its position or by feeling the motor for vibration when the skew knob is set at its end positions. Some actuators have timers which shut the motor off in 5 seconds so this test may require two people, one at the receiver and one at the antenna. Both the IF gain and skew fine tuning are usually accomplished by rear or bottom panel adjustments.

Read the manufacturer's instructions carefully to ascertain that all that can be done has been done. These instructions are invaluable during the next step, programming the receiver and actuator.

At this point, if pictures are not as perfect as expected, check that all connections to the satellite receiver are secure. Another source of difficulty could be a customer's television set which will occasionally require fine tuning. Its channel frequency should precisely match the satellite receiver's modulator output. This television channel is rarely used for over-the-air programming but may have been used if the customer was a cable TV subscriber or used a VCR. In addition, to guarantee that the user's TV set is in good health, tune to a strong over-the-air broadcast or use a VCR or a portable signal simulator which generates test patterns on the modulated channel. If the television is in good shape, excellent pictures should be observed.

K. WATERPROOFING

Water in the form of vapor, rain, ice or snow is the chief enemy of a home satellite system. It can corrode connectors, accumulate in feedhorns or weigh down antennas. Once the system has been tested and is working perfectly, the waterproofing team gets on the job. This task should be saved until the end when fine tuning is completed. If a component must be replaced, water seals would not have to be broken and then redone.

All connectors and junctions where water could enter should be sealed with silicon grease to prevent corrosion. Note that when caulks are used, some types are made with ammonia or lactic acid which can corrode metals; others harden, become brittle and crack. A sealant which stays soft like Coaxial-Seal™, a hand-molded, non-corrosive rubbery material, is recommended.

The equipment at the antenna focus should be protected with an LNB cover, a small expense and excellent insurance. In addition, drip loops should be used whenever applicable.

When a ground mount is used in climates with snowy winters, the antenna should be installed high enough so that it will not be pushing snow as it tracks the arc. In addition, if at all possible, leave adequate space under the bottom rim so a lawnmower will not run into and damage the reflector.

Be sure that water cannot accumulate anywhere on the mechanical system. There have been cases where water has collected at the bottom of mounts which were open at the top but closed at the bottom: the water froze, split the tubing and weakened the mount.

In the case of ground mounts, the cable trench can be covered up and the sod replaced after waterproofing is completed. At this point, the installation is finished.

L. PROGRAMMING THE RECEIVER AND ACTUATOR

Each brand of programmable receiver and actuator is different and the manufacturer's instructions should be carefully followed in programming these units. Realize that some remote controls require that standard 9-volt batteries be added by the user.

Making the System "User Friendly"

It is generally a good policy to ensure that a user either fully understands any complex programming procedures for a system or is guided to perceive the system as user-friendly.

Technicians must also be familiar and comfortable with any components that are installed as they may require service in the future. For those who install numerous systems using a common receiver type, it is wise practice to maintain consistency in programming this particular component. For example, if possible, ensure the preset channels are identical in all receivers or that satellites are stored with similar abbreviations or mnemonics on all actuator controllers. If such continuity is maintained, a set of simple instructions for in-field work can be created.

A copy of the chart for each receiver along with a detailed setup procedure can then be issued to each installer. The installer's copy should have a diagram of the remote control. It also should be sufficiently detailed to enable the technician to talk the user through the procedure over the telephone. If it would be necessary to travel long distances in order to re-program a receiver, the cost of this should be included inthe price of the system or provided for in a service contract.

Many receivers have a back panel hardware switch that enables the programming and re-programming procedure. If this switch is set in the locked position, it prevents the receiver from being inadvertently re-programmed by the remote control. This switch should always be in the locked position after a receiver is programmed.

The actuator set-up procedure includes parameters used in programming a channel into memory. Such steps which are generally receiver specific. The installer chart should also include the battery type used in the remote control for the particular receiver.

Sample forms for customers and technicians are illustrated in Tables 5-11a and b.

TABLE 5-11a. SAMPLE CUSTOMER CHART

Satellite Name:
Satellite Abbreviation:

Service	Channel Number	
Channel 01	....	....
Channel 02	....	....
| |	|	|
Channel nn	....	....

Satellite Name:
Satellite Abbreviation:

Service	Channel Number	
Channel 01	....	....
Channel 02	....	....
| |	|	|
Channel nn	....	....

TABLE 5-11b. SAMPLE INSTALLER CHART

Satellite Receiver Type:

Positioner Type:

Satellite:

Satellite Abbreviation:

Channel	Frequency	Service	Bandwidth	Audio
Channel 01................				
Channel 02................				
| |	| |	| |		
Channel nn................				

Detailed receiver setup procedure:

Battery type in remote control:

Detailed positioner setup procedure:

Include parameters used in programming a channel into memory

M. CONNECTING STEREO PROCESSORS AND OTHER ACCESSORIES

Every installation will have slightly different requirements for in-house equipment. Extra TVs, stereo processors, modulators, or even large screen or projection TVs may be needed. But installing peripheral equipment is usually an easy task if the design has been carefully planned.

Extra TVs

One satellite receiver can power two or three nearby TVs without amplification or any number of televisions with insertion of the necessary line amplifiers. All that is needed are splitters. These divide the video signal into two, three, four or even more equal signals. The placement and type of splitters used depends upon the location of each television. For example, assume that there is one set adjacent to the satellite receiver and three at the far end of the house. A 2-way splitter would be used at the receiver output to drive the nearby television and a trunk line having a 10 to 20 dB line amplifier would be connected to a location near the other sets. There, a 3-way splitter would divide the signal. This would certainly be a much better strategy than using a 4-way splitter at the receiver and running three long cables.

The type of cable selected depends upon the distance traveled and number of televisions connected. For short distances, RG-59 is adequate. RG-6 can be used for lines up to 100 meters. RG-11 can be used for longer trunk lines. However, for example, if 30 televisions were in close proximity to the receiver, it would still be necessary to use lower loss RG-6 or RG-11 since the original signal would be split or tapped so many times. The rule is to use common sense and minimize cable runs.

Line amplifiers which are easily inserted via F-connectors into a trunk line system may be required for extensive distribution systems.

If a multiple receiver system is being installed, splitters or taps rated for 950 to 1450 MHz are required if the coax from the LNB is divided before entering each receiver. This could be accomplished in a variety of ways. More details about these configurations are presented in Chapter VII.

Stereo Processors

Many satellite receivers have built-in stereo processors. All that is necessary to hook them to a home stereo system are two leads connected by RCA jacks, one for the left and one for the right sound channel.

The few stand-alone stereo processors that are in use are driven by the composite baseband output from a satellite receiver. This output is most often via an RCA jack. The receiver instruction manual should clearly specify which output is dedicated to a stereo processor.

Modulators

Occasionally it will be necessary to modulate the receiver composite audio and video signal onto channels other than the standard. Using an American system as an example, three receivers may be used to send three different channels along one trunk line to many televisions. In order not to interfere with over-the-air channels, it may be necessary to modulate onto UHF channels 20, 22 and 24 or onto VHF channels 5, 11 and 13. In general, it is best to modulate onto VHF channels since cabling losses are much higher in the UHF frequency range.

Modulators ranging in quality from those built into satellite receivers to commercial brands providing 30 to 60 dB of amplification or more and excellent channel isolation are available. They generally require both an audio and video input and have one RF output. Such modulators also usually have both audio and video level adjustments which are used to fine tune a well-designed distribution system.

N. TOOLS OF THE TRADE

The old boy scout adage "be prepared" applies to a satellite TV installer. Having to make an extra round trip of 60 km for an F-connector coupling or a 75 to 300 ohm transformer is certainly a waste of valuable time and money.

Tools of the trade can be carried in a small pickup or a thoroughly equipped lorry. The type of equipment needed depends upon what type of antennas are being installed, the climate and how a dealer wants to organize his installations. For example, one crew can be used to plant the pole, assemble the antenna and run the cable and then another to do all the electrical and alignment work. Or a single crew can complete the job in one visit.

Heavy Machinery

Heavy machinery includes vehicles needed to transport poles and antennas and all the equipment required for excavation and concrete work. Although some 3 meter mesh antennas can be packaged to fit into a small car, spun or stamped reflectors come in one piece and require at least a small pickup truck for transportation. At very least, a small pickup truck or a single or tandem axle trailer which can be pulled behind a car should be owned by any serious dealer.

Excavating tools include a wheelbarrow or a portable cement mixer, a pick ax, breaker bar, hoe, rake, hole digger and shovel. Others may be required in different situations.

Major Tools

This category includes all those tools required for every installation. This includes an inclinometer and compass, extension cord, wire strippers and carpenter's level.

Occasionally Needed Tools

Tools on this list are needed occasionally in special situations. For example, if no power is available at the site, a portable generator may be required. Or for long pole or roof mounts, an extension ladder may be needed.

Connectors, Cables and Other Accessories

This category includes all those small items that are most often required for an installation. These should always be on hand for both installing and troubleshooting. The lack of a small but essential item could be a source of great frustration.

The installer's equipment checklist in Table 5-12 is provided for convenience. Modifications to this list can be made to suit a dealer's tastes. However, using a checklist before leaving the shop to install or troubleshoot a system is a highly recommended practice.

TABLE 5-12. INSTALLER'S EQUIPMENT CHECK LIST

Heavy Machinery

___Pickup Truck or Trailer
___Wheel Barrow or Cement Mixer
___Post Hole Digger or Power Auger
___Shovel
___Pick Ax
___Rake and Hoe
___Bucket and Hose
___Breaker Bar
___Trencher

Major Tools

___Inclinometer
___Compass
___Signal Strength Meter
___30 Meter Extension Cord with 3 or 4-Way Adapter
___Extra TV or Monitor
___Multimeter (volts, ohms, etc.)
___Hex Crimp Tool for RG-6 and RG-59 Connectors
___Wire Strippers/Exacto Knife
___Tape Measure (at least 30 meters); or Ultrasonic "Tape Measure"
___Soldering Gun (20 and 40 watt) and Solder; or Gas Powered Iron
___Carpenter's Level
___Flashlight and/or Fluorescent Light
___12 Millimeter (1/2 inch) Heavy Duty Drill, Hammer Operation Switchable
___10 Millimeter (3/8 inch) Variable Speed Drill and Bits
___1 Kilo (2 pound) Hammer
___Hack Saw and Blades
___Large Diagonal Pliers
___Large Long Nose Pliers
___Small Long Nose Pliers
___Vise-Grips
___Assorted Screw Drivers
___10 to 30 Millimeter (3/8 to 1-1/8 inch) Set of Opened Ended Spanners
___Set of Socket Spanners
___Small Set of Metric or SAE Sockets
___Large and Small Crescent Wrenches (30 to 40 cm, 12 to 15 inches)
___Awl or Center Punch (for Aligning Holes in Antenna Petals)
___Set of Allen Wrenches
___A Short Ladder
___String and Tape
___4 to 5 cm (1/5 to 2 inch) Electrical Conduit, Solvent and Sweeps
___Metal File and Assorted Sandpaper
___16 mm by 45 cm (5/8 by 18 inch) Wood Drill Bit
___16 mm by 30 to 45 cm (5/8 by 12 to 18 inch) Masonry Bit

Often-Needed Tools

___Extension Ladder
___Generator
___Fish Tape (for feeding cable inside walls)

TABLE 5-12. INSTALLER'S EQUIPMENT CHECK LIST (continued)

Connectors, Cables and Other Accessories
___F-Connectors
___F-to-F Bullets (F-81 Barrels)
___N-Connectors (UG-57 and UG-29)
___Right Angle and Straight F and N-Connector Adapters
___RCA Jacks, MolyPlugs, Lugs, Scotch Locks and Assorted Adapters
___Video Connectors Kit with Cables
___Rolls of Direct Burial, RG-6 and RG-59 Cable
___RG-213 or 214 Pigtails
___75 to 300 Ohm Transformers
___A/B Switches
___2-, 3-, and 4-Way Splitters
___75 Ohm Terminators
___Spare Fuses and Voltage Regulators
___Electrical Tape
___Box of Assorted Wire Nuts
___Box of Romex Staples
___Can of Touch Up Spray Paint
___Coaxial Sealant
___Non-Conducting and Non-Corrosive Sealant
___A Spare Polarotor Servo Motor
___Surge Protector with Multiple Outlets
___10 and 20 dB Amplifiers
___Ground Rods and Heavy Gauge Grounding Wire
___T-25 Arrow Roundtop Staple Gun

O. COLD WEATHER INSTALLATIONS

Doing an installation in very cold weather can be uncomfortable. However, there are many factors to consider other than discomfort. Things that are taken for granted do not necessarily happen as expected. For example, electrical tape which normally sticks perfectly well becomes brittle and useless. It is necessary to buy a good brand and keep it in a warm place or a pocket until it is needed. Low power soldering guns cannot produce enough heat to melt solder. A very large gun is needed at -30 °C. In these cases, crimp-on connectors, not solder, may be chosen.

When it is sub-freezing, fiberglass becomes brittle and can shatter (see Figures 5-88 and 5-89). Many fiberglass antennas have been damaged in these conditions by over-tightening the bolts that hold the reflector to the mount. The answer to this problem is either to be very careful or, if possible, to assemble antennas indoors, out of the cold weather. If this option is not possible, let the unassembled pieces sit in the sun under a plastic covering, until they are warm.

Concrete will not set properly in very cold weather unless the right precautions are taken (see Figures 5-90 to 5-92). Two percent calcium chloride must be added to prevent freezing and the surface must be covered with a blanket, straw or another insulating material until it sets. Other binding materials dependent upon chemical reactions, like Dish Set™, set more slowly when cold. A simple practice to aid the hardening of either this material or a fast setting cement is to place the pole indoors or under the exhaust of a running vehicle to warm it up just prior to planting it.

Figure 5-88. Assembling a Fiberglass Antenna. *In this case the shipping crate was taken apart and laid down on the snow to keep water from freezing between the antenna panels as bolts were tightened. It was important to be sure that the panels were lined up without any gaps between them. (Courtesy of Brent Gale)*

Figure 5-90. Assembling an Antenna Directly on the Mount. *In this case, the antenna had to be assembled directly on the mount. This was somewhat more difficult than a typical installation.(Courtesy of Brent Gale)*

Figure 5-89. Penetrating the Frost Line. *This photo shows the use of a digging bar to penetrate the frost line so that a power auger could subsequently be used. Note that snow cover does have the the advantage of acting as an insulator and keeping the ground from freezing. In contrast, those areas of the country that have extreme cold but little or no snow can have frost penetrations of over 3 feet. In such cases, a barrel filled with burning wood or charcoal can be placed over the site of the pole; this will thaw the ground in a matter of hours. (Courtesy of Brent Gale)*

Figure 5-91. Use of a Power Auger in Cold Weather. *This two-man power auger is a very useful tool that allows rapid excavation of holes in cold climates. (Courtesy of Brent Gale)*

Figure 5-92. Use of Leveling Stakes. *This photo was taken shortly after the concrete was poured. The mount was installed on the pole only after curing was complete. (Courtesy of Brent Gale)*

Some brands of silicone sealants will not harden. Make sure that a silicone designed specifically for cold weather is used. Other products behave in surprising ways when cold. For example, one of the authors purchased two thousand 30 cm wire ties one summer expecting that they would have a long service life. When winter came, the ties were so brittle that they snapped when used. When warmed up they were fine, but at below freezing temperatures they were useless. Another brand manufactured specifically for cold weather had to be ordered. Similarly, most non-colored nylon wire ties will fail after only one year's exposure to sunlight. Black, blue, brown or specially UV- stabilized brands will, however, last years longer.

Tools can become very brittle in the cold, especially at temperatures below -25 °C. For example, ends of screw drivers snap off and crescent wrenches break. A solution is to use oversized wrenches which have greater leverage and need less torque, such as 50 cm instead of 30 cm crescents.

Plastic components are especially susceptible to damage in such extreme conditions. If TVs are treated roughly in the cold, their plastic containers easily break. Knobs also snap off receivers and TVs. In addition, some brands of satellite receivers do not work properly below a certain temperature. A recommended method is to keep the receiver and television in a warm car or truck during testing. Many types of cable will not unbend when cold and should be kept warm until needed. Some cables are made especially for such use. Center conductors of some coaxial cables will snap in two when cold. Therefore, care should be taken when uncoiling coax in cold weather.

Cold metal can be dangerous. For example, one dealer was working on a system when the chill factor was -50 °C. He went inside to check a picture and ran back to raise the elevation adjustment without his gloves. A crescent wrench which had been outside froze to his hand and could only be removed by running water over it. It is just as easy to pull skin off a hand by tightly grasping a very cold mount. Always wear a good pair of gloves.

Many problems can be traced to freezing of water which enters components either by condensation or leakage. Water left inside a waveguide can easily freeze and crack the housing. Even small amounts of condensation can cause problems in polarizer motors in extremely cold climates. Ferrite devices having no moving parts are recommended in such situations instead of mechanically rotated probes. There have been cases of water collecting in mounts or actuator motors and cracking the surrounding metal. Using drain holes, effective sealants, LNB/feedhorn covers and other protective measures is especially recommended in such climates (see Figure 5-93).

Figure 5-93. Weather Sealing. *A thorough weather sealing including the use of drip loops and coax seal is especially important in cold climates. (Courtesy of Brent Gale)*

A properly equipped dealer should be able to install in any weather, be it rain, snow or blizzard. For example, in very cold weather prefab concrete bases poured in a shop can be trucked to an installation site. But there are times when it is simply too cold or wet or snowy to warrant the effort.

P. CUSTOMER RELATIONS

Customer relations are common sense behavior, athough common sense must sometimes be learned, often at a high price. The most important piece of common sense dictates that a dealer or installer be punctual and reliable.

Some customers have an amazing amount of technical know-how and can teach most any dealer some interesting tricks and methods. Often a dealer who knows how to spot such people will make a sale where others have failed.

Some customers may want to participate in an installation by simply helping to assemble the antenna, plant the pole or run cable lines, or by actually completing the whole job except for the final wiring and fine tuning. Ninety percent of the time they can do an excellent job with the proper guidance and may become a source of many referrals. Other customers are pleased to do the entire installation knowing that a competent dealer is ready waiting in the wings when needed. Such installations are often examples of fine workmanship.

Whether or not a customer wants to participate, every installation should be be treated as if it were the dealer's own. It should be a source of pride and referrals and should last for years. Before leaving a site, a professional will take one last look around the premises in order to see if anything else can be done more perfectly.

Contracts and Warranties

Never make a sale or do an installation without a signed contract. Customers should be informed of all warranties and special conditions of this legal document. Each section of the contract should be explained so that there are no hidden expectations or unwanted surprises. Contracts protect both the dealer and the customer. Whenever a situation deteriorates to the point that a lawyer must be called in, both parties usually have already lost.

Q. DOCUMENTATION

It is important that records be kept of each installation in case any problems arise in the future. Documentation for both the customer's and installer's records should include the following:

- Date of Installation
- Original Site Map
- Antenna Type
- Mount Type
- Pad or Pole Support
- LNB Temperature and Make
- Receiver Make
- Actuator Make
- Type and Length of Cable
- Number of TV Sets
- Serial Numbers of All Components
- Any Special Installation Details

During the installation, a table outlining the measurements taken during final adjustments should be created. This will prove very useful if performance comparisons are required at a later time (see Table 5- 13). Remember, however, that as satellites age their transponder power will decrease and these numbers will slowly fall.

Signal strengths can be measured using equipment such as the Squawker and or other signal strength meters.

Business Documents

A variety of proposal, quotation and other business documents are used in the everyday running of a satellite TV dealership. Documents kindly provided by Ron Long and presented on the next four pages provide excellent samples.

TABLE 5-13. SIGNAL STRENGTH COMPARISON TABLE

Date: March 10, 1990
Location: R.F. Jones, St. Laurent, Quebec.

Satellite	Transponder Number	Signal Strength
Satcom K2	14	5.0
Anik C3	12	5.5
GSTAR 1	19	4.7
Spacenet II	13	6.4

INSTALLATION

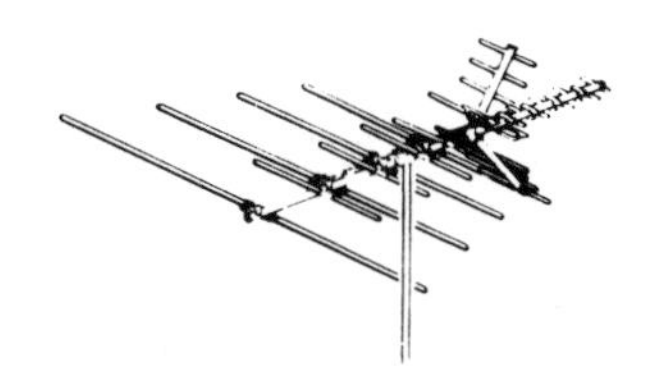

LONGFELLOWS
ANTENNA SERVICE

532 Dawson Drive, Suite 4-N
Camarillo, California 93010
(805) 484-2432

№ 1103

DATE

NAME

ADDRESS

CITY PHONE

NATURE OF SERVICE REQUEST

QTY	DESCRIPTION	PRICE	AMOUNT
	Antenna		
	Antenna		
	Antenna		
	Amplifier		
	Amplifier		
	Coupler		
	Coupler		
	Cable		
	Connectors		
	Base plate		
	Outlets		
	Outlets		
	Wall plates		
	Wire		
	Matching xfmr.		
	Eyebolts		
	Screw eyes		
	Seperators		
	Guy—Wire		
	Mast		
	Misc. Items		

Service Details:		
	TOTAL PARTS	
	TAX	
	TRIP CHARGE	
	TECHNICAL CHARGE	
	TOTAL	

CUSTOMER SIGNATURE ______________________ DATE ______________

MATRIX PRINTING

LONGFELLOWS
ANTENNA SERVICE

532 Dawson Drive, Suite 4-N
Camarillo, California 93010
(805) 484-2432

№ 1068

DATE ____________

NAME

ADDRESS

CITY PHONE

NATURE OF SERVICE REQUEST

QTY	DESCRIPTION	PRICE	AMOUNT
	Dish		
	Feed		
	LNB C		
	LNB C		
	LNB KU		
	Actuator ACME/BALL		
	Receiver		
	Receiver		
	Satellite Cable		
	Coaxial Cable		
	2-way Splitters		
	Coax Connectors F/BNC		
	Concrete		
	Ground Pipe		
	PVC Pipe and Accessories		
	Satellite TV Week Subscription		
	Misc.		

Service Details:		
	TOTAL PARTS	
	TAX	
	TRIP CHARGE	
	TECHNICAL CHARGE	
	TOTAL	

CUSTOMER SIGNATURE ____________________ DATE ____________

MATRIX PRINTING CENTER '89

QUOTATION

LONGFELLOWS
ANTENNA SERVICE
532 Dawson Drive, Suite 4-N
Camarillo, California 93010
(805) 388-5453

INQUIRY NO. ____________

DATE ____________

TERMS ____________

DELIVERY ____________

PRICES QUOTED ARE F.O.B.:

____

TO ____________

____

____

____

WE ARE PLEASED TO QUOTE ON YOUR INQUIRY AS FOLLOWS:

QUANTITY	DESCRIPTION	PRICE	AMOUNT

QUOTED BY:

MATRIX PRINTING

LONGFELLOWS
ANTENNA SERVICE
532 Dawson Drive, Suite 4-N
Camarillo, California 93010
(805) 388-5453

No.____________________
Date____________________
____

PROPOSAL

Proposal Submitted To:

Name____________________
Street____________________
City____________________
State____________________
Phone____________________

Work To Be Performed At:

____
Street____________________
City____________________ State____________________
Date of Plans____________________
____

We hereby propose to furnish the materials and perform the labor necessary for the completion of

____

All material is guaranteed to be as specified, and the above work to be performed in accordance with any drawings or specifications required for above work and completed in a substantial workmanlike manner for the sum of

Dollars ($).

with payments to be made as follows:____________________

Any alteration or deviation from above specifications involving extra costs, will be executed only upon written orders, and will become an extra charge over and above the estimate. We are not responsible for damages due to acts of God.

Respectfully submitted____________________

Per____________________

Note—This proposal may be withdrawn by us if not accepted within ________ days.

ACCEPTANCE OF PROPOSAL

The above prices, specifications and conditions are satisfactory and are hereby accepted. You are authorized to do the work as specified. Payment will be made as outlined above.

Signature____________________

Date____________________ Signature____________________

MATRIX PRINTING

VI. UPGRADING A SATELLITE TV SYSTEM

In many instances a service call will evolve into a need for a system upgrade. For example, the actuator tube may be bent and parts for this particular actuator may not be available, the receiver or LNB has given up the ghost, on older systems the LNA or downconverter could have failed and parts are no longer available, or the signal quality is poor to terrible because the cable runs need replacement. Occasionally, a customer orders a new receiver that is incompatible with the previous actuator counting mechanisms or polarizer control, or someone purchased a home with an existing satellite system that is not functioning. In some cases, a person could have purchased a home with a functioning satellite system and have requested help in learning to operate it only to realize that an upgrade is desired. It is also not uncommon for a casual conversation during a simple troubleshooting visit to turn into a request for an upgrade.

However the opportunity arises, system upgrades can tax the ingenuity of even a seasoned professional. During the short evolution of the satellite industry, numerous engineering approaches have come and gone. For example, it is difficult to impossible to obtain parts for some potentiometer and optical sensors used as actuator counting mechanisms. The challenge is to salvage as much of the old equipment as is reasonably possible when upgrading the system to be compatible with modern technologies.

A. COMPONENT EVALUATION

LNAs and LNBs

Although many LNA/downconverter systems are still functioning, some of the original manufacturers are out of business and, in any case, parts are rarely available. Furthermore, the 80 to 120°K noise temperatures typical of older systems are inadequate today.

Whenever a new block downconversion satellite receiver is purchased, the LNA and downconverter must be replaced with a new LNB. This may be true even if an LNB was included in the previous system because many older receivers were mated with units having uncommon output frequency ranges such as 430 to 930 MHz, 900 to 1400 MHz or 930 to 1430 MHz.

Feeds

Earlier satellite television systems incorporated feeds and polarizers that used techniques that are phased out today. Products included feeds with ferrite polarizers or low noise feeds (LNFs) that were LNAs with built-in polarizers. Many older feeds accepted from 3 to as high as 15 Vdc for driving dc polarizer motors. The polarizer voltages for three wire servomotors ranged from 3 to 8 Vdc. There were dc motors requiring both two and three-wire hook-ups. Some motors had just a 100° range of rotation while others could rotate a full 180°. Note that the autotune feature on some modern IRDs could burn out the motor of some of these limited range polarizers by running them up against a limit. Trying to match any of these product features to current equipment could be as difficult as trying to find parts for antique automobiles.

The current standard feed is a polarizer with a servomotor that rotates 270° and that can withstand approximately 8 Vdc without possible motor damage.

Actuators

Pioneer actuators were in many respects crudely designed and manufactured. A variety of unusual count mechanisms were employed. Although the potentiometer and optical counting mechanisms are now historical curiosities, there are still such units in the field. Some earlier actuators did not even have counters or limits but were moved by simply depressing a "west" or "east" movement button on the controller. Some actuators had circuit boards attached to the mount that was specially designed for a particular antenna. Today, such boards are normally located in the satellite receiver or actuator controller. In these cases the only possible upgrade is to replace both the dish and actuator.

Actuator motors have been powered by 12 Vdc, 24 Vdc and 36 Vdc. The latter voltage is the standard today. Actuators used on older systems often are not capable of moving much past Satcom F4 or Galaxy 1. This range of motion is not sufficient today.

Cable and Connectors

Some of the coaxial cable manufactured and installed during the infancy of the satellite industry was substandard. It was generally not necessary to sweep test this cable because 70 MHz frequencies were relayed in many cases. Remember that transmission at 950 to 1450 MHz is much more difficult than at 70 MHz. The extremely long runs of RG-59 cable installed in many older systems are simply inadequate to relay signals from an LNB. Some installations that use RG-58, RG-62 or RG-92 coax are still on-line. These cables have characteristic impedances that do not match the 75 ohms required with satellite receiving equipment. While this mistake could be tolerated at 70 MHz, where almost any type of cable will function properly, such cable is a disaster at 950 to 1450 MHz.

Some cable used for earlier satellite systems that was white or beige in color would not tolerate ultraviolet radiation. The dielectric around the center conductor would shrink, causing the characteristic impedance to change. The end result is deterioration in picture quality. This cable damage shows up as an outer jacket that is tacky or sticky to the touch and has blisters. Moisture could enter at the blisters further altering the characteristic impedance.

Some earlier all-in-one cables were enclosed in a single round jacket. In order to separate the actuator, polarizer and LNA/downconverter or LNB runs, the outer jacket had to be removed. This exposed the wires to the degrading effects of ultraviolet rays and moisture. Most installers did not even attempt to tape these awkward configurations to reduce damage.

Cable that was installed years ago without a conduit could easily have been damaged by the action of rodents or water. Similarly, connectors may have degraded to the point that substantial signal losses occur.

It is interesting to note that ribbon cables manufactured in the past used the second extra coax line to transmit LNB power and the IF signal on separate lines. Additional wires may also be unused if an actuator was not installed. These extra lines may have frayed or degraded over time.

In these situations as well as others where seemingly excellent cables were installed, it is crucial to be certain that no damage has occurred. If any damage is detected, all cables should be replaced.

Satellite Receivers

The satellite receiver is the system controller. It must be properly mated with the LNB, actuator and polarizer. Voltages, size and number of wires and electronic control circuitry must be compatible.

If the existing receiver has a TI filter attached, it is a wise policy to replace the existing feed with a TI rejection type and the receiver with one having built-in filters or, at least, an IF loop. In the latter case, select a receiver that has an IF loop frequency for which filters are available. Notch filters with 70 MHz IF are readily available.

General Considerations

Any new component must be compatible with any other with which it interacts. Unfortunately, many upgrades require replacing almost everything other than the antenna.

In an upgrade, there are many concerns similar to those that arise when troubleshooting a system. For example, the system might need an alignment, connectors may have to be replaced or a mount may have to be stabilized. In some cases, installing an entirely new system may be the simplest solution.

B. UPGRADING TO KU-BAND

The majority of C-band antennas can be retrofitted to receive Ku-band broadcasts. Either a dual-band feedhorn or one dedicated to Ku-band reception can be installed. C and Ku-band LNBs have outputs in the 950 to 1450 MHz band range and most modern receivers have dual-band capability. Another option is to install a second smaller Ku-band antenna in parallel with the C-band system. An A/B switch, either built into or external to the satellite receiver, is used to select between formats. Note that many older satellite receivers do not have dual-band capability because only C-band channel frequencies can be tuned.

Although a Ku-band antenna requires a more accurate reflective surface for efficient reception, the satellite EIRP is generally high enough so that even inefficient C-band dishes suffice. Ku-band dishes can typically be smaller than C-band antennas and the extra reflector area can be technically very forgiving. Nevertheless, many older mesh dishes have relatively large openings and are quite inefficient at Ku- band. A general rule of thumb to follow is that holes in the mesh should be less than one tenth of the microwave wavelength (1 mm at Ku- band since the wavelength at 12 GHz is approximately 1 cm). Holes less than this size prevent the microwave energy from passing through the openings and nearly all the incident energy is reflected.

Antennas must also be pointed with greater accuracy to properly track Ku-band satellites because the beamwidth of any antenna decreases in inverse proportion to microwave frequency. Therefore, for a given antenna, the beamwidth at Ku-band (at triple the frequency of C-band) is one third that at C-band. This need for tracking accuracy is similar to that encountered in out-of-footprint installations that require very large antennas with narrow beamwidths (see Chapter IX).

VII. MULTIPLE RECEIVER SATELLITE TV AND DISTRIBUTION SYSTEMS

Satellite TV signals can be combined with off-air broadcasts, VCRs, video games or even closed circuit television messages and relayed to any number of television sets. There is really no limit to the design possibilities once a sufficiently powerful satellite signal has been intercepted and processed.

This chapter is organized into three sections: (1) a survey of the basic components of a multiple receiver system, (2) an examination of the different configurations ranging from a simple one receiver/24 channel system to a more complex multiple receiver/dual feedhorn/24 channel design and (3) a brief study of signal distribution systems. A wide range of designs are possible but all are really quite simple to understand once the different pieces of this jigsaw puzzle are known.

A. THE BASIC COMPONENTS

The most complex configurations of satellite reception equipment are pieced together from basic components including a dish, feedhorn, LNB, cables and connectors, receivers and televisions. These are then combined with splitters, line amplifiers, attenuation pads, terminators, DC power blocks, A/B switches, combiners and, in some cases, coaxial relays to create any combination desired.

Televisions

Television receivers function best when their input signal level is between 0 and 3 dBmv (1 to 2 millivolts), although the optimal levels can vary between different brands. When signal levels are above approximately 10 dBmv some older televisions may be overdriven so that pictures could be distorted. However, nearly all modern TV sets have AGCs (automatic gain controls) which compensate for excessively strong signals. This feature allows inputs ranging from -20 dBmv to as high as 50 dBmv to be managed, to a point. Nevertheless, it is safe practice to limit the input signal level to at least the 0 to 40 dBmv and, if at all possible, to the 0 to 10 dBmv range for best results.

It is much easier to manage signals that are too powerful than too weak. If the ratio of signal to noise power has been allowed to drop too low on its way to a television set or is "in the mud," no level of amplification will improve the situation because as much noise as signal would be amplified. In contrast, when signal levels exceed the upper limits, attenuation pads can easily be inserted to reduce the power and restore normal operation. Therefore, too much power is never really a problem but allowing signal strength to deteriorate can cause serious impairments in reception.

Note that many older televisions often have a 300 ohm, two-screw VHF input where the familiar flat TV leads can be attached. Signals from satellite receiving equipment as well as from other devices are transmitted by 75 ohm coax and require F-connector inputs. So a 75 to 300 ohm matching transformer, known as a balun (derived from balanced-unbalanced), must often be used in the transition from the coax F- connector to the television VHF terminals (see Figure 7-1). The transformer matches the impedance between the coaxial cable and the television input to allow maximum signal power to be transferred.

Cable, Connectors and Splitters

Cable, connectors and splitters are the conduits for carrying signals between any number of locations. Each type has characteristic power losses which depend upon frequency. These must be accounted for when designing systems (see Figures 7-2 and 7-3).

Attenuation or signal loss in standard 75 ohm cable such as RG-6, RG-11 and RG-59 have been

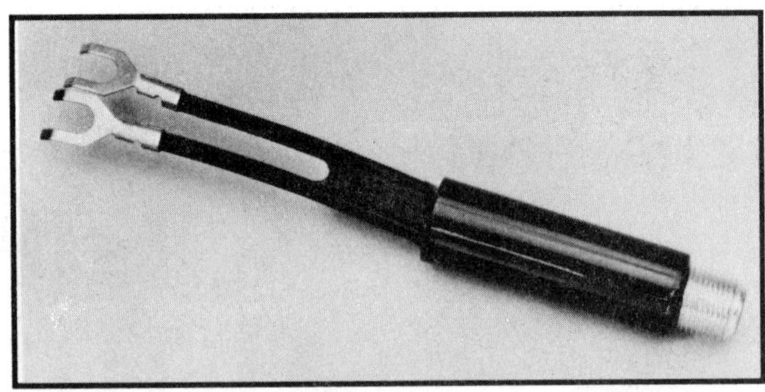

Figure 7-1. Matching Transformers. *This device is used with televisions that require 300 ohm twin-lead cable inputs. Today most sets are manufactured with 75 ohm F-connector VHF inputs. (Courtesy of MACOM Industries)*

discussed in Chapter 2. A good ballpark loss figure to use for RG-59 cable, the standard type for distributing television signals in small systems, is 2 dB per 100 feet at 50 MHz and 4 dB per 100 feet at 200 MHz. Note that 50 MHz is about the same frequency as channel 2 and 200 MHz is just below channel 13 (see Tables 2-7 and 2-8 and the cable attenuation graph, Figure 2-60, in Chapter II).

Coaxial cables can be purchased in any lengths usually up to a maximum of 1000 feet. Different colors such as black, beige or white are available and can be chosen to improve the look of any installation or to help conceal the cable.

Splitters do just what the name implies, divide the signal into two or more branches (see Figures 7-4, 7-5, and 7-6). They may also be used to combine signals in the correct situation. Splitters are designed to handle a specified frequency range. For

Figure 7-2. Drop Cables. *Drop cables are packaged in a variety of easy-to-use containers. These can be dispensed from "easy-out" boxes or reusable plastic cases to facilitate pre-wiring. (Courtesy of Comm/Scope MA/COM)*

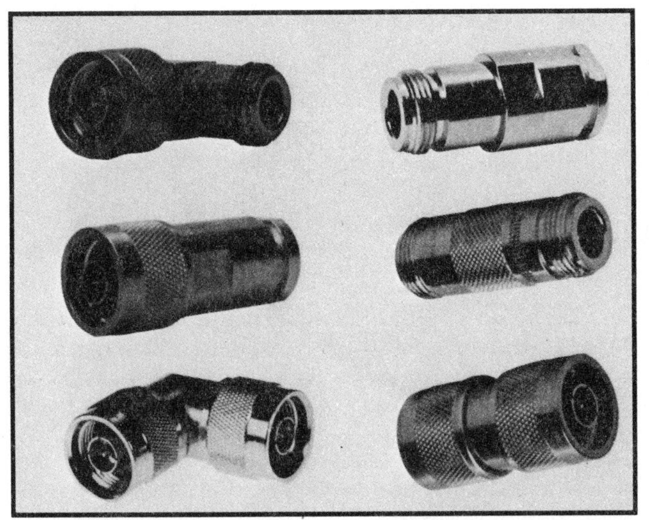

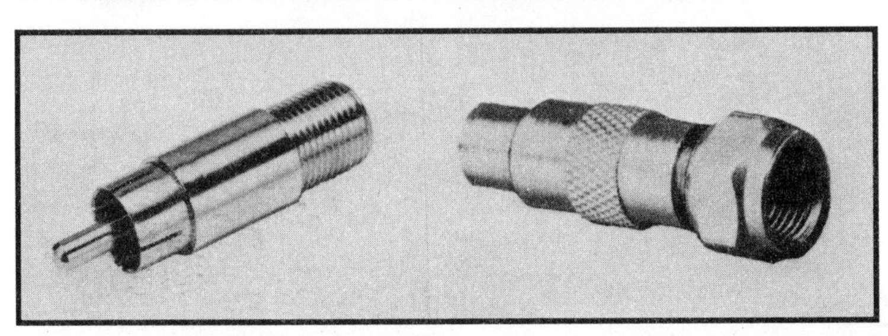

Figure 7-3. Cable Connectors. *F-connectors are commonly used with RG-59, RG-6 or RG-11 coax. RCA male-to-female F-connectors are used to connect an RCA plug to a coax line. RCA female-to-male F-connectors are used to convert coax to an RCA plug. F-81 barrels are occasionally used to splice two F-connector type coaxial cables together. N-connectors, most commonly encountered for attaching to an LNA output, are used on the outputs of some European LNBs.*

example, when dividing a 950 to 1450 MHz block downconversion satellite signal, a splitter rated up to 1450 MHz must be used or losses would be too high. Some brands of splitters also have built-in bandpass filters so that frequencies below and sometimes above the designed range are attenuated. Limiting the bandwidth can protect cables from ingress interference. Typical insertion losses for two, three, four, eight, and sixteen-way splitters are listed below in Table 7-1.

A 2-way splitter cuts the signal a little more than in half (since a 3 dB reduction is a halving of power). A 3-way splitter has one -3.5 dB and two -7 dB legs. For example, if a signal of 3 dBmv is divided in a 4-way splitter each output leg would have 3 dBmv less 7 dB or -4 dBmv of signal output, which is not really enough to produce a studio quality television picture. If this signal was then relayed down 100 feet of RG-59 cable, additional losses of 4 dB at 200 MHz would result in a final signal of -8 dBmv at the television input. Therefore, amplification would be needed before the 0 dBmv level was reached. Line amplifiers should be used before cable losses cause a situation where noise power becomes too large compared to signal power.

TABLE 7-1 SPLITTER LOSSES PER LEG

Type of Splitter	Loss/Leg (dB)
2-way	3.5
3-way	3.5 and 7
4-way	7
8-way	10.5
16-way	14

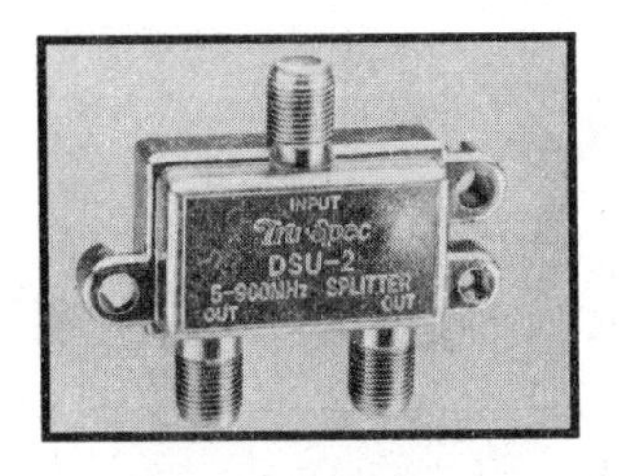

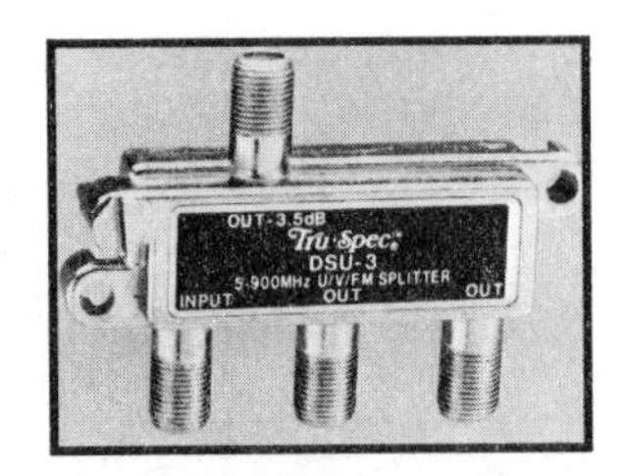

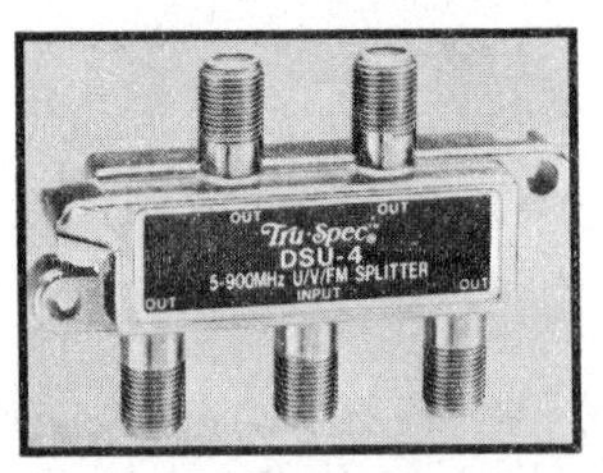

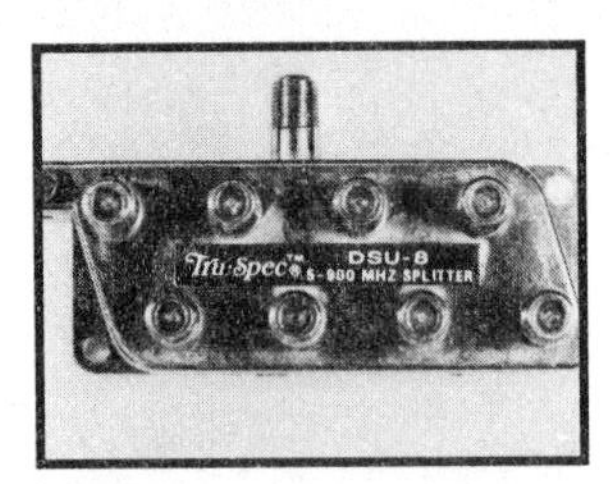

Figure 7-4. MATV Splitters. *2, 3, 4 and 8-way splitters are most commonly employed when dividing signals in MATV applications. The usable frequency range of this brand ranges from 5 to 900 MHz which extends past the UHF channel frequency band.*

Figure 7-6. High Frequency Splitters. *Quitar models HFS-2 and 4 operate in the 900 to 1500 MHz band, have a 20 dB port-to-port isolation and pass dc power through one port. Models HFS-2P and 4P pass power through all ports. (Courtesy of Qintar, Inc.)*

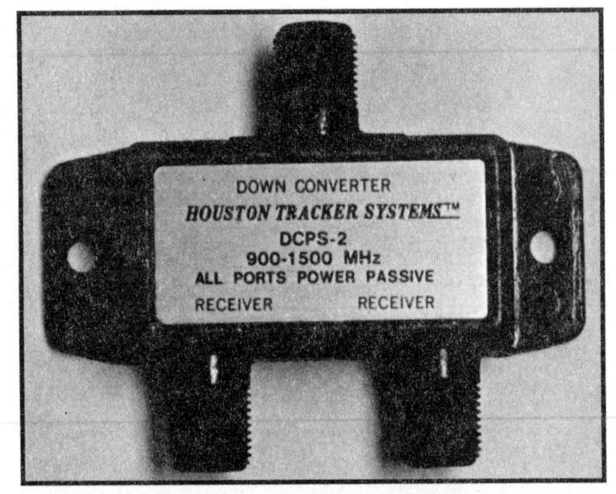

Figure 7-5. DPS-2 and DPS-4 Splitters. *These splitters pass dc power through all ports, so that any receiver in the system can power the LNB. (Courtesy of Houston Tracker System)*

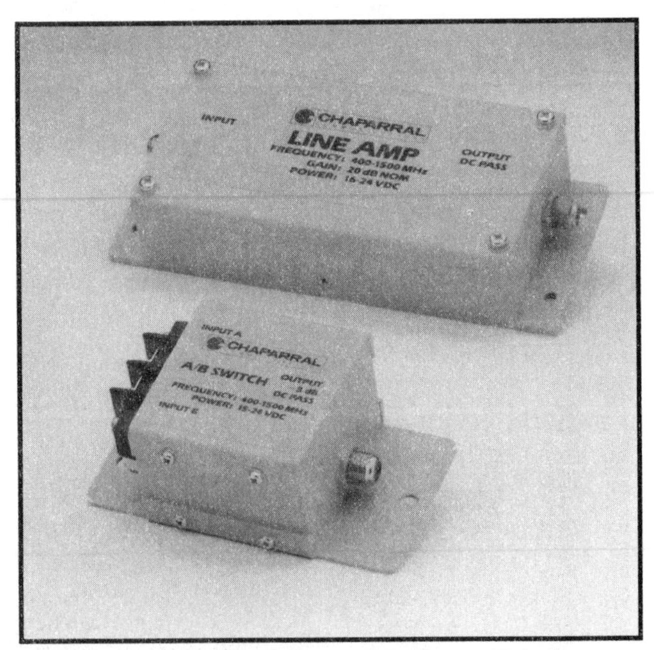

Figure 7-7. Line Amplifier and Splitter. *This line amplifier is designed for a frequency range of 400 to 1500 MHz. The A/B switch in the bottom photo operates in the same frequency range. (Courtesy of Chaparral Communications)*

Line Amplifiers and Tilt

Line amplifiers which are inserted via F-connectors directly into a coax line are available for boosting signals from either a single channel or a whole range of channels. Most of the simpler units are powered from the dc voltage relayed down the coax (see Figures 7-8, 7-9 and 7-10). More costly commercial amplifiers draw current from a regular wall outlet and convert it to dc power. Preamps on off-air antenna masts for boosting signals operate in a similar fashion.

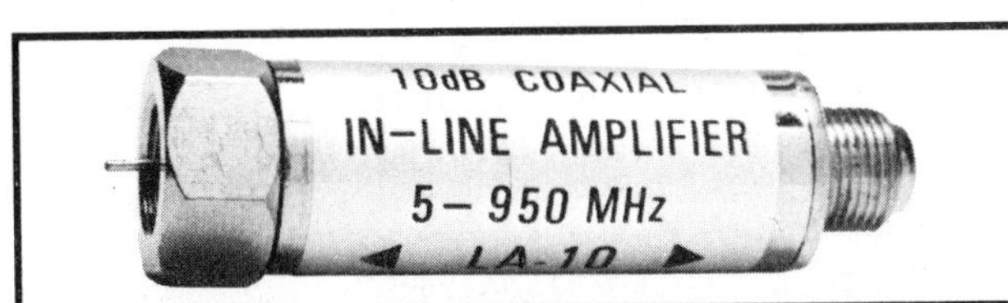

Figure 7-8. Bullet Amplifier. *The LA-10 Bullet line amplifier is rated for frequencies from 5 to 950 MHz and is normally used in SMATV distribution systems or in conjunction with an LNB having a 400 to 950 MHz output. It can be inserted as a line amplifier between a pre-amp output and a channel processor input in long cable runs, but must receive line power to operate. (Courtesy of Pico Products)*

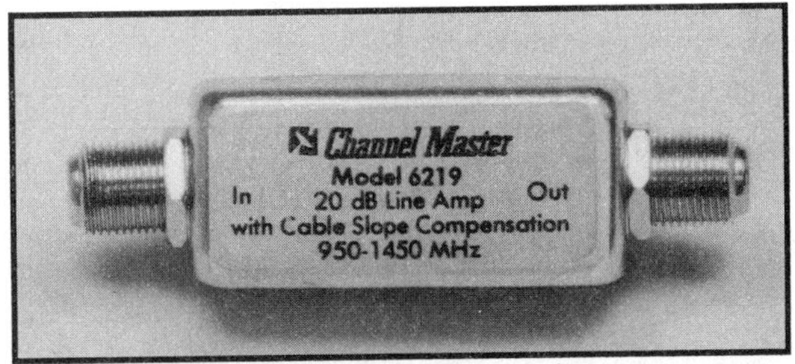

Figure 7-9. 20 dB Line Amplifier. *This 20 dB line amp operates within a frequency range of 950 to 1450 MHz and has cable slope compensation of 6 dB between its low and high frequency ranges. (Courtesy of Channel Master)*

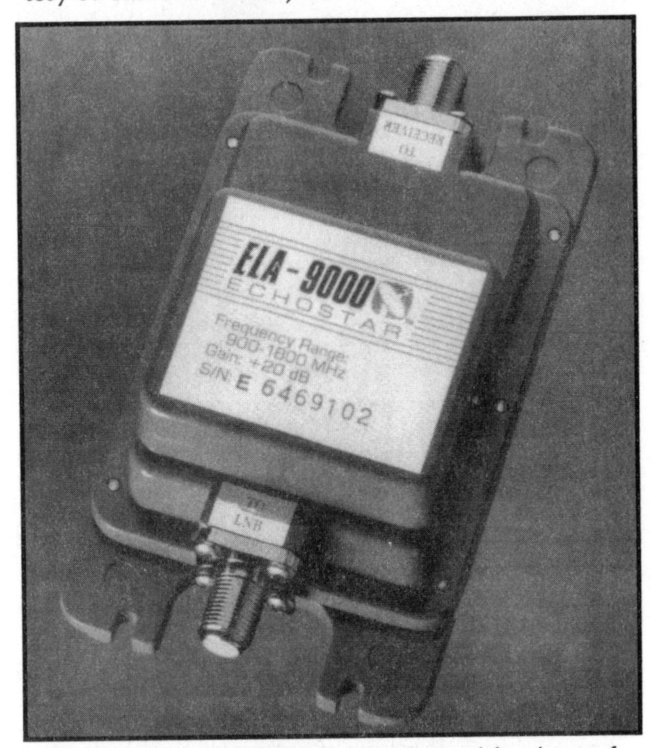

Figure 7-10. ELA-9000. *This line amplifier has a frequency response of 900 to 1800 MHz, a gain of 20 dB and a slope of 4 dB across this band. It has F-type connectors and operates in the IF line on a +15 to +24 Vdc. (Courtesy of Echosphere Corporation)*

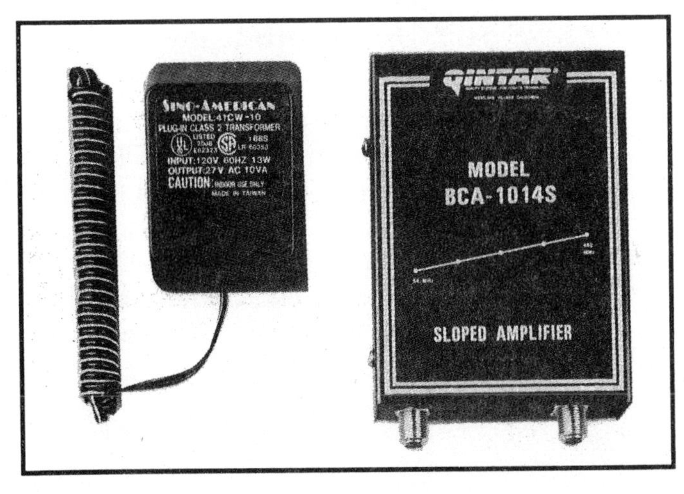

Figure 7-11. CATV Amplifier. *This 54 channel, push-pull amplifier covers the 54 to 440 MHz frequency range. It has a gain of 10 dB at 54 MHz and a gain of 14 dB at 440 MHz to compensate for tilt. (Courtesy of Qintar, Inc.)*

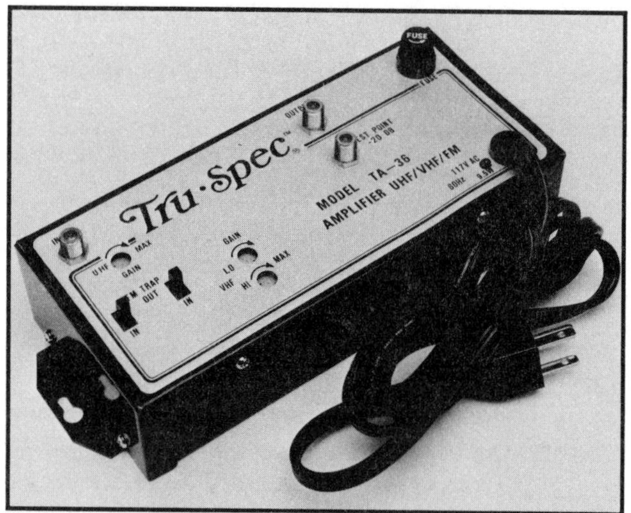

Figure 7-12. Pico TA-36 UHF/VHF/FM Distribution Amplifier. *This amplifier is used in an SMATV distribution system providing amplification to feeder lines throughout a building. Frequency bandwidth ranges are: lo-VHF 54 MHz to 108 MHz; high-VHF 174 to 216 MHz; and UHF 470 to 812 MHz. Gain at VHF is 31 dB and at UHF is 36 dB. (Courtesy of Pico Products)*

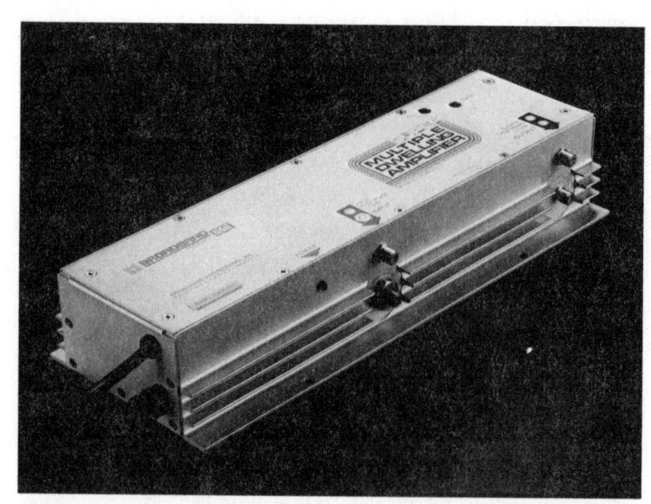

Figure 7-13. Commercial Trunk and Distribution Amplifier for SMATV. *This top of the line amplifier meets most requirements for apartment, condominium, hotel, motel, and hospital distribution systems with a 60 channel loading capability. This product is available for use in both North America and Europe. (Courtesy of Broadband Engineering)*

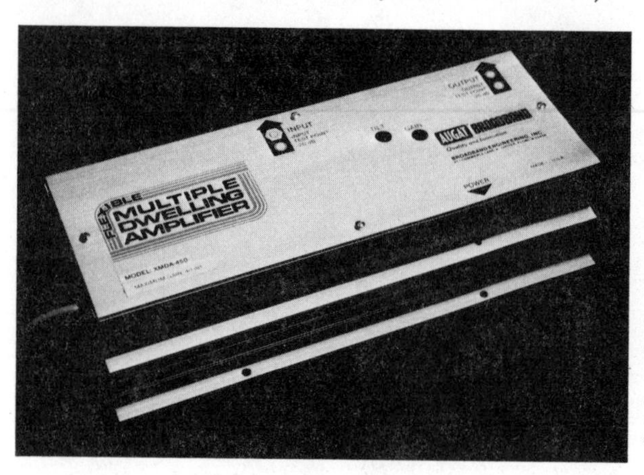

Figure 7-14. Multiple-Dwelling Amplifier. *This commercial MATV distribution amplifier is used in SMATV distribution systems. It operates in a frequency range of 50 to 450 MHz and has a gain of 20 to 50 dB. It is available with variable gain and tilt controls and can be tailored for operation in locations where a line voltage of 220 Vac is standard. (Courtesy of Broadband Engineering Incorporated)*

Amplifiers can have either fixed or adjustable gain. Some amplifiers designed for longer runs have gains which increase with frequency to compensate for the increased attenuation of signal at these higher frequencies. This is called tilt (see Figures 7-11, 7-12, 7-13 and 7-14). For example, if RG-59 is used for especially long runs, a tilt of about 2 dB per hundred feet between 50 and 200 MHz (channels 2 and 13) should be used. This is based on the difference between the 4 dB and 2 dB loss per hundred feet that this coax has at these two frequencies. For example, if a cable run of 300 feet were used, a line amplifier delivering 26 dB at 200 MHz and 20 dB at 50 Mhz would properly compensate for the 6 dB difference in cable losses between these frequencies. Then the power levels after 300 feet of cable would be flat across all frequencies.

In-Line Attenuators

An in-line attenuator, also known as an attenuator pad or simply a pad, is used to reduce the strength of an excessively powerful signal. Pads are small, inexpensive devices which screw directly into a coax line via F-connector fittings (see Figures 7-15 and 7-16). They are available with either fixed or variable attenuations and occasionally with built-in tilt compensation. Most pads are not designed to pass dc power so they cannot be used in the line between receivers and LNBs.

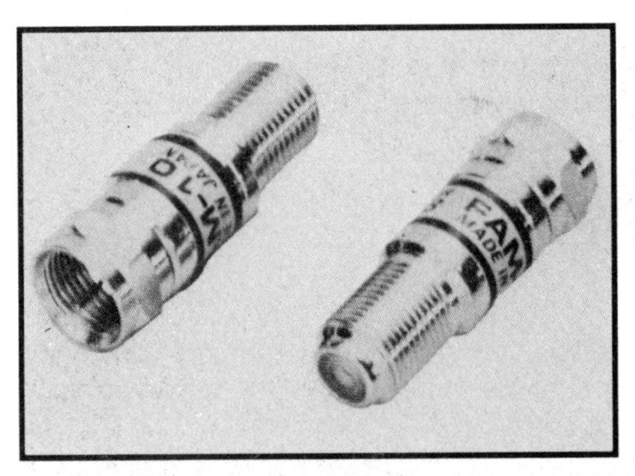

Figure 7-15. Attenuator Pads. *These pads have one female and one male port for insertion into a coax line using F-connectors. These models have a variety of rated attenuations including 3, 6, 8, 10, 12, 16 and 20 dB. Their bandwidth ranges from dc to 1000 MHz. They are generally used in MATV systems to decrease signal levels when balancing channels. This pad does not pass power and therefore cannot be used in a satellite block distribution system. (Courtesy of Pico Products)*

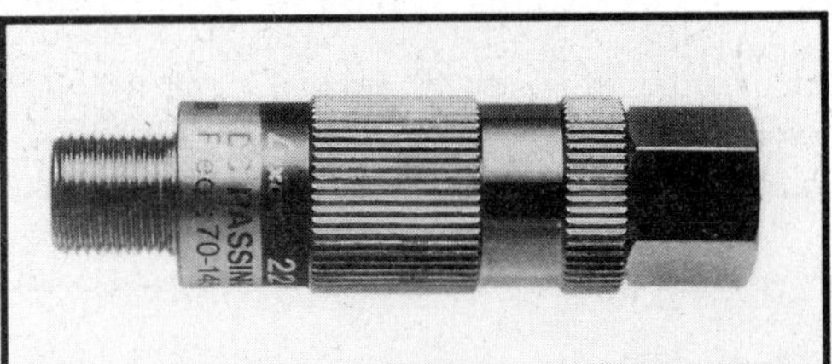

Figure 7-16. Attenuation Pad. *This pad has one male and one female port for insertion into a standard RG-59 or RG-6 coaxial line. It provides 10 dB of signal attenuation and passes dc power. (Courtesy of Luxor Corporation)*

Terminators

Any output port on a television distribution system must end in an appropriate device such as a television set, a satellite receiver or a terminator. If not, the opened port could pick up interference or cause signals to be reflected back into the system as "ghosts." A terminator for 75 ohm coax is a resistor in the form of a cap that screws onto an F-connector (see Figure 7-17). It has a 75 ohm impedance that matches the coax impedance. As a result, the cable does not "see" an opened end or discontinuity at the unused port; it appears that the cable never ends.

Figure 7-17. Terminator. *Terminators are installed on an unused splitter port or at the end of a distribution line in order to provide a 75 ohm impedance match. Doing so eliminates both signal reflections and ingress interference. (Courtesy of Pico Products)*

dc Power Blocks

A dc power block is a simple device which allows higher frequency television signals to pass unattenuated but which blocks the passage of any direct current (see Figure 7-18). Therefore, dc blocks are used to isolate various components in a satellite system from dc voltages. This is accomplished by a circuit element called a capacitor which passes higher frequency signals but which blocks direct current.

Figure 7-18. Voltage Blocking Coupler. *VBCs are used to block any dc or low frequency ac power which might be present on a coaxial cable. They can be used to protect equipment such as televisions or some types of satellite block distribution system relays from damaging power surges. (Courtesy of Pico Products)*

A/B Switches and Combiners

An A/B switch is used to select between either one of two input signals (see Figures 7-19 and 7-20). High quality A/B switches have at least 40 or 50 dB isolation between ports. This means that if a 10 dBmv signal is present on the disconnected input port, this signal would be reduced to by 40 dB to -30 dBmv (10 dBmv minus 40 dB which equals -30 dBmv) at the output.

A signal combiner is used to combine and balance the powers of satellite and off-air signals (see Figure 7-21). It can be used in place of an A/B switch. For example, a simple unit might take an input from the satellite receiver channel 3 modulator and another line from an off- air antenna. Note that if two adjacent channels are being combined there is always the possibility that they will interfere with each other when a lower quality signal combiner is used. In any case, it is always important to balance the power levels of both the satellite and off-air signals before they enter a combiner. In order to eliminate interference between channels, it may be necessary to use a bandpass filter which cuts out the lower sideband and restricts the satellite TV input to a selected narrow range of frequencies. More details about equalizing and separating adjacent channels are discussed below.

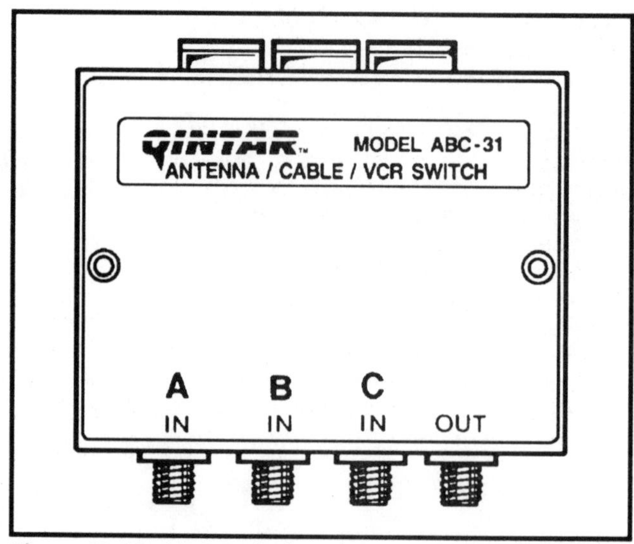

Figure 7-19. ABC Switch. *The Qintar model ABC-31 switch has three inputs and one output. Isolation between ports is more than 70 dB. (Courtesy of Qintar, Inc.)*

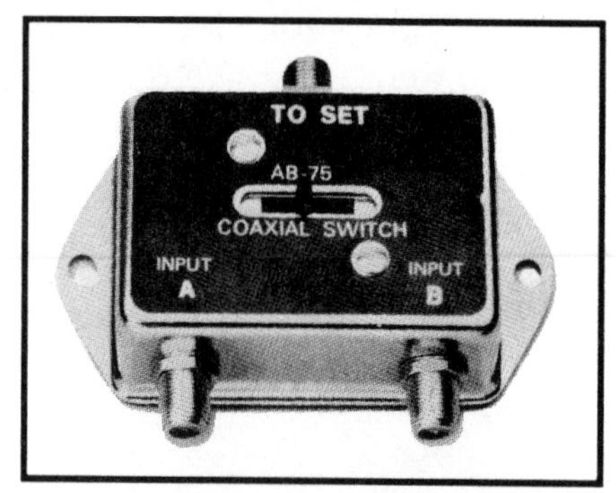

Figure 7-20. Coaxial A/B Switch. *These switches come in a variety of models with isolations ranging from 30 to 90 dB. They can be used to switch between inputs of either horizontally or vertically polarized satellite signals and an over-the-air broadcast. (Courtesy of Pico Products)*

Figure 7-21. Inexpensive Signal Combiner. *This combiner consists of a high quality, six-stage bandpass/bandstop filter with adjustable balancing attenuators. It allows 3, 4 or 6 output signals from a videotape recorder, MDS receiver, TVRO or video game to be combined on a distribution system carrying adjacent channels. Properly balancing signal levels in an MATV system eliminates the need for an A/B switch.*

Coaxial Relays

A coaxial relay is a switch which passes signal is only one direction at a time. Typically, 12 volts dc from the rear of a receiver or power supply is used to open or close a relay. Coaxial relays do not pass dc power and must be protected from excessive currents. A coaxial relay is nothing more than an electrically operated A/B switch (see Figures 7-22, 7-23 and 7-24). Coaxial relays are also referred to as H/V switches, C/Ku switches or coax switches.

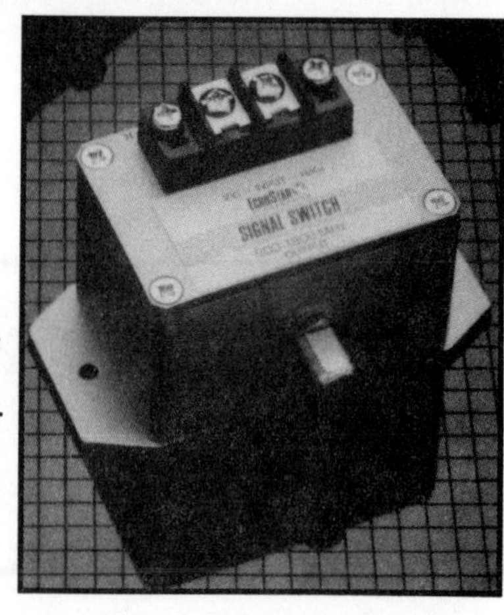

Figure 7-22. Coaxial Relay. *This coaxial relay operates on switching voltages of +15 and 0 Vdc. (Courtesy of Echosphere Corporation)*

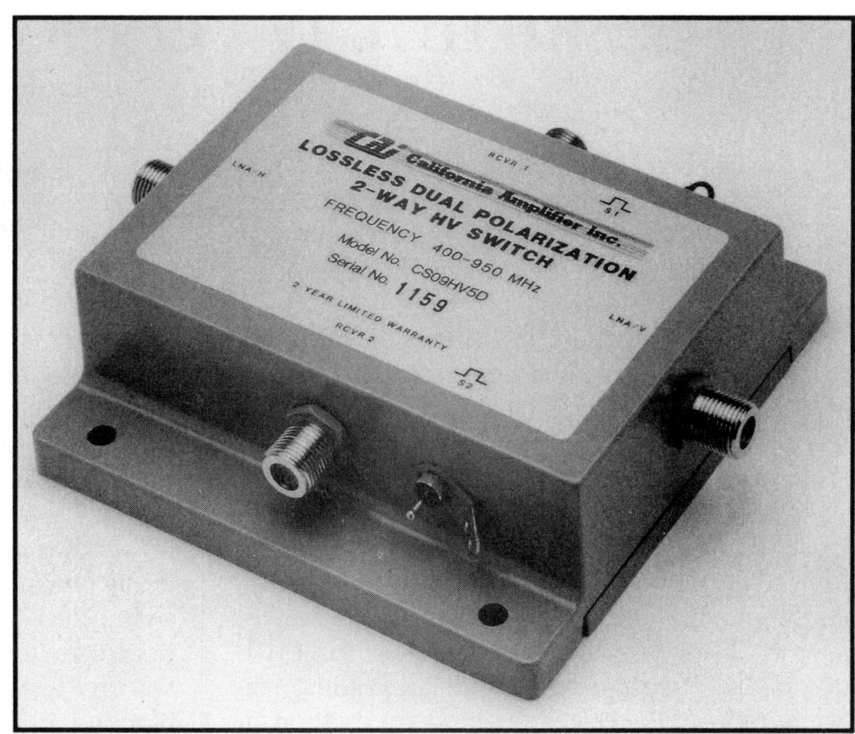

Figure 7-23. Model CS09HV5D Dual Polarization H/V Switch. *This lossless dual polarization 2-way H/V switch replaces a host of components. It is rated to operate at 400 to 950 MHz and is connected to two LNBs or block downconverters. Horizontal or vertical signal outputs are chosen internally by pulses from satellite receivers. This device has at least 55 dB isolation between receivers and 20 dB isolation between H/V ports. It passes power directly to the downconverters/LNAs and has a built-in amplifier to provide lossless operation. (Courtesy of California Amplifier, Inc.)*

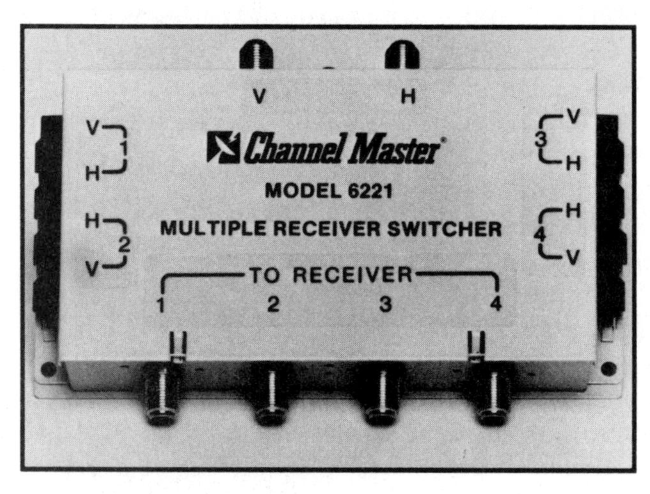

Figure 7-24. Channel Master Multiple Receiver Switcher. *This switcher/divider is designed to take a horizontal and vertical LNB input, divide this signal and then select a path to any of four satellite receivers giving each independent polarity control. (Courtesy of Channel Master)*

B. SATELLITE TV CONFIGURATIONS

Connecting more than one receiver to a single satellite dish is rather straightforward. This situation differs dramatically from just five ago when much more complex arrangements were required.

There are certain limitations in multiple receiver systems. Just one satellite can be viewed at one time for each dish installed and only one receiver can control dish movement. In the United States and other nations where programming is scrambled, a separate decoder is required at each location to view a different subscription program. Otherwise, only the unscrambled, non-subscription programming can be viewed at the second location. The user must pay a fee for both descramblers which each have different authorization numbers. Programmers have no method of knowing if more than one decoder is located in the same home.

Just a few years ago, low cost slave receivers were available. However, today most satellite receivers are IRDs and the auxiliary unit costs almost the same amount as the master receiver. When a UHF remote is used, each auxiliary receiver must be tuned to just one remote, if possible so all receivers would not be affected at the same time. An alternative is to use infrared line-of-sight remotes.

The representative examples described below are diagrammed in the accompanying figures.

Basic Off-Air/Satellite TV Pioneer System

This system is presented for comparison with more current off-air/satellite systems that are most commonly encountered in the home satellite TV market. The one diagrammed here consists of a single dish and satellite receiver combined with off-air broadcasts (see Figure 7-25). Either an A/B switch or a 3/4 channel combiner is required to avoid interference between satellite and off-air signals. (Today most satellite receivers have a built-in A/B switch). Note that an optional VCR could have been used as an A/B switch. When its controls are set on "tuner" position, the satellite signal is blocked. In the "camera" or "audio/video" position the composite baseband signal feeds into the VCR. When the satellite audio/video signal feeds directly into the VCR and bypasses the receiver's modulator, recordings will be of higher quality than those from conventional broadcasts which must be demodulated, remodulated and then demodulated for viewing.

One line of RG-6 or RG-59 coax carries the LNB power and RF signal from the antenna to the satellite receiver. A power supply for a VHF or UHF preamplifier is required in the regular over-the-air television line if the signal is relayed from a distant source or if there is a long cable run from the conventional antenna. This amplifier boosts power levels to at least an acceptable 0 to 3 dBmv range as input to the A/B switch or channel combiner.

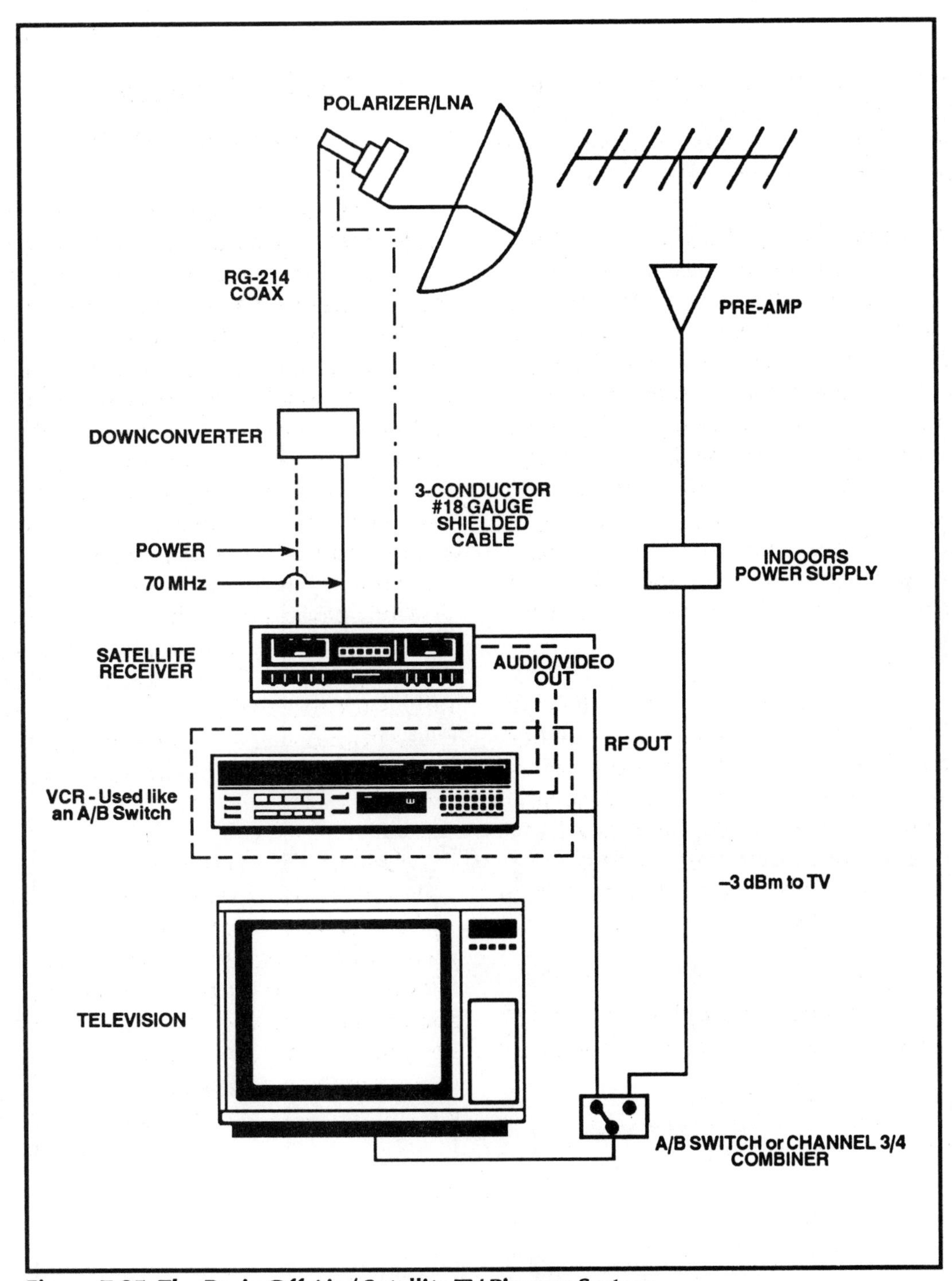

Figure 7-25. The Basic Off-Air / Satellite TV Pioneer System.

Basic Off-Air/Satellite TV System

In the current off-air/satellite TV system, an LNB relays the signal directly to a satellite receiver. The off-air side of the system again uses a pre-amp and power supply but also feeds to the rear panel of receiver. The receiver has an internal A/B switch that is used to select between off-air and satellite television signals (see Figure 7-26).

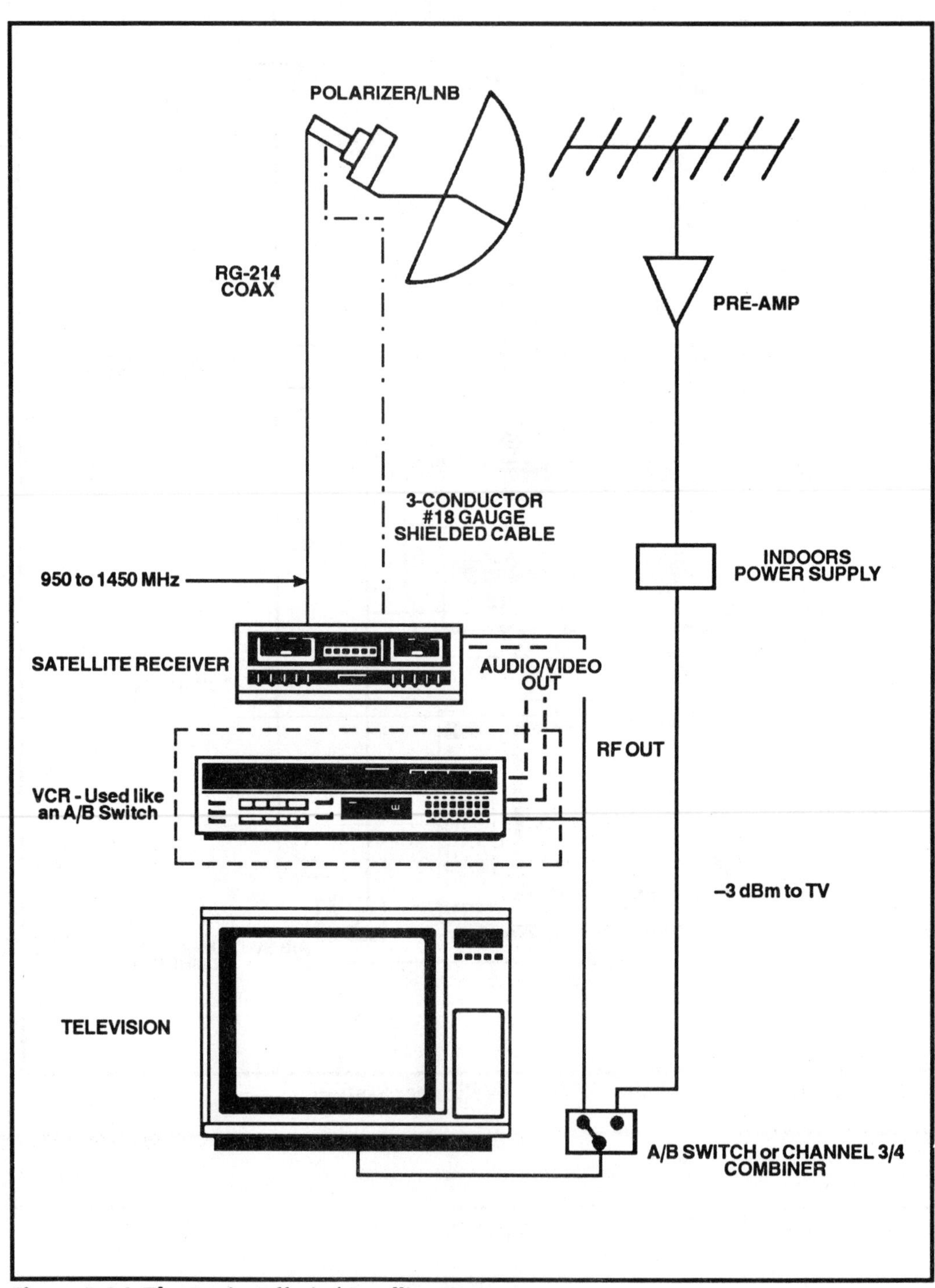

Figure 7-26. The Basic Off-Air / Satellite TV System.

Single Receiver, Two TVs and Extra Remote Control Headend

This design is similar to the basic system except that an extra TV has been added (see Figure 7-27). The main television is located near the satellite receiver which is assumed to have an infrared, line-of-sight remote control. The satellite receiver can also be controlled from the second infrared transmitter, which is located in another room some distance away but usually in the same building, by use of an extra remote control which communicates via the coax to the receiver. Using such a "remote- remote" or an "extra-link" can be very convenient for controlling the receiver from a bedroom or family room. This system can be controlled from up to four rooms.

Some brands of receivers feature a UHF radio frequency, hand-held remote. The satellite receiver can then be controlled from any location within a radius of about 60 meters (about 200 feet) without the need for extra wires. Even though this remote operates on UHF frequencies in the vicinity of automatic garage door opener range, it is rare that these two different systems interfere with each other. When extra range is needed on such a UHF remote, additional antennas can be installed by running coax from the antenna output port to remote locations via standard cable and splitters.

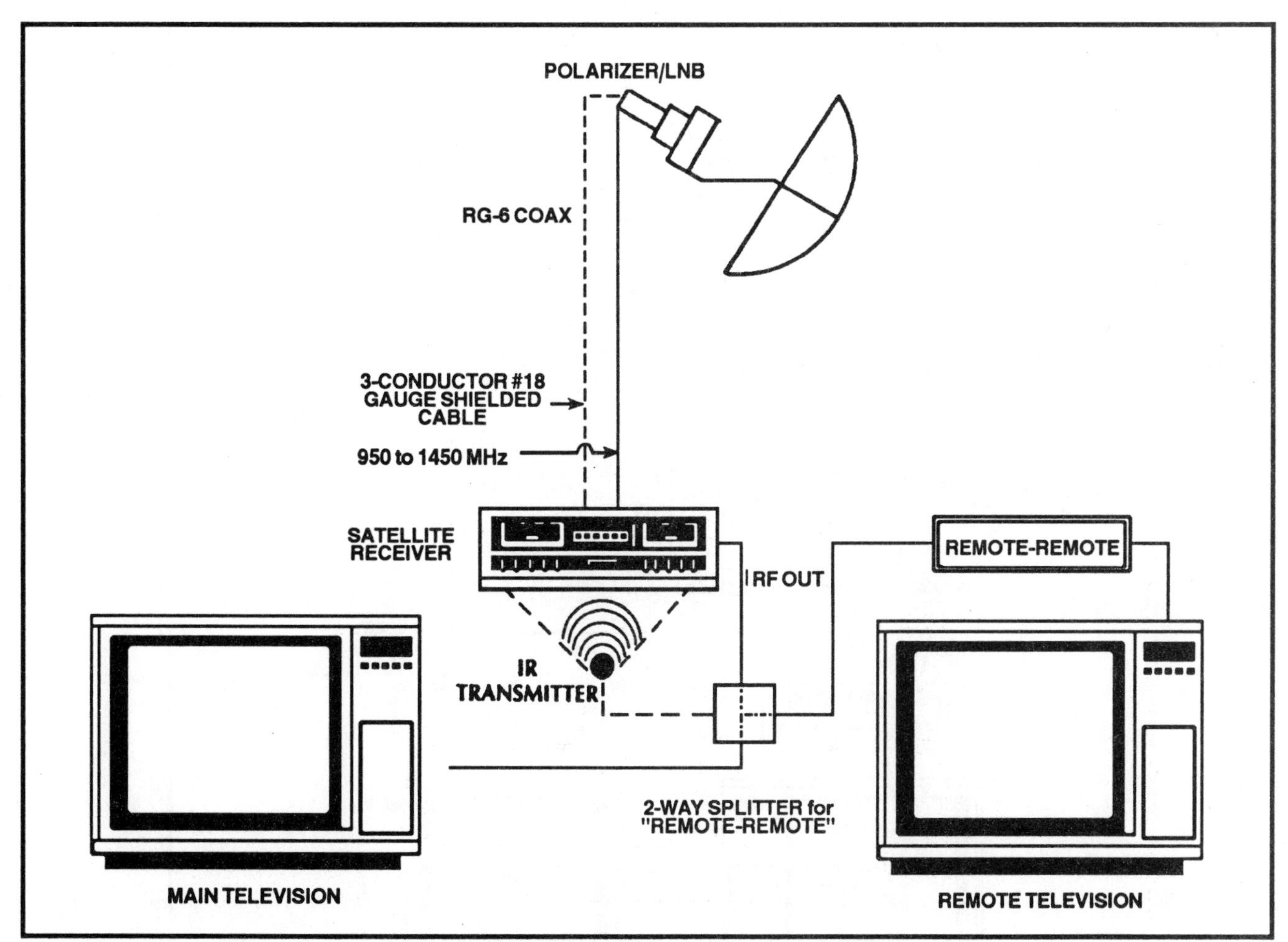

Figure 7-27. Single Receiver, Two TVs and Extra Remote Control

Dual Receivers, Single Polarity Headend

This system uses a splitter rated for an LNB output range to feed two receivers (see Figure 7-28). Usually, one side of this splitter is dc blocked so that only the master receiver passes dc current to power the LNB. Either master or slave receiver can independently select any of the channels available on the downconverted block. Only the master receiver controls the polarizer probe to select from either horizontal or vertical, or LHCP or RHCP polarizations. The slave unit is tuned to only those polarities selected by the master.

Similar systems can be designed using any number of slave receivers. All that is required is more splitters and, when necessary, a line amplifier rated for these high frequencies to boost signal strength. The master receiver must always be turned on unless an independent power supply and power insertion blocks are used. Otherwise, switching it off would interrupt power to the LNB.

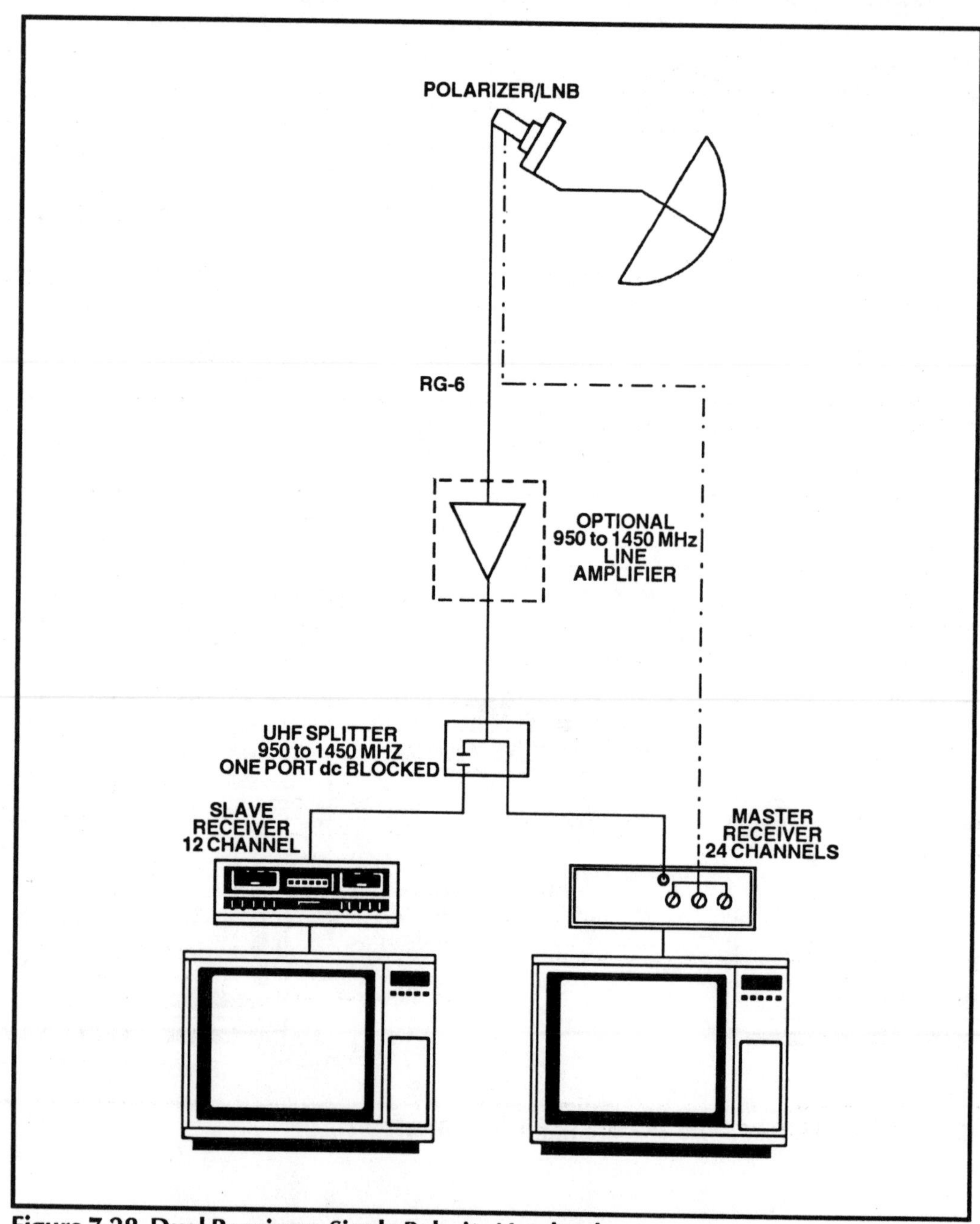

Figure 7-28. Dual Receivers, Single Polarity Headend.

Two or Four Receivers – 24 Channels Reception

An orthomode feed that simultaneously receives both vertical and horizontal signal polarities can be used to feed two or more satellite receivers. Four representative systems are illustrated here. In Figure 7-29 both LNBs are driven by external power supplies that are dc blocked from the vertical/horizontal (V/H) 2-way switch. (A 4-way switch could also have been used to drive four receivers). Each receiver relays a control voltage to the switch. Note that a V/H is the same component as a C/Ku switch.

Figure 7-30 illustrates an easier method to drive a 2 or 4-receiver system. Again a C/Ku (V/H) switch is controlled by voltages from the receivers.

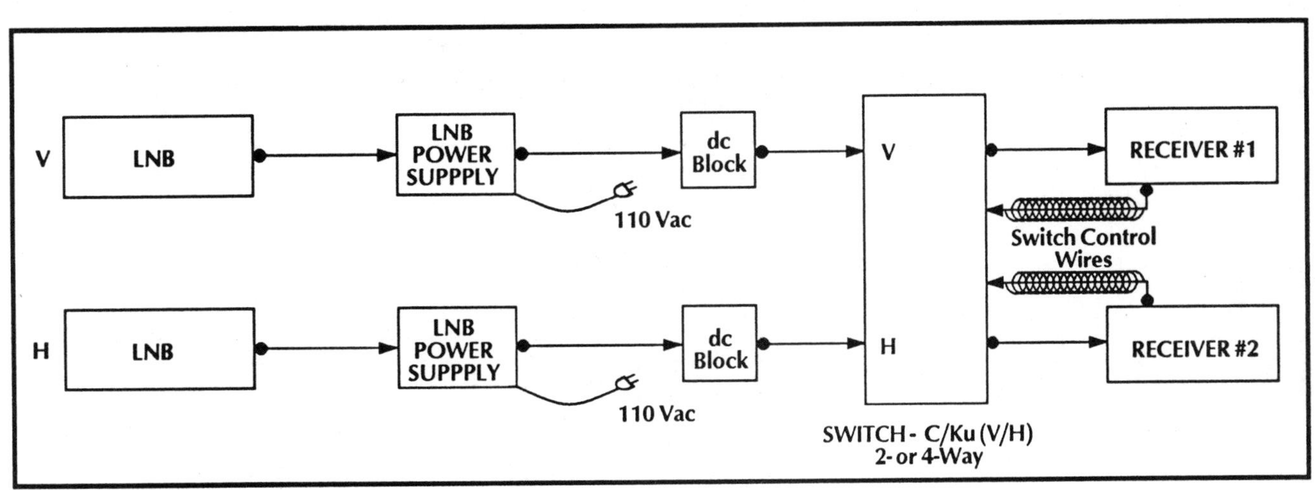

Figure 7-29. dc Blocked Two-Receiver System.

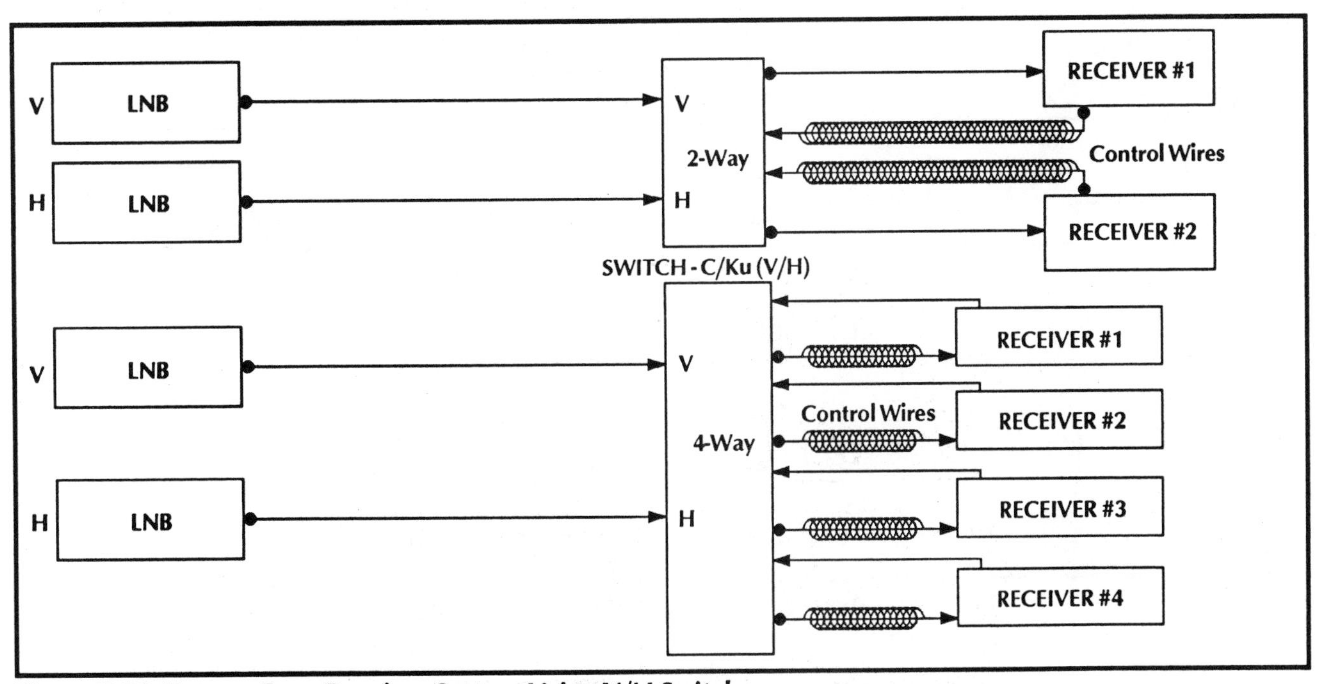

Figure 7-30. Two- or Four-Receiver System Using V/H Switches.

Figure 7-31 illustrates the use of high frequency 2-way splitters. As before, 4-way splitters could be used to drive four independent receivers. In this design, each receiver can internally switch between vertically and horizontally polarized signals.

A fourth method to control multiple receivers is the same as the third except that the H/V switches are not assumed to be integrated into the receivers (see Figure 7-32).

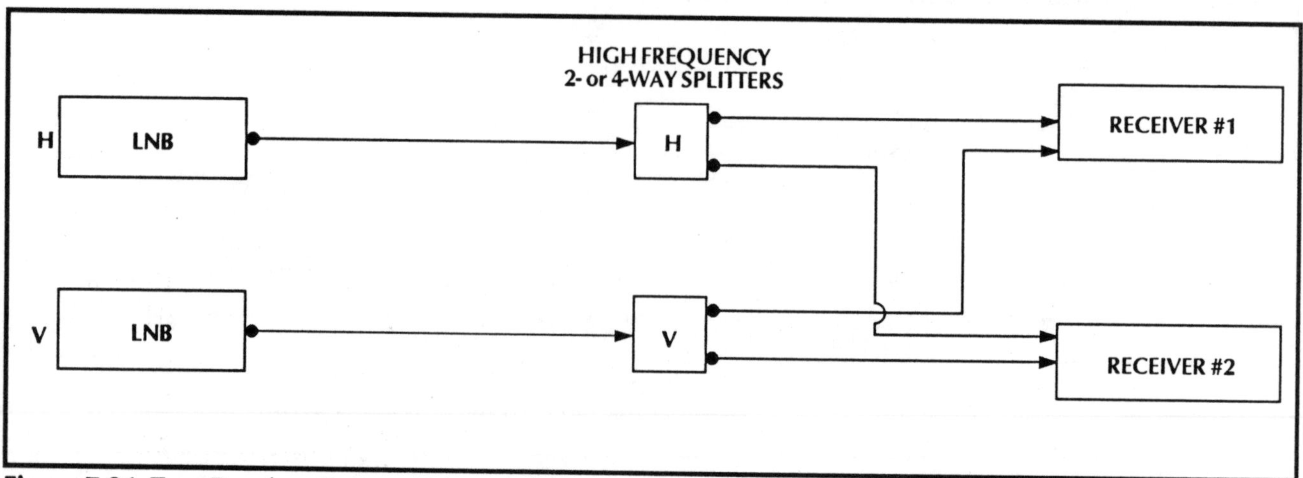

Figure 7-31. Two-Receiver System Using High Frequency Splitters.

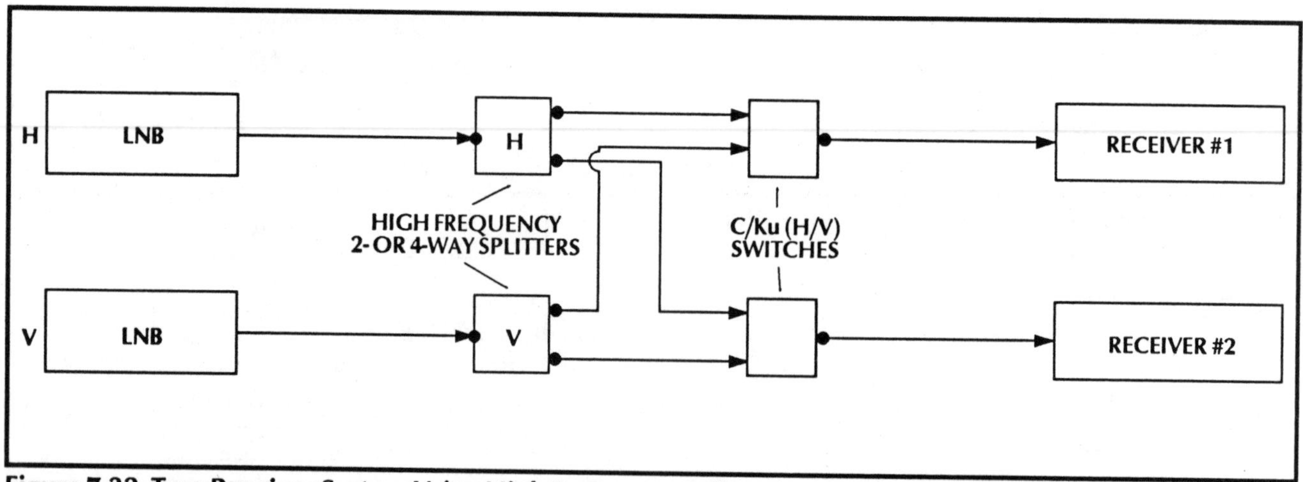

Figure 7-32. Two-Receiver System Using High Frequency Splitters and C/ Ku Switches.

Single Receiver – Dual Frequency Bands

This system allows the user to view either C-band or Ku-band broadcasts over the same receiver and TV system. Since both the C and Ku-band LNBs convert to the 950 to 1450 MHz range, just one receiver can process both signals. Here, a power passing A/B switch activates either LNB. Many modern receivers accomplish the switching internally (see Figure 7-33).

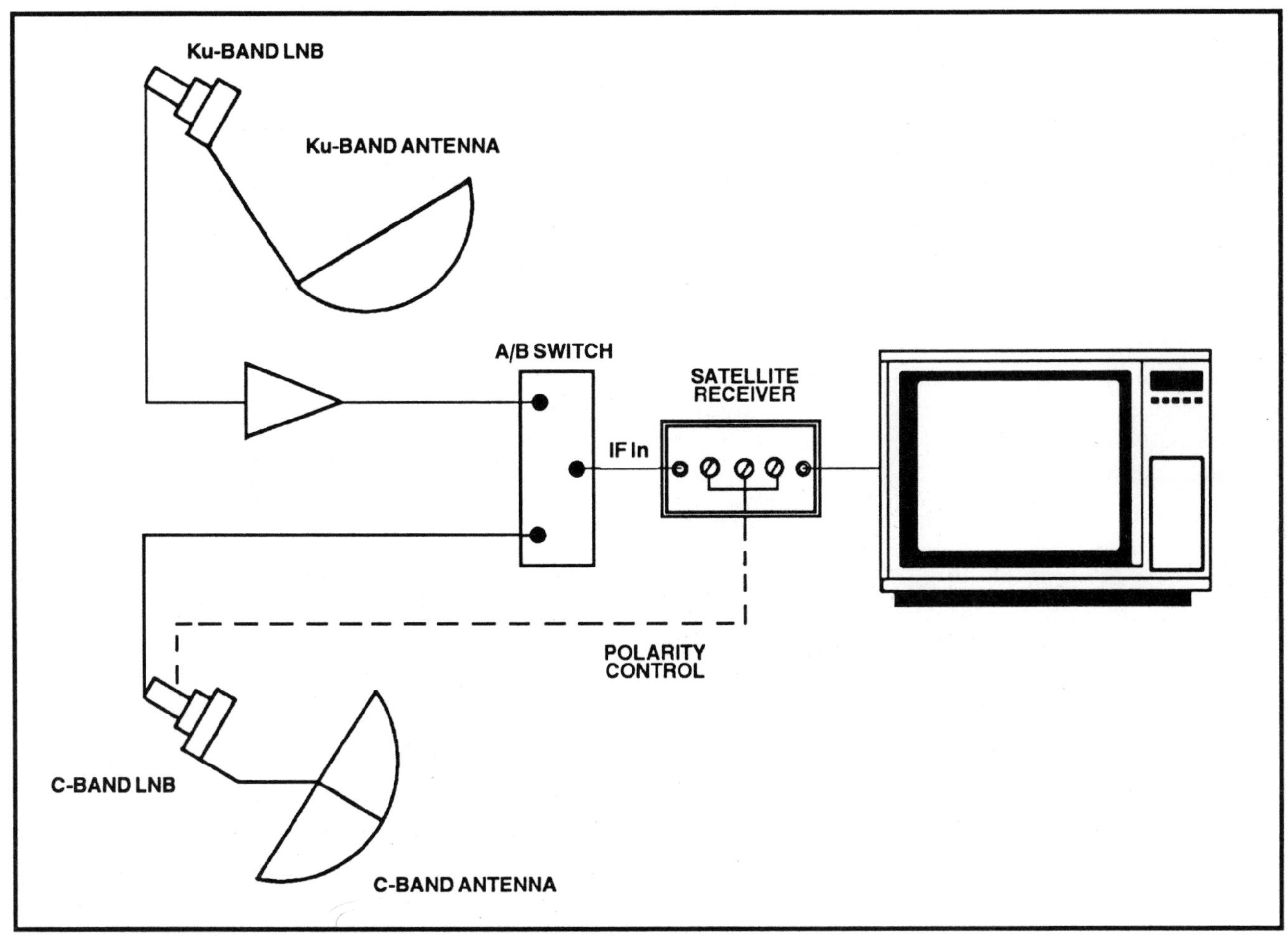

Figure 7-33. Single Receiver - Dual Frequency Bands.

C. SMATV SYSTEMS

The Headend

The objective in designing an SMATV system is to provide each television set on the network with one or more channels which have sufficient power for clear reception and which do not interfere with each other. At the "headend," signals are detected from the various broadcast sources and processed so that all channels have nearly equal power when combined on the output line. Outputs from each receiver configuration outlined above can be input to an SMATV headend.

Signals can be captured from any combination of sources including off-air antennas, VCRs or single or multiple focus satellite dishes. Commercial networks often use larger, fixed satellite antennas to deliver a signal many decibels stronger than the receiver's threshold to the headend so there is little possibility of "down time." Commercial systems are usually designed having C/Ns in excess of 15 dB while C/Ns of 11 dB are considered more than adequate for home satellite TV systems.

The Distribution Network

The distribution network is designed to take the combined output from the headend and deliver a balanced, sufficiently strong signal to each television. Master antenna television (MATV) or satellite master antenna systems (SMATV) can be compared to water distribution systems. With insufficient pressure in the system, water will just trickle out of shower heads; with too much pressure, pipes can burst and water knocks develop. When receiving television broadcasts, snow can appear if the signal is too weak and picture flutter or jitter may be seen if the signal is overdriven.

The Basic Components

An SMATV system is composed of a number of basic components including those devices outlined in section A as well as modulators, bandpass filters and taps.

Modulators, Processors & Bandpass Filters

The purpose of a modulator is to "rebroadcast" a television signal from one frequency range/format to another. For example, those modulators which are built into home satellite receivers take the baseband signal and translate it into an AM format, usually tuned to either VHF channel 3 or 4. A commercial grade modulator (see Figure 7-34) might convert the satellite TV baseband signal to VHF channel 13 or to UHF channel 36. Note that devices called channel converters take one channel and re-modulate it onto another. So, for example, a converter might take VHF channel 5 and translate in onto UHF channel 38. These are usually crystal controlled.

Amplitude modulation was originally chosen for off-air broadcasts and later for cable TV relays because signal powers were relatively high as required for AM broadcasts, and also because the transmission bandwidth must be maintained at a relatively narrow 6 MHz to permit a maximum number of channels per cable line. This system is different from that used in satellite FM broadcasts which can manage low powers and where the required wide bandwidths are necessary.

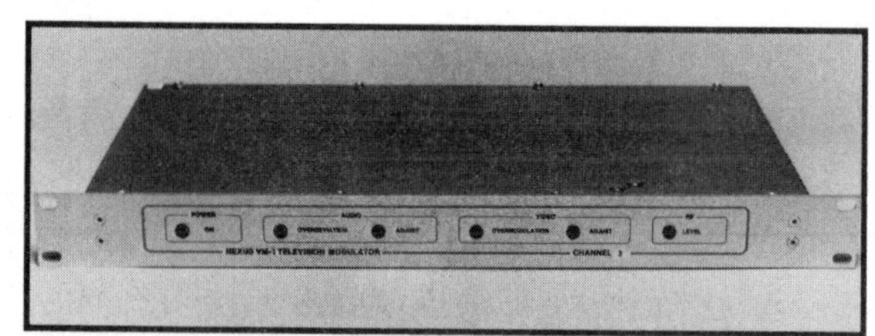

Figure 7-34. Nexus VM-1 Modulator. *This high performance television modulator with a 54 dBmv output is specially designed for small cable systems. The SAW (surface acoustic wave) filter design makes it easy to mix adjacent channels. (Courtesy of Nexus Engineering)*

The baseband output signal from a satellite receiver can be modulated onto any chosen VHF or UHF television channel. In those cases where some channels are occupied by local off-air broadcasts, satellite signals can be modulated onto any other desired frequency. The FCC allocation of all 58 TV frequency slots in the United States is listed in Table 7-2.

TABLE 7-2. VHF AND UHF TELEVISION CHANNELS

Designation	Channels	Number of Channels	Frequency Range (MHz)
Lo-VHF	2 to 6	5	54 to 88
FM Radio			88 to 108
Aeronautica	I		108 to 120
Midband	A to I	9	120 to 174
Hi-VHF	7 to 13	7	174 to 216
Superband	J to W	14	216 to 300
Hyperband	AA to WW	23	300 to 438

Modulators fall into three broad classes. "Home style" modulators which have evolved from home VCR technology usually operate on channels 3 or 4 and have low output powers, typically on the order of 0 to 3 dBmv but always lower than 10 dBmv. They do not have built-in bandpass filters to limit their output to a narrow frequency range centered on the targeted channel. This type of modulator is inadequate for commercial systems.

Commercial modulators have much higher outputs, with typically 20 dB or more of gain (see Figures 7-35 and 7-36). A distinction is drawn between those brands with and without built-in bandpass or SAW (surface acoustical wave) filters. Those not having bandpass or SAW filters can generate a signal which may interfere with an upper or lower adjacent channel (see below for a discussion of this topic).

Crystal stabilized modulators are permanently set on one channel. Many modulators used in home satellite receivers are crystal stabilized. Usually home satellite TV modulators can be switched between channels 3 or 4. Other brands also can often be switched between a different set of channels. Those modulators which are LC (an abbreviation for inductor/capacitor) driven can be tuned by a small set screw across a range of channels. Note that when this adjustment is made (for example, on the Anderson line of receivers) a non-metallic screwdriver is required since metal affects the circuit elements and can make fine tuning onto the selected channel difficult.

Figure 7-35. Maspro MDA-300 Modulator. *This microprocessor-controlled, frequency-agile, commercial modulator can be programmed for all the North American NTSC channels as well as for channels having European PAL formats. (Courtesy of USS/Maspro)*

Commercial modulators have adjustable audio and video levels as well as RF output level controls. The audio and video settings control the "percentage of modulation" or how much of these signals are added to the carrier. Higher video output makes for a brighter television picture but too

Figure 7-36. Channel Plus Multiplex Modulators. *These units are manufactured to transmit any channel between 14 and 60 in the UHF frequency band. A front panel LED displays the selected channel number. A ROM inside the unit's logic module contains the specifications for video and audio carrier selections for each UHF channel. A second model is also available which spans the entire NTSC hyperband spectrum and which can be configured for PAL broadcasts. (Courtesy of Multiplex Technology)*

high a level causes a buzz in the audio and washed out pictures. The RF adjustment sets the output power level.

Each satellite receiver in an SMATV system is set to a given satellite transponder and has a dedicated external modulator tuned to any chosen UHF or VHF television channel. These modulators are of higher quality and are more expensive than those built in to home receivers. The signals are combined at the "headend" and relayed to all the subscribers on a common cable. Then customers can select desired channels on their conventional TV sets by turning knobs or buttons.

A cable network is designed somewhat differently. Each receiver has a dedicated modulator as in a commercial system. But channel selection is accomplished by a separate device called a converter which sits next to the TV set. These channels are usually transmitted at mid or super-band and the converter modulates them to channel 3 or 4.

Channel processors also have outputs which can be combined with signals from modulators. However, processors take already modulated signals usually from off-air broadcasts and "clean up" the signal. They do this by demodulating down to baseband and amplifying and filtering both the input and output signal to eliminate the unwanted sideband. Then the output is remodulated onto the same channel as the input signal (see Figure 7-37).

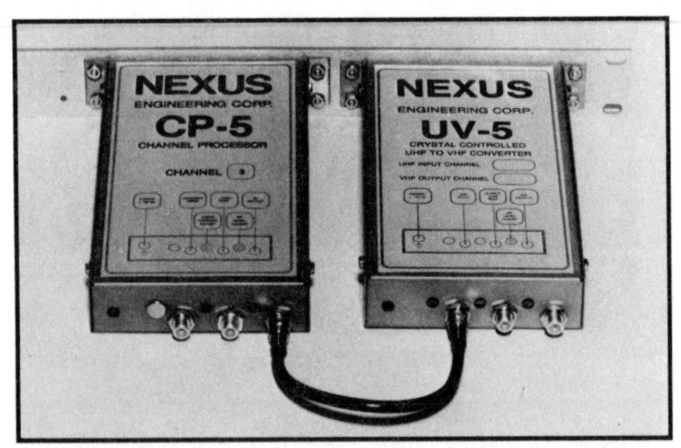

Figure 7-37. Nexus Channel Processor and Channel Converter. *The UV-5 is a low cost crystal controlled UHF-to-VHF converter used to translate UHF over-the-air television signals to the VHF band. Its output is then fed into a channel processor which filters and amplifies the signal to be combined into the mixing system for distribution. (Courtesy of Nexus Engineering)*

Taps

A tap, also known as a directional coupler or directional tap, extracts a specified portion of an incoming signal while allowing most power to pass through to its output (see Figures 7-38, 7-39, 7-40 and 7-41). For example, a 24 dB tap would take a 30 dBmv signal, siphon off 6 dBmv (30 less 24 dBmv) and pass the remaining portion less a small insertion loss of about 0.5 dB to its output leg. Table 7- 3 shows that the lower the tap value, i.e. the more signal extracted, the higher its insertion loss.

To illustrate, a 40 dBmv signal would be reduced to 39.5 dBmv after passing through a 24 dB tap but to 37.8 dBmv following a 6 dB tap. The first tap would extract 16 dBmv of signal; the second 34 dBm. Note that a tap differs from a splitter which divides a signal equally into two or more output legs. But both devices are used to accomplish the same function. Taps are usually installed to pull signals off a main feeder line since the throughput losses are much lower. Then splitters are used for local distribution.

TABLE 7- 3. TAPS and TYPICAL INSERTION LOSSES

Rated Tap Value (dB)	Insertion Loss (dB)
30	0.5
27	0.5
24	0.5
20	0.5
16	0.8
12	1.0
9	1.5
6	2.2

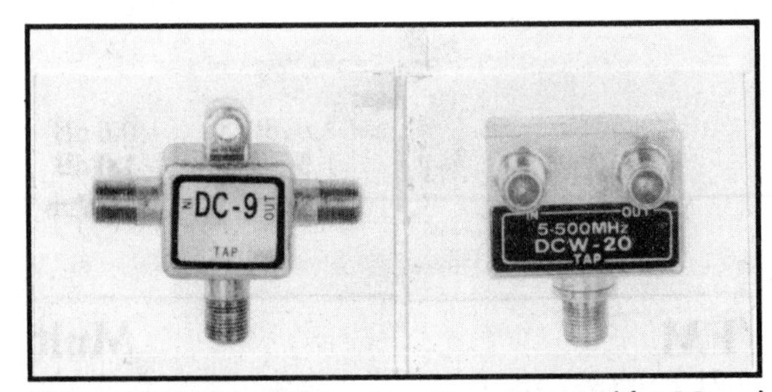

Figure 7-38. Taps. *These two taps are rated for 20 and 9 dB and have maximum insertion losses of 1 and 0.5 dB, respectively. (Courtesy of M/A COM Industries)*

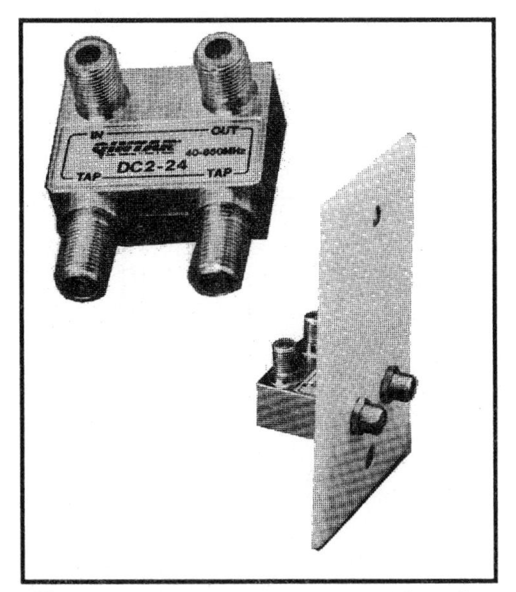

Figure 7-39. DC2 Two-Port Direction Coupler. *This tap operates in the 5 to 900 MHz range and is available in 12, 16, 20, 24, 27 and 30 dB values. (Courtesy of Qintar, Inc.)*

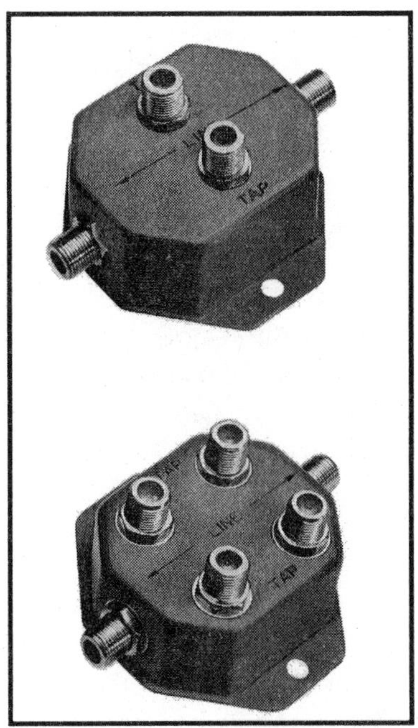

Figure 7-40. Multiport UHF/VHF/FM Tap. *This tap operates in the 5 to 890 MHz frequency range and is ac/dc isolated on its output ports. It is available is 12, 17 and 23 dB tap values with insertion losses at VHF frequencies of 1.5, 1.3 and 1.0 dB, respectively. (Courtesy of Qintar, Inc.)*

Figure 7-41. Indoor/Outdoor Direction Coupler. *The Model DC-4 directional coupler operates in the 5 to 900 MHz band and is available in six tap values. (Courtesy of Qintar, Inc.)*

Television Channels and Balancing Signals

Whenever signals are modulated onto two adjacent channels there is always the possibility that interference will occur. In the United States the FCC wisely allocated sets of non-adjacent, off-air broadcast channels in each region or metropolitan area of the country. Thus, for example, New York could have channels 2,4,6,7 and 9 in use while the Philadelphia area TV stations broadcast over channels 3,6,8,10 and 12. (See below to understand why channels 6 and 7 are not adjacent to each other.)

Today, SMATV systems may be designed to relay, for example, 5 off-air channels along with 4 satellite TV broadcasts onto the VHF television band. Headends therefore process and transmit adjacent channels and must be designed so each one is received without interference from all others. The UHF band can also be chosen as a target for modulation but cabling losses are higher than in the lower frequency VHF band. Therefore, extensive distribution systems will often be designed to use the VHF range.

Each television broadcast has its video, color and audio information organized as shown in Figure 7-42. If the audio signal power is too high, it can bleed into the upper adjacent channel video and cause a cross hatch or pattern over the picture. If the video bandwidth has not been cut off or filtered below the lower channel edge, it can bleed into the lower adjacent channel and cause scratching sounds and "cross-talk" on its audio output. Therefore, whenever adjacent channel modulation is used, two rules must be followed. The signal must be restricted by a bandpass or SAW filter to protect the lower adjacent channel and the audio level must be 15 dB below the upper adjacent video channel.

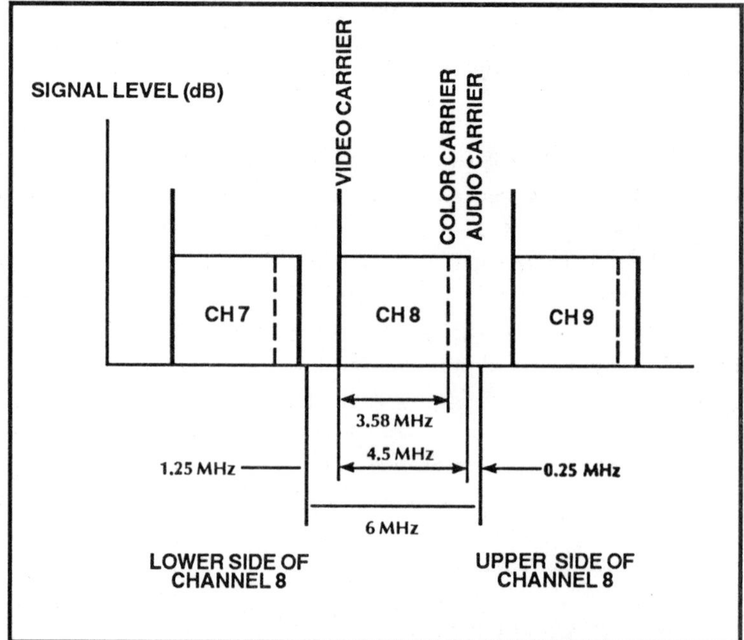

Figure 7-42. Layout of Audio and Video Channels. *Each television channel is organized into a video subcarrier centered 1.25 MHz above the lower side of the channel, a color signal centered 3.58 MHz above the video carrier and an audio subcarrier at 0.25 MHz below the upper edge of the channel.*

Table 7-4 shows which channels should be considered adjacent in the VHF band. Since there is a large space between the VHF low band (2 through 6) and VHF upper band channels (7 through 13), channel 6 does not have an upper adjacent and channel 7 does not have a lower adjacent. Note that fortunately for system designers there is also a frequency space between channels 4 and 5, a 4 MHz guard band which is occupied by marine radio communications.

TABLE 7-4. VHF TV CHANNEL ALLOCATIONS

Channel Number	Frequency Range (MHz)	Upper Adjacent	Lower Adjacent
2	54-60	3	none
3	60-66	4	2
4	66-72	none	3
5	76-82	6	none
6	82-88	none	5
FM Band	88-108		
7	174-180	8	none
8	180-186	9	7
9	186-192	10	8
10	192-198	11	9
11	198-204	12	10
12	204-210	13	11
13	210-216	none	12

This table has valuable information for satellite TV technicians. For example, assume that a small motel wants to send two satellite channels to each room along with its off-air broadcasts. If channels 2,4,7 and 9 are occupied, satellite TV could be modulated onto channels 11 and 13 with no trouble. But what if an off-air station is already relaying a broadcast on channel 11. A good choice would be to modulate the satellite signal onto channel 6 which has no upper adjacent channel as well as onto the free channel 13. This sensible choice would save time and money in designing a headend. Less expensive modulators and signal balancing techniques could be used since non-adjacent channels were chosen instead of channels 5 and 6.

Note that a precise balancing of signal levels is not nearly as important in systems which do not use adjacent channels. Systems should be designed, if possible, with non-adjacent channels as targets for modulation. When adjacent channels are used, however, high quality modulators which "clean up" the audio and video signal and which provide filtering are a must.

Combining and Distributing the Signal

Once each output from each modulator has been equalized in power level by use of any necessary line amplifiers or by lowering the RF power level and the necessary bandpass filtering and processing has been accomplished, signals can be combined. Note that the signal levels can easily be measured by using a field strength meter. This is an essential tool for any SMATV installer. Levels can also be calculated in advance of installation. For example, if a modulator with 40 dB output is drawing a -3 dBmv baseband signal from a satellite receiver, it will provide 37 dBmv to the distribution system. If an adjacent channel is being relayed at 32 dBmv, in order to set the level of the first to an equal level either a 5 dB pad could be inserted or the RF level adjustment could be lowered by the appropriate amount.

Figure 7-44. Scientific Atlanta Headend. *This photo shows four satellite receivers, four B-MAC decoders, four channel modulators and one spare. This configuration receives GStar 1 broadcasts for the Holiday Inn High-Net system. (Courtesy of Scientific Atlanta)*

Note that rooms which house headend equipment must be maintained at a relatively constant temperature (see Figures 7-43 and 7-44). Modulator outputs can increase by more than 10 dB if the ambient temperature drops from summer to winter. Humidity must also be controlled since excessive amounts of water vapor will cause electronic components to degrade. Signals can be combined by using either splitters or taps in reverse. Thus, for example, a 3-way splitter could take three separate signals and combine them into one stream. Or a network of taps could take signals and add them together. Insertion losses will be the same regardless of whether these splitters are used in a reverse or forward direction. Devices called combiners are specially designed for this purpose.

Figure 7-43. Nexus Compact 12 Channel SMATV Headend. *The pre-packaged headend of 12 channels include six satellite receivers, six modulators, and six channel processors. It requires a total of just 12 by 19 inches (30 by 48 cm) of space. (Courtesy of Nexus Engineering)*

Similarly, two types of distribution systems can be designed using either taps or splitters. In general, splitters are used for smaller systems. Designs ranging from larger SMATV systems to enormous cable TV networks use the trunk line and feeder cable with taps. A main trunk line is never tapped but uses only splitters and amplifiers so that the distribution system will be well protected from ingress interference. Then taps are used in feeder lines. For example, if a hotel having two small 5-room buildings was being cabled, a 2-way splitter might be used to feed each building. Then either a splitter or a tap network could be used in each building to feed each of the five rooms.

The objective in designing a distribution system is to provide from 0 to 3 dBmv of signal to each television. Once the signal power leaving the headend is known, it is simply a matter of subtraction to calculate the losses at each tap, splitter or pad in each cable run to ascertain that adequate signal reaches each set.

Example of SMATV Headends

Three examples of SMATV headends are illustrated below:

4 Channel Headend

This SMATV headend may be located at a hotel in an area where no off- air TV is available. Four satellite TV channels, two horizontally and two vertically polarized, are being distributed. Note that the even and odd channels from the block downconversion receivers are combined in reverse 2-way splitters before being added together on one line. The system is designed this way to protect adjacent channels from interfering with each other. Note that since channels 4 and 5 are not adjacent in frequency, lower cost modulators without bandpass filters could be used, but this is not advised.

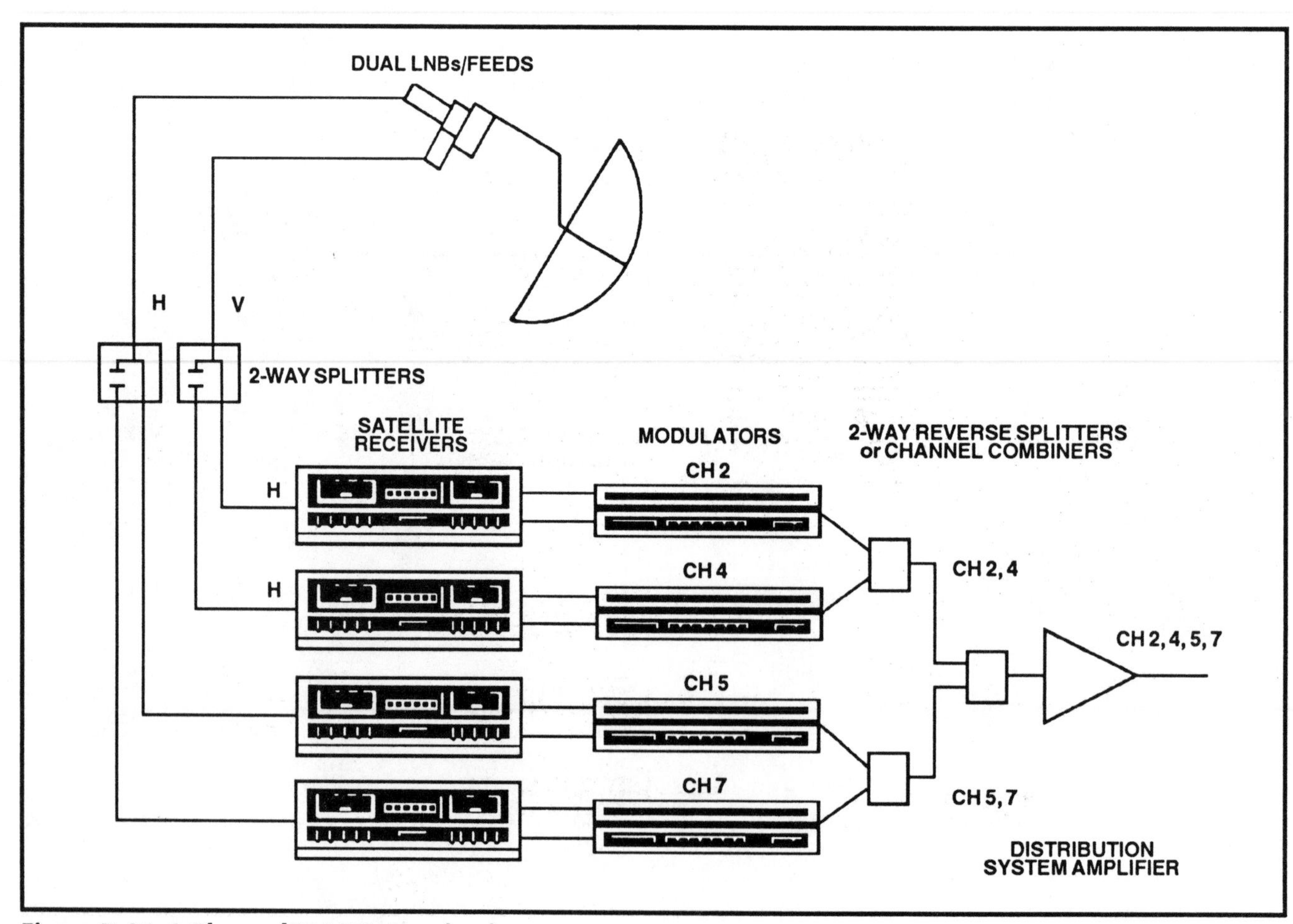

Figure 7-45. 4-Channel SMATV Headend

7 Channel Headend With Off-Air TV

This 7 channel headend combines four satellite TV channels with 3 off-air stations into one distribution network. In this case, care has to be taken in protecting channels 8 and 9 from interfering with each other by properly balancing signal levels and by using a bandpass filter either as an extra component or built into the channel 9 modulator.

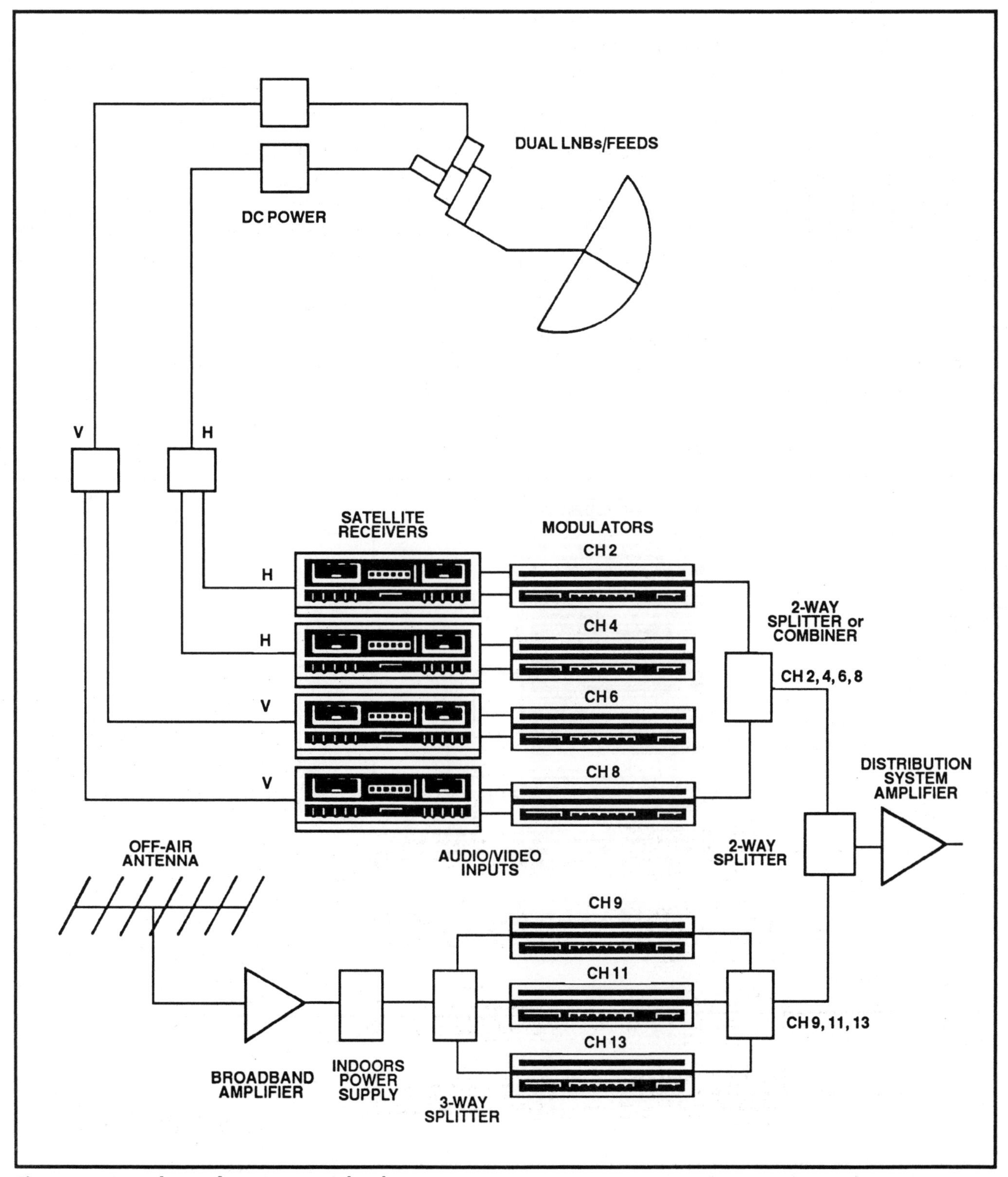

Figure 7-46. 7-Channel SMATV Headend

12 Channel Headend

This 12 channel headend combines all the available VHF channels. Three 4-way splitters used in reverse, sometimes also labeled as combiners, feed their outputs into a 3-way combiner and subsequently into the distribution network.

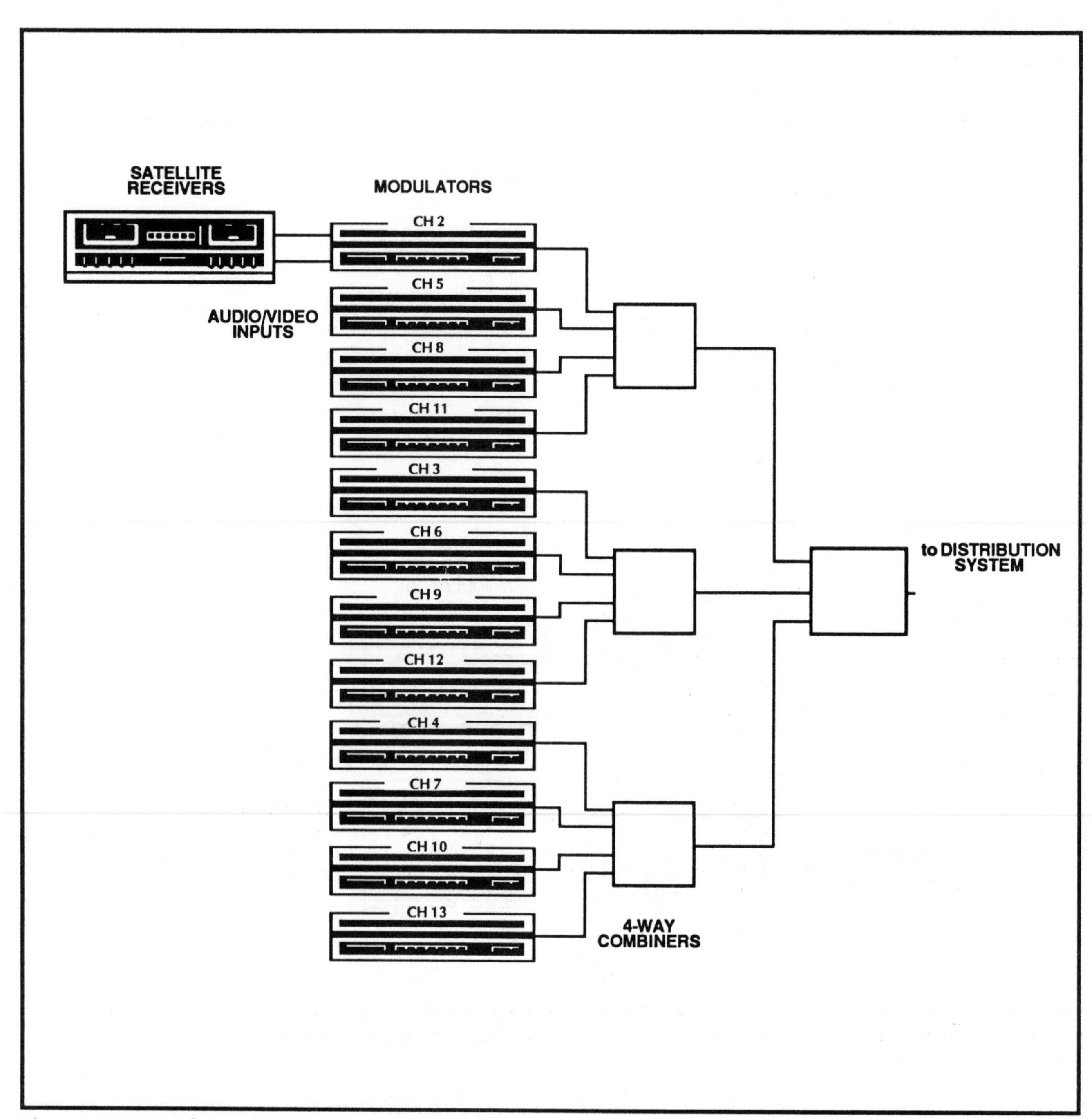

Figure 7-47. 12-Channel SMATV Headend

Examples of Distribution Systems

Two simple distribution systems, one using taps and one using splitters are shown below. These diagrams clearly demonstrate how power levels can be traced to their final destination.

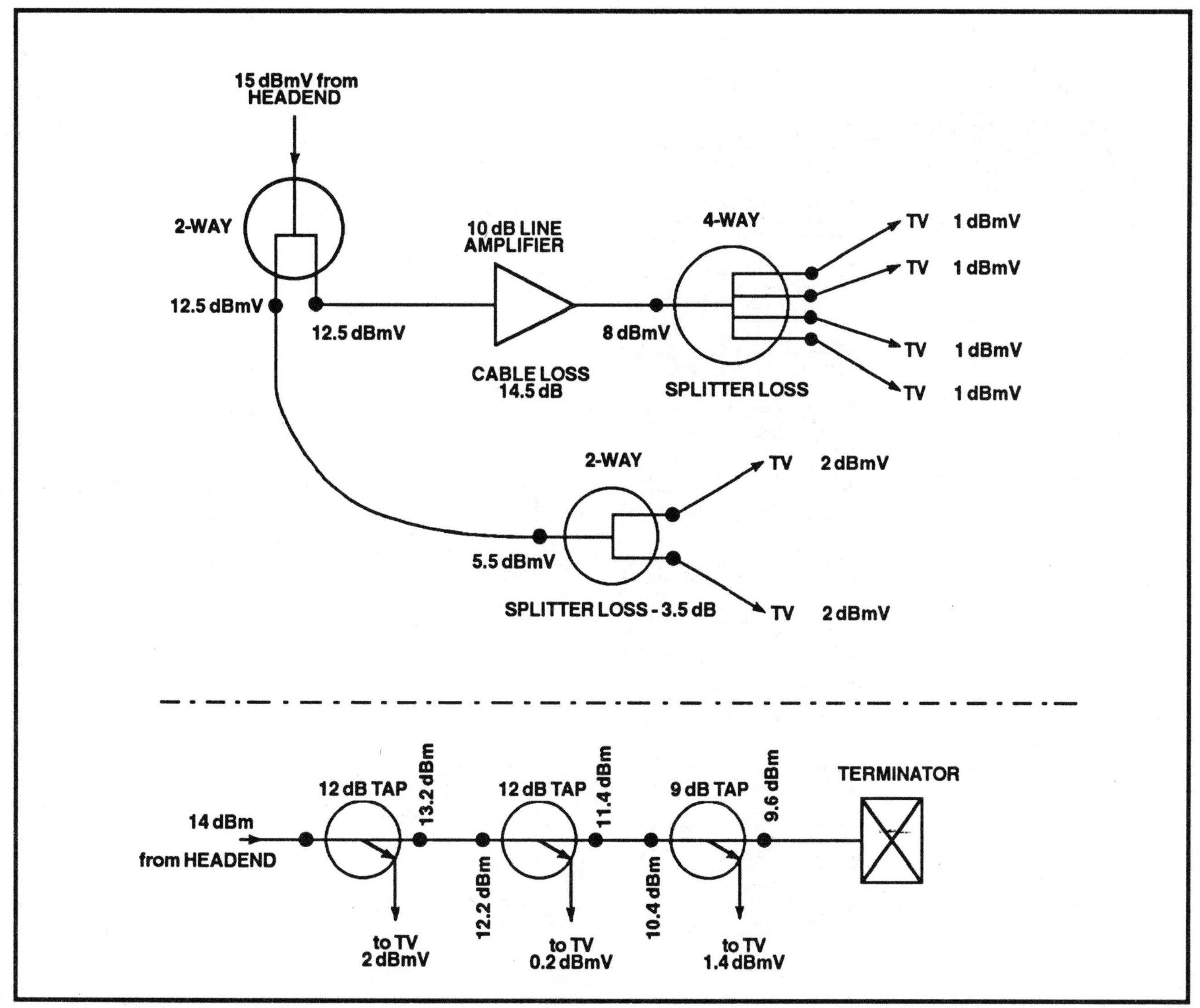

Figure 7-48. Sample Distribution Systems. *These two examples illustrate how power levels are calculated for both tap and splitter distribution networks.*

VIII. TROUBLESHOOTING & REPAIR

A. INTRODUCTION

It is inevitable that every home satellite system develops operational problems of one sort or another. When this occur depends to a large extent upon whether equipment was properly selected and installed. For example, poor weatherproofing is the most common installation error. Although it is clear that no man-made device lasts forever, all problems can be solved by thoughtful troubleshooting. The purpose of this chapter is to provide a consistent method of solving any difficulties that will eventually arise. Finding out what is wrong, the diagnostic procedure, can in itself be an interesting and rewarding exercise if a correct approach is followed.

A competent technician should expect breakdowns and factor service and repair calls into the sales price. It is important to realize that the cost for a service call including travel time, wear and tear on the service vehicle, fuel plus a normal hourly rate can be substantial. Clearly it is wise to have a clear and effective approach to troubleshooting and repair.

Troubleshooting can be relatively easy when only one component of a system goes bad. This can often happen a short time after an installation has been completed. Then diagnosis and repair can generally be quite simple, especially if the service call is made by the same technician who installed the system. The installation documentation should facilitate locating the fault. Unfortunately, if another person installed the system, diagnosis may be more difficult. Some outrageous mistakes have been discovered in servicing home satellite systems which were poorly installed.

However, most problems often arise over a longer period of time when several unrelated, minor faults that result in poor pictures simultaneously develop. For example, connectors can corrode, concrete may settle and throw a pole slightly off from a vertical orientation, a center conductor in a coax may slightly contract and not make perfect electrical contact, the antenna may have sagged slightly or shifted from its north/south alignment, or the feed may have moved noticeably over time. Eventually a customer may telephone to complain about poor reception. As a result, a check of antenna tracking, feed alignment and all connections would be required.

Troubleshooting during an installation can be much more difficult, especially if proper procedures have not been followed from the start. These "proper procedures" are very important in order to avoid future headaches. To illustrate, assume that the components were not tested before installation and the LNB was faulty and that a cable which had already been buried also had an internal short. The problem with the LNB would have been easily located if it had been tested before installation. But once the installation is completed, trying to separate the two problems without going back to square one could be difficult.

It is clear that as the number of installed systems continues to grow and as equipment in the field ages, the troubleshooter's role can only become more important and more profitable. In fact, during the brief but dramatic turn-down in the home satellite television business in North America during 1986 following the initial scrambling of signals by HBO, many dealers survived by offering a professional repair service. The hallmark of a successful troubleshooter has been and will continued to be an approach that combines technical knowledge with a systematic and methodical process for diagnosing, isolating and repairing TVROs.

Troubleshooting can be interesting. Every seasoned expert has some unusual and often amusing stories to relate. For example, a satellite dealer had a customer who recently had purchased a home with an existing mesh dish. This person was a collector of exotic birds. He consulted with another dealer who supported with the idea of building a wire cage over and around the dish to house the birds. Not only did reception disappear, but the birds virtually demolished the mesh dish.

B. A SYSTEMATIC APPROACH TO TROUBLESHOOTING

Diagnosing an operational problem in a satellite TV system should follow a logical path to a rapid solution. The strategy is to take the simplest route having the highest probability of solving the problem before proceeding with more complicated steps. However, even these more complicated steps can be approached with a similar perspective of tackling the most quickly solved problems.

The first step in a systematic approach to troubleshooting and repair is an in-depth interview, preferably over the telephone, in order to identify the problem. Of course, this interview can be successful only if the underlying theory is clearly understood. A technique known as the subsystems approach can be quite useful in isolating problems in any satellite system. When necessary, the next step would be a visit to the installation site for a visual inspection. If this simple inspection does not isolate the problem, a series of technical troubleshooting procedures can be followed to eliminate the fault.

None of these procedures is magical or complicated. All that is required is knowledge of how a satellite TV system works and a clear, level-headed approach. No problem is too difficult to solve and, with experience, solutions come more quickly.

The Subsystems Approach

Every satellite TV system can be broken down into three subsystems: the mechanical, consisting of the mount, antenna, feed support and feedhorn, the electromechanical consisting of the actuator and the polarizer, and the RF consisting of the LNB, satellite receiver and TV (see Table 8-1). The modulator that is incorporated in a home satellite receiver is most often a separate component in a commercial system. Each of these subsystems can be diagnosed somewhat independently even though symptoms may be similar in nature. Of course, the only reason for defining and isolating these three subsystems is to simplify troubleshooting and repair procedures.

TABLE 8-1. SATELLITE TV SUBSYSTEMS

Mechanical	Electromechanical	RF
Mount	Actuator	LNB
Antenna	Polarizer	Receiver
Feed Support		Television Set
Feedhorn		

The Diagnostic Interview Identifying the Problem

A telephone conversation between the customer and an experienced technician can often solve problems and eliminate the need for many service calls. This is especially true during the initial few weeks or months of ownership when customers are not familiar with many aspects of their systems. Troubles can stem from a television set that is incorrectly tuned to the modulated channel or from a customer's inexperience with a handheld remote or the operation of the parental lockout feature on a receiver.

In particular, a new owner generally has expectations about how the pictures should appear and would certainly notice any problems or irregularities. Complaints about poor picture quality are most typical.

The interview can also identify symptoms of a problem that may require a service call. Such information can prove extremely useful during the initial diagnosis as well as in assembling the proper equipment to isolate and repair the system. For example, if the customer was out tilling his garden when the satellite system failed, certainly bring some extra cable and connectors. There is nothing worse than driving out to a site and then having to return to the shop for one small forgotten component such as an extra fuse or voltage regulator. If adequate documentation about an installation is on file, questions can be guided by knowledge of the particular system. The documentation should include information about antenna size and make, the location of cable runs, the type of electronics installed and the other factors outlined in Chapter V.

A satellite reception system can have three general categories of symptoms: completely inoperative; intermittent where the operation alternates between normal and bad; and unusual operation where out-of- the-ordinary behavior is exhibited. The questions can be guided with these categories in mind.

In summary, the interview is an important first step before anything else is done. Some sample questions are:

- When did the trouble first start?
- Was it a gradual change or a sudden occurrence?
- Describe the problems.
- Are these difficulties constant or do they occur at regular or irregular periods during the day or night? Perhaps the difficulties are simply solar outages (see Figure 8-1).
- Was the equipment recently moved or has any component been replaced?
- Which channels and how many channels are coming in clearly?
- Can you receive all the satellites?
- Does the antenna move easily from satellite to satellite and does it make any unusual noises in doing so?
- Have you possibly cut or disconnected any cables by mistake?
- Does picture quality vary from channel to channel or from satellite to satellite?
- Is the audio or video most affected?
- Has there been any recent construction in front of or near the antenna?
- Did the problem first occur after a storm and did it persist when the weather cleared up?
- Do your problems occur more during rainstorms than on clear days or more during hot than during cold weather?
- Is it possible that a fuse has been blown in the receiver? Do the lights come on? Could a circuit breaker in the house providing power to the receiver have been tripped?
- Have there been any children playing around the antenna?
- Did you recently dig a garden or do any other excavation work?
- When the wind blows can you hear any part of the antenna rattling?

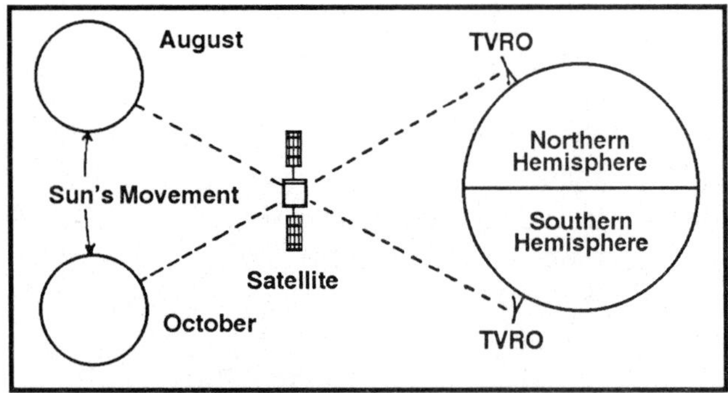

Figure 8-1. Solar Outages. *The timing of solar outages is directly related to the site latitude. They occur when the sun lines up directly along the antenna boresight. Solar outages always take place within 3 1/2 weeks of the equinoxes on March 21 and September 21, when the sun crosses the equator during its annual north/south journey. Solar outage in February, March and April begins at the northernmost latitudes and moves southward until satellite systems on the equator have outages at equinox time. Then earth stations in the southern hemisphere begin having outages until those at southernmost latitudes experience them 3 1/2 weeks after the equinox. In August, September and October, the entire pattern is reversed because the sun is moving in the opposite direction. Outage season lasts for a little over a week in any given location. During this period, there will be an outage daily on each satellite. It appears first as small amounts of video noise, rapidly becomes annoying interference and, on the days at the center of the outage season, peaks out as a total loss of incoming signal. Antennas should be moved away from this satellite during outages in order to avoid heat build-up on the feedhorn and LNB. Spun aluminum and shiny surfaced fiberglass antennas can raise the temperature at the focal point sufficiently to cause severe damage and possibly destroy components.*

Can your television receive over-the-air channels?

A competent troubleshooter is like a doctor. The patient often knows something is wrong but simply does not have the vocabulary to express it. A well conducted interview will gather useful and important information. With a little experience, a complete list of questions can be created so that such an interview will proceed smoothly.

Many problems can be quickly solved by this technique. For example, a customer may complain that pictures are perfect except during heavy rain or snow storms. This might be caused by rain fading because the system was under-designed. Or a wet connector or leaky gasket between the feedhorn and LNB may be at fault. Also, for example, if every transponder went bad after a severe lightning storm, the cause could likely be an LNB with a blown stage or simply a blown fuse or circuit breaker.

Then there are occasional satellite problems or sun outages that happen infrequently enough that people forget that they are even possible. Sun outages occur twice a year during the spring and fall equinoxes when the sun lines up directly behind the targeted satellite and its RF radiation completely overpowers the broadcast for twenty to thirty minutes. Some technicians are simply not aware of these occurrences (see Figure 8-1). This underlines the need to have first hand acquaintance with satellite TV by owning an operational system for comparison purposes.

C. VISUAL INSPECTION

Once the customer interview and a thoughtful evaluation of the assembled facts are completed, a visual inspection may be required. This is especially important if the system was installed by another technician.

The visual inspection can be guided by symptoms of the problem. For example, if the customer reports that he can receive television from only one satellite, the first step would be to check the actuator and its control box to see if the antenna is stuck in one position. If all channels are received with crystal clarity on some satellites but poorly on others, the first visual check would be to determine if the reflector is warped, if the feed is moving off the focal point as satellites are being tracked, or if the north/south alignment has been disturbed.

Figure 8-2. Antenna with an Obstructed View. *This antenna was installed in the middle of dense foliage. Although it may not have been completely screened during the winter when the surrounding trees had no leaves, chances are excellent that few satellites in the arc can be clearly received in the summer.*

It is important to realize that many symptoms can arise from the same problem. It is advisable to conduct a visual inspection at first when a not-so-obvious symptom is encountered than to jump into a more technical troubleshooting approach or to immediately start exchanging components which might create additional problems. Symptoms will usually give an indication of where to start the visual inspection. For example, if picture quality is poor, it would certainly make more sense to first try moving the antenna east and west with the actuator or a hand crank than to start measuring voltages. A second sensible step would be to check for loose or corroded connectors and a third one would be to check that the feed is properly centered and the antenna is correctly aligned. Perhaps the mount has rotated in a strong wind because the bolts on the collar were not tightly secured or pinned. Sight with a compass to be sure that the polar mount is still properly oriented in the north/south plane.

The basic questions below show the thought process underlying a simple visual inspection. Many more questions that may arise are presented in the discussion of techncial details in the following setions.

RF Subsystem

In general, the visual inspection should begin at the system control center, the satellite receiver/television set. Turn on the TV and what do you see?

1. Does the television work properly on over-the-air channels?

2. Is the picture completely snowy or covered with "bug-fight" sparklies? If there is a weak picture in the background perhaps the satellite is targeted onto a side lobe.

3. Is the receiver turned on and are all the indicator lights working? Maybe a rear panel or internal fuse has blown or a 120 or 220 Vac circuit breaker in the house has been tripped.

4. Is the television tuned to the correct channel that matches the modulator output?

5. Have any cables behind the receiver been disturbed? Perhaps the coax was improperly crimped and the cable is simply disconnected.

6. Are any other wires loose or disconnected? If this is a new installation make sure that all the wires are well attached to the proper place before powering on.

7. Is there a high signal strength meter indication but no picture? This would suggest that the modulator may be bad or that the TV set is not tuned to the right channel.

8. Is the receiver shutting off because it is overheating? Most receivers have ventilation slots that allow cooling by natural air flow. If heat is not carried away, the voltage regulator that generates substantial amounts of heat might be internally compensating and shutting down.

9. Are all the cable runs intact? Check all the cable routes from indoors to the antenna. Perhaps some over-enthusiastic gardener has severed the cable with a weed-eater, shovel or lawnmower. Maybe a dog or rodent has chewed through the cable.

10. Are all the connections at the antenna secure? Make sure that all connectors are tight and that no water has leaked in. Are drip loops installed in the correct places? Maybe a connector on the LNB has corroded. When a cable is used to carry power, water entry can cause corrosion similar to the build-up that can be seen on a car battery terminal. Cut the cable back and install a new connector past the point where the corrosion ends. Make sure that coaxial sealant has been used. Clean any suspected connectors with a solvent like carbon tetrachloride, dry them thoroughly and re-seal. Also, if a defective connector has a slightly short center pin, it may perform well until cold weather arrives. The cold could cause this lead to shrink and break contact. This phenomena is appropriately known as center connector suck-out. If the center conductor is too long, it may split the shroud on the center pin when inserted into an LNB or receiver and cause it to short out against the inner connector wall.

11. If an LNB/feed cover has not been used, check to see if any water is collecting in one of the waveguides. If a polarizer has been used with the right angle waveguide fitting, water may also have collected and a cure might be to drill a small 1 or 2 mm drain hole on its underside. If the gaskets between the various flanges have been omitted this problem may be worsened.

12. Does the system work intermittently? Sometimes a bad LNB functions only when it is cooled down to the point where its gain increases. A test for this malfunction is to use a hair drier to warm it up on a cold day or to use a can of freeze spray on a warm day to bring it back to life.

Electromechanical Subsystem

1. Does the antenna respond to the actuator? Maybe a voltage transient from a nearby lightning hit fried the Hall effect or reed switch sensor. If the sensor leads were not shielded, the actuator controller may be miscounting and aiming the antenna incorrectly. Possibly water has entered the arm and has frozen or rusted the internal assembly so that it does not move.

2. Make sure that the motor is pointed up so that the drain holes are pointed down. If there are no drain holes, drill some. Maybe the gasket between the motor cover and housing has been omitted.

3. Is the actuator tube bent or is there wear or scoring on the inner tube? Perhaps the original installer did not use ball joints on the actuator arm supports in order to minimize lateral force on the tube or did not set both mechanical and electrical limits to prevent bending.

4. Is the polarity adjustment having any effect on picture quality? If it is a mechanical device, is the probe moving when channels are changed or when the skew adjustment is altered? Check to see if obstructions caused by moisture collection, birds or even wasps are interfering with signal reception in the throat of the feedhorn.

5. Is the skew adjustment having little or no effect? Maybe the probe has been bent or mechanically distorted by freezing and thawing water. Make sure that the waterproofing was ensured by correctly installed gaskets.

6. Is the probe motor bad? Have a spare handheld controller because the 555 timer in the actuator controller or receiver may be at fault, not the motor. The 555 timer chip is a basic component of most actuator controllers.

7. Is the motor servohunting? This means that the motor is driving the probe back and forth un-

successfully looking for the maximum signal. This can cause hum bars visible as wide horizontal regions of disturbances rolling across the screen. Have a 10 Vdc, 1000 microfarad electrolytic capacitor on hand and install it on the polarizer motor at the antenna. Also, check that the wire feeding the motor is of sufficient gauge and thus thick enough to avoid this problem. When using two servo motors in a parallel dual-band configuration, a capacitor should be used to provide an extra current when simultaneously switching both probes.

Mechanical Subsystem

1. Does the antenna look as if it is mounted correctly and that all the angles are still set properly? If the reflector is facing north, the problem is obvious. The collar was not properly tightened, drilled and bolted.

2. Has the concrete shifted or settled under the pole or pad so that the pole is no longer level or plumb?

3. When using fiberglass antennas, perhaps the surface of the antenna has delaminated or is blistering resulting in poor reception.

4. Has the owner painted the antenna with a roughly applied metallic paint that has affected its reflective surface? Or maybe he has applied a highly reflective paint that is concentrating sunlight enough to overheat the components at the focus.

5. Shake the antenna from the outer rim for maximum leverage in order to check for loose nuts and bolts. Maybe the installer did not use lock washers or lock-tight and the wind has gradually loosened up the assembly.

6. Are any nuts or bolts missing?

7. Is there excessive wear at any attachment points? Scoring on the support pipe caused by such movement may be a telltale sign. If this is the case, set a bolt through the outer to the inner pipe when all the re-adjustments are completed. This will maintain a true north/south tracking axis.

8. Is the antenna warped? When sighting along one lip does the other side line up perfectly parallel? Do the two or three string test, if necessary, to check for warping. Maybe the mount support was not designed properly and the reflective surface has twisted.

9. Is the feed system secured and centered? This can be checked with a focal finder or with a tape measure. A loose feed could possibly move in the wind or be jarred out of place as the antenna rotates across the arc. Guy wires are recommended.

10. Is the feed opening free of bugs or wasp nests? A quick visual inspection will eliminate this possibility.

11. Does the actuator jack have complete freedom of movement across the entire belt of satellites?

D. TECHNICAL TROUBLESHOOTING

If the initial visual inspection does not provide a solution to the problem, then more technical approaches can be followed. There are always a number of ways to arrive at the same conclusion. It is important to understand that the intuitive approach is a valid one, not to be diminished in value. Troubleshooting certainly has its share of personal touch.

The tools used in technical troubleshooting are test instruments. A voltmeter, spectrum analyzer, portable signal simulator or portable receiver can greatly simplify and speed up the diagnostic process. Switching components is included in this section as one of the technical troubleshooting methods because it involves more than casual observations of TVRO problems.

Test Instruments

Voltmeters and Multimeters

A voltmeter is by far the simplest, lowest cost and potentially most effective test instrument. Most electrical breakdowns occur in power supplies that provide energy to operate the LNB, actuator and all other signal processing circuitry. A competent technician can isolate power supply and other voltage related problems in minutes with the proper use of a voltmeter (see Figure 8-3).

Voltmeters, like most test instruments, have variable sensitivities. The maximum range for each setting is typically 200 millivolts, 2 volts, 20 volts, 200 volts and 1000 volts dc. Testing should always begin in a range which can manage the maximum voltage expected so the indicator will not be driven off the upper end of its scale.

Most voltmeters are designed to read resistance and ac voltage as well as dc voltage. These capabilities are extremely useful in testing for shorted wires, in checking line voltages or in troubleshooting other areas.

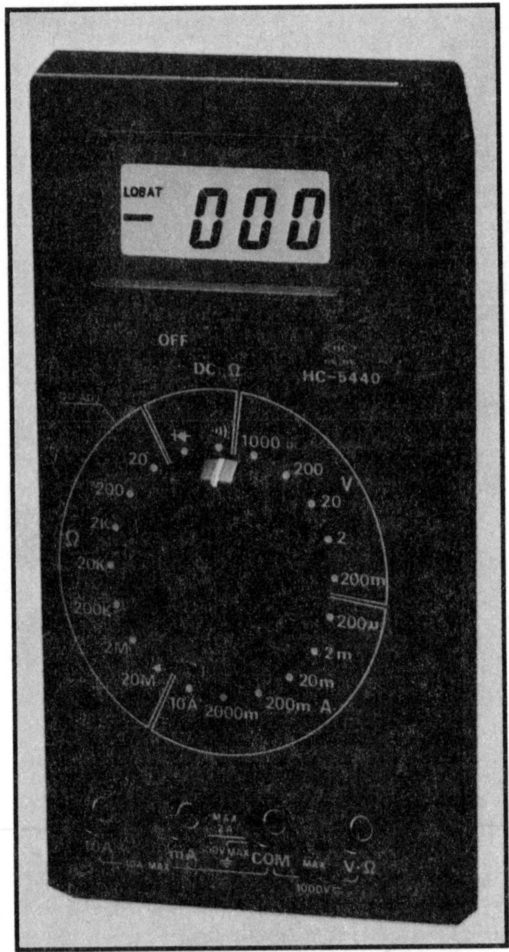

Figure 8-3. Multimeter. *One of the most frequently used pieces of test equipment either in the field or on a test bench is a multimeter. It has the capability of reading dc and ac voltages or currents as well as taking resistance measurements for checking cable continuity.*

Spectrum Analyzers

Spectrum analyzers are powerful diagnostic tools designed so that incoming signals in a range of frequency bands can be observed and measured (see Figures 8-4 and 8-5). This instrument essentially plots a graph of signal power versus frequency. Frequency increases from left to right on a CRT screen while signal amplitude is presented as a series of peaks and valleys. Incoming microwaves located anywhere within the entire band of satellite communication frequencies can be observed at once or fine details of the frequency and amplitude structure of any single carrier can be studied. The spectrum analyzer is to an RF designer what an oscilloscope is to an electrical engineer. A spectrum analyzer displays the signals plotted versus frequency whereas an oscilloscope displays signals plotted against time.

Portable spectrum analyzers have long been used for cable television applications. Many analyzer manufacturers have marketed add-on modules designed to allow coverage of the satellite television IF block frequency ranges. Such analyzers can be used to observe signals at any point in a satellite system. When the signal from an LNB, or preferably an LNB and feedhorn assembly, is fed into this instrument, microwaves of any chosen polarity can be detected and measured. By scanning a feedhorn or an antenna across the sky and observing the variation in detected power, the source can also be pinpointed. This capability makes a spectrum analyzer invaluable in aiming an antenna and in peaking a feed.

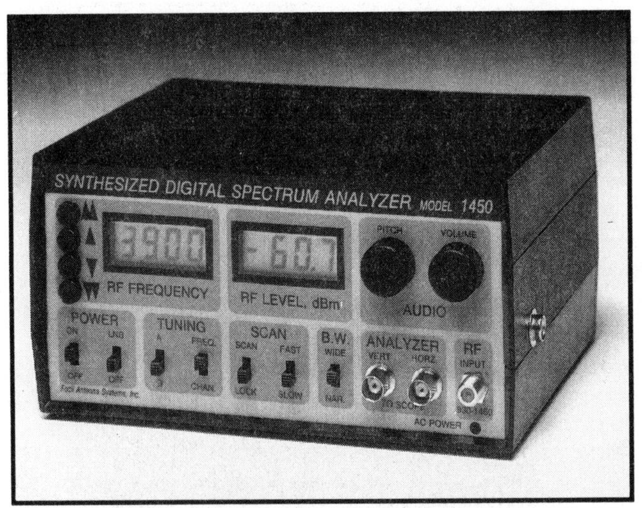

Figure 8-4. Focii Digital Spectrum Analyzer. *This battery-powered unit accepts input signals in the range from 950 to 1450 MHz and has both digital, audio and oscilloscope outputs. Center frequency is adjusted in steps of 1 or 20 MHz. (Courtesy of Focii Manufacturing Company)*

An LNB typically generates a signal having powers ranging from -60 to -80 dBm while the highest level signal at the modulator output is on the order of -40 dBm. One common portable, relatively low cost instrument, the Avcom PSA-35, has a sensitivity spanning the range of zero to -90 dBm and therefore can respond to RF signals produced at any point in a typical TVRO.

Technicians at uplink facilities use on-line analyzers to monitor the power, frequency, stability and composition of signals relayed to communication spacecraft. Powers are generally equalized across all transponders on-board a broadcast satellite. This is seen as a series of equal peaks spread across the entire C or Ku-band frequency range. A spectrum analyzer monitoring the output of an LNB should display the same, equal distribution of transponder power.

The ability of an analyzer to create a visual display of signals over a wide range of frequencies makes it a valuable troubleshooting instrument which can pinpoint system or component problems (see Figure 8-6). This tool can also be directly connected into the output of an LNB, the output of the cable leading from the LNB to the satellite receiver, the output of the satellite receiver or any other point in the RF system. Signals can therefore be traced enroute from the antenna to the television set in order to identify any faulty components or cable breaks.

Two common symptoms observed on an analyzer are tilt and wipe-out. If transponder power appears to decrease across the satellite band on the instrument's display, the feedhorn or LNB may be faulty. If the problem is a bad component, it might be identified by replacing it with a healthy one. Tilt can also be a result of attenuation in coax-

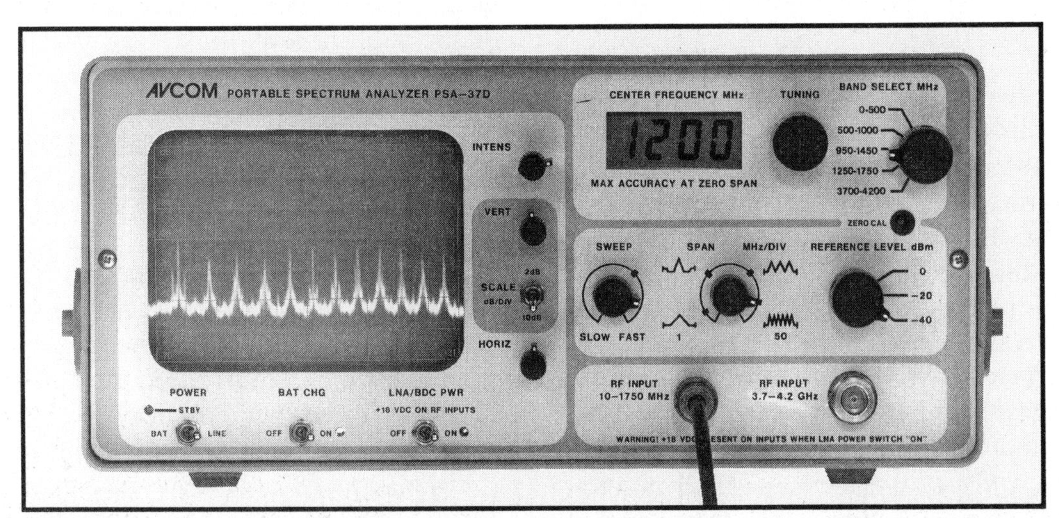

Figure 8-5. Avcom PSA-37D Spectrum Analyzer. *This battery-operated analyzer accepts input signals from 10 to 1750 MHz and from 3.7 to 4.2 GHz in five bands. It has both an LCD digital and CRT output. A built-in supply powers LNAs and LNBs. (Courtesy of Avcom of Virginia)*

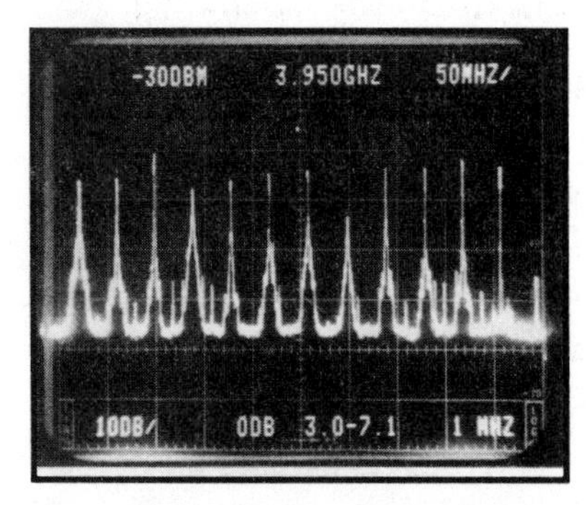

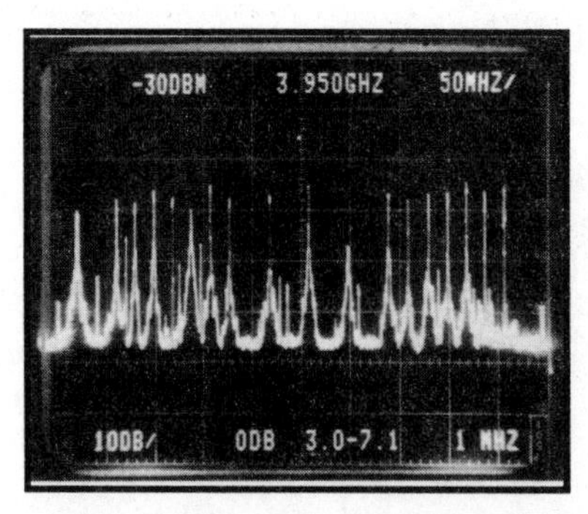

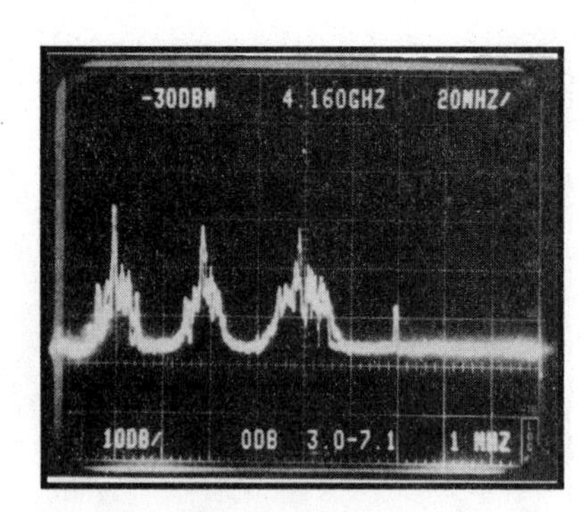

Figure 8-6. Spectrum Analyzer Displays. *These three illustrations display different views of transponders received from satellite. The left shows a view of 12 transponders and a control beacon on Galaxy I. The middle displays all 24 cross-polarized transponders. The right photo shows three transponders and Beacon. Its amplitude is approximately one third that of the satellite broadcast signals. (Courtesy of Tektronix)*

ial cables that have losses which increase with frequency. An analyzer can be used to locate the source of high frequency roll-off in block downconversion distribution systems. Once the source has been isolated a solution might be to use a line amplifier with tilt compensation or to somehow shorten the cable runs. Another common problem, transponder wipe-out, which can also be traced with an analyzer, appears as a sharp reduction in power of one or a group of transponders. One source of this problem may be an impedance mismatch that causes signals in a rather narrow frequency band to be reflected back to the antenna. Something as simple as a missing terminator on an unused splitter port or a faulty connector could be the cause.

The potential uses of an analyzer are limited only by the creativity of the technician (see Table 8-2). Some unusual tasks can be created. For example, this instrument can even be used to test a hand-held remote which emits RF signals. As buttons are pushed on the remote, spikes should appear on the analyzer's screen. Probably one of the most important uses of an analyzer is to locate faint Ku-band satellites on narrow beamwidth antennas during alignment. An analyzer can also allow visualization of weaker signals that would not normally be seen with a satellite receiver and monitor. Thus an installer can detect those satellites that may not yet be transmitting television signals by observing their telemetry beacons.

Portable Signal Simulators

A portable signal simulator can be used to quickly diagnose faults in a satellite system. This device transmits a signal which mimics the television color bar test pattern received from a satellite. The Newton Electronics GBS 1600 (GBS stands for ground-based-satellite) is a good example of a simulator that was designed for testing C-band TVROs. The one designed by Satellite Test Equipment, the ET-15 Simulator, generates Ku-band as well as C-band test signals. These simulators also generate the same block of frequencies as the LNB output as well as some typical IFs and an AM modulated signal. These can therefore be used as diagnostic tools at any point in a satellite receiving system. For example, the ET-15 transmits color bar signals having IFs of 70, 130 and 612 MHz. This unit also provides both normal and inverted video and has an output to test continuity of polarizers. Both units are battery powered thus eliminating the need for a bulky extension cord trailing behind during the troubleshooting procedure. The battery in the GBS-1600 lasts about two hours before recharging is required. The ET-15 has the added advantage of being capable of generating test patterns in NTSC, PAL or SECAM formats.

If the output of the ET-15 is aimed into the antenna and the TVRO is functioning properly, clear color bars should be seen on the television screen, assuming that the TV channel has been correctly tuned to match the modulator output frequency. The satellite receiver must be set to 1190 MHz in

TABLE 8-2. SPECTRUM ANALYZER ANALYSIS REPORT

Customer Name____Date_______________

Address____Telephone_________________________

____

City____State_______________ZIP________

System Description

Antenna: Manufacturer________________Diameter___________Model_______________

Polarization/Feed: Manufacturer_____________________________Model_______________

LNB: Manufacturer____________________Max. Noise Temperature________Gain________

Coaxial Cable: Length from LNB to Receiver__________________________Type_______________

Receiver(Main): Manufacturer_________________________________Model_______________

Serial Number______________________Location_______________________

Receiver(Slave): Manufacturer_________________________________Model_______________

Serial Number______________________Location_______________________

System Performance Statistical Data

Satellite-Targeted____Transponder#_______________

Output Levels: LNB________dBm @ 3.7 GHz________dBm @ 3.9 GHz______dBm @ 4.2 GHz

Coax_______dBm @____MHz________dBm @____MHz_______dBm @____MHz

Terrestrial Interference: Frequency______MHz Level________dBm Bandwidth_______MHz

Notes

____

____

the 950 to 1450 block downconversion range as this instrument broadcasts on 1190 MHz. This test shows whether or not a satellite is targeted properly and can conclusively prove that the RF system is functioning. If color bars are seen, the RF system is functioning well and it is probable that the antenna is not tracking properly or has its vision obstructed.

If color bars are not seen, the LNB can be bypassed by connecting the 950 to 1450 MHz F-connector output from the simulator directly into the coaxial cable leading to the receiver. If color bars are observed, the LNB is defective or is not receiving power. The signal simulator can also be used to ensure that the necessary 15 Vdc power is being received by the LNB. A light on its front panel would illuminate if there were adequate voltage at the LNB.

If the test pattern is still not detected on the TV, the simulator can be connected directly into the satellite receiver. This test is used to determine if the cable between the receiver and LNB is faulty. If this is not the problem, then the channel 3 (or 4) output of the ET-15 can be connected directly into the television receiver. This may show that the channel 3 (or 4) input or the set itself is faulty or the TV channel has not been properly fine tuned. If the television is functioning properly, then there is a problem with the satellite receiver.

If all these components are healthy, the problem may be something as simple as a tree screening the antenna and blocking the satellite signal, a non-functioning polarizer or even a noisy, partially corroded connector that operates with the high levels of power generated by the test unit but may not be capable of detecting very weak satellite signals.

Another much simpler diagnostic tool, also a source of microwave energy and in a sense a portable signal simulator, is a portable fluorescent light (see Figure 8-7). If it is waved in front of a functioning LNB some change in noise level should be seen on the television and the signal strength meter should respond. If this does not occur, the LNB may not be receiving adequate power or it may be defective. Similarly, the human hand is a source of 300°K noise. When placed in front of a feedhorn, a sensitive meter should indicate an increase in detected power, assuming that the LNB is operating. Of course, this action will blank out any picture which might have been present if a satellite was targeted.

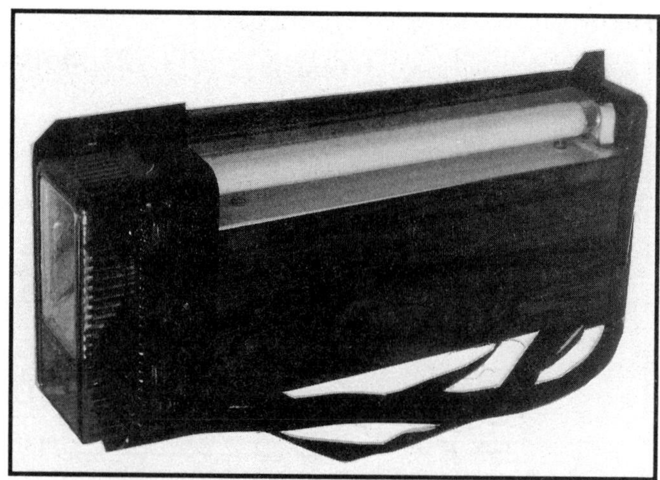

Figure 8-7. A Portable Fluorescent Light. *This portable fluorescent light can be used as a simple tool to diagnose the condition of an LNB. In addition to generating light, it emits microwaves covering a broad frequency band. When waved in front of a healthy LNB, the noise level on the television screen should change and the signal level meter should indicate full scale reading.*

Portable Receivers

Portable test signal simulators are valuable, but will not help in aligning an antenna onto the satellite arc. This job is best accomplished with a portable receiver or other sensitive signal detection instruments described above and in Chapter V. While most installers simply connect the customer's receiver or an extra unit from the shop, using a self- contained portable satellite receiver avoids the difficulty of hooking up three or four components as well as lugging a TV monitor or receiver out to the site each time a test is conducted.

Most portable receivers today are battery operated and are equipped with either a color or a black and white monitor all packaged in a relatively lightweight container (see Figure 8-8). The only extra equipment required to receive a satellite broadcast is an antenna and LNB/feed. This portable tool eliminates the risk of forgetting some necessary component back at the shop or of dropping a customer's prized set in the mud during testing.

Testing with a portable receiver proceeds similarly but in an opposite sense to that used with a signal simulator. The LNB output is first connected to the portable receiver 950 to 1450 MHz or 950 to 1700 MHz input. If an adequate picture is seen on the small TV monitor, the LNB is fine and any problem would lie downstream towards the customer's TV set. Next, the output of the coaxial cable can be connected to the portable test unit. In this fashion, by working from the antenna towards the television, the faulty component can ultimately be identified.

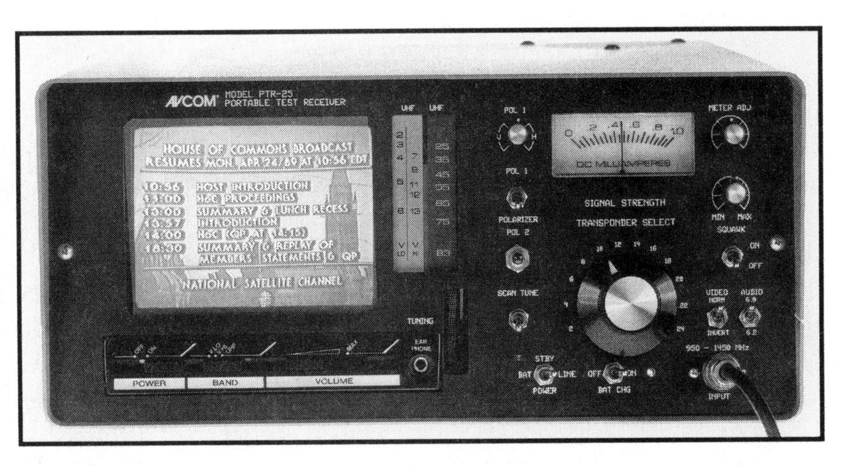

Figure 8-8. Avcom PTR-25 Portable Test Receiver. *Acvom's COM-2 and COM-3R satellite receiver circuits as well as PSD controls are included in this battery-operated unit. Outputs include connections to drive TV monitors, video recorders and audio amplifiers, an audible tone and a signal strength meter. It accepts an input frequency of 950 to 1450 MHz at a -70 to -25 dBm level. (Courtesy of Avcom of Virginia)*

Testing with both portable signal simulators and receivers can also follow along the lines of a process called "half-split" troubleshooting. Instead of checking each component sequentially, the diagnosis begins somewhere in the middle of the system. This might save time if the input components seemed to be working fine but nothing was happening at the output, the television set.

Switching Components

One very simple method of troubleshooting a satellite system is to switch out components, one by one. Thus, if the LNB is suspected of being faulty, substitute a healthy one and see what happens. Remember, in this and all other similar situations when any component is switched, always be sure that the power is turned off and the receiver is unplugged to avoid causing further damage. Some brands of receivers continue to send voltages to the system even when the power switch is off. This diagnostic method is direct but can be somewhat time-consuming. Also, if switching components is attempted during a new installation and more than one component happens to be bad, it may prove to be a frustrating endeavor.

Interchanging or shotgunning components is used mainly to diagnose problems in the RF system. It is easy to check if the actuator or polarizer is not functioning properly by visual inspection and then to find the problem by voltage checks and other troubleshooting methods.

Power Supplies

Most electronic malfunctions can be traced to satellite receiver and actuator power supplies. Unless adequate measures have been taken to suppress voltage spikes, it is common to have a fuse blow or a voltage regulator fail. Some brands of equipment have external, screw-in and internal fuses so both must be checked. If nothing happens when a receiver or actuator is connected, the unit should be unplugged and the fuses examined. To be absolutely certain that fuses are blown, if this is not obvious by a visual inspection, check their continuity with an ohmmeter. When replacing fuses it makes good sense to install spike suppressers so this problem does not reoccur.

If fuses are repeatedly blown, a very inexpensive current limiter can be constructed with a 40 or 50 watt light bulb and a 1 ohm resistor connector in series with the power line into the receiver. The ac voltage is measured across the resistor. If there is a direct short in the receiver, the bulb will glow brightly as it limits current to the receiver and protects the fuses. Make sure that this test setup is well insulated or severe shocks could occur.

Integrated circuit (IC) voltage regulators are often not adequately protected and can malfunction. Excess voltages can damage the receiver, actuator or LNB. In unusual cases, the antenna could even be damaged by being driven into the ground by a runaway actuator. However, if a regulator does fail, identification of the faulty part is easy with the voltage tracing methods outlined below. Replacement is just as simple.

The 7800 and LM340 series of positive regulators are capable of supplying up to 1 amp of positive current and are often mounted directly onto the receiver chassis. The 7900 and LM320 series are the complementary negative components but must be isolated from the chassis and therefore mounted by use of a mica insulator and insulated screw. The output voltage is identified by the last two digits of the 7800 and 7900 regulators. For example, a 7818 regulator provides a positive 18 Vdc. Two other adjustable regulators, the LM317T and LM723, are also often used in satellite receivers.

Voltage Checking

Tracing voltages with a voltmeter is an effective method to rapidly identify a faulty component. When examining an RF problem, all voltage checking should begin at the receiver. It is the brains of a satellite system and it (1) powers the LNB, polarizer and modulator, (2) selects channels by sending a correct voltage to an internal VTO and (3) communicates polarity, channel selection and audio format information to the user.

The satellite receiver relays from 15 to 24 Vdc to the LNB and typically 5 to 6 Vdc to the polarizer. If this power does not reach its destination for one reason or another, the system will not function.

Voltage checking should begin at the receiver output. The voltmeter is set on an appropriate range, from 20 to 200 Vdc. An output of 36 Vdc is common on the actuator circuit so the 200 Vdc scale would be used here. If the range is set too low, the indicator will be pushed off scale. In general, the voltmeter red lead is connected to power and the black lead to ground. Voltage is measured between the central conductor and ground at one port of a two-way, power passing splitter. The input coax is connected to the second port. It is important not to leave the splitter in the line because 3.5 dB will be lost by its insertion. An open port is also an invitation to ingress interference and causes signal losses because of reflection at this discontinuity. If no voltage or a low voltage is measured at the receiver IF input, turn the power off, unplug the receiver and disconnect the cable connected to the LNB. If a second reading obtained after powering back on indicates the correct voltage, there is a short in the system upstream towards the antenna. If the voltmeter still reads zero or near zero, 99% of the time an internal or external fuse or a voltage regulator is blown in the receiver.

If the disconnected receiver was delivering full power, but when connected to the cable the voltage was too low, there is probably a short or a bad component further upstream towards the antenna. Unplug the receiver and disconnect the coax at the LNB input. Then re-plug the receiver and turn it back on. If full voltage is measured across the disconnected cable, the fault lies in the LNB. If zero voltage is measured, then there is certainly a shorted connector or coax. It is possible that the cable could be crushed somewhere along its length.

Always carry extra fuses and voltage regulators. For pocket change, standard 7812, 7815, 7912 or 7915 regulators for the receiver or 7805 regulators for the polarizer control circuit can be purchased at electronic supply stores. Also, have an extra 555 timer chip used to send timing pulses to many brands of polarizers. These can easily be soldered into a receiver in the field, saving many hours of travel time. 555 timers are 8-pin chips which look like a small centipede. They control the position of the polarizer probe.

Whenever leads are disconnected or connected it is important that the satellite receiver be unplugged in order to avoid inadvertently destroying a good component during the troubleshooting procedure. Just because the receiver is turned off does not mean that power is not being sent to the various components.

It is also wise to have an extra receiver, actuator and LNB, as well as appropriate connectors on hand so a bad one can be temporarily replaced. Most users appreciate the loaned equipment during the time their faulty ones are being repaired.

Similar techniques can be used to ensure that the proper voltages are reaching the polarizer as well as the actuator motor. The path can be traced from the controller or, if voltage is present at the peripheral equipment, it can be followed in reverse from the antenna to either the receiver or actuator control box. When checking voltages on the actuator motor two people are required, one to activate the motor via the control box and one to take the reading at the motor.

The voltmeter can also be used to check both continuity and shorting of cables. If a coax or other wire is suspected of having a break, follow this procedure. Set the volt/ohm meter on the ohms x 10 scale. Then use alligator clips to join the central conductor and ground together at one end. Make sure that the other end is disconnected from power and then attach the meter at this end. If the meter reads zero ohms, then the circuit is continuous; otherwise the cable is broken. If the alligator clips are not attached and a resistance reading is taken on the disconnected cable, a zero resistance indicates a short between the central conductor and the ground shield.

When a connector is shorted, look first to see if there is any mechanical cause. Sometimes water is soaked into a cable at a leaky connector just as wax is drawn up a wick on a candle. This "wicking" can pull water into a cable as far as 12 inches or more. If this is the case, the sign is usually a discoloration of the center conductor. If the cable is made from aluminum, it can be severely corroded. Cut back until the discoloration disappears and reattach a new connector.

Grounding Problems and Hum Bars

Occasionally problems arise that cannot be traced to a single faulty component but are related to how all the components and wiring fit together. Diagnosing and servicing such difficulties can be somewhat more time consuming but also more rewarding than simply replacing a bad LNB or finding a shorted cable.

Ground loops are a good example of this type of system problem. A ground loop results from having two grounding points which offer slightly different resistances to ground within a TVRO. In most installations all the electrical components are grounded to a common point through the wall outlet. However, if the cable run between the antenna and the indoors equipment is long enough, the resistance between the ground potential at the earth near the antenna and the coax's ground resistance can be substantially different. When a ground loop occurs, it may be necessary to "float" the indoors components from the wall outlet and use only a single common ground at the antenna.

Ground loops can create a series of problems including "hum bars" on the screen, difficulties with polarity control when a pin diode controller has been installed, actuator positioning errors or a low- level hum in the audio. Hum bars appear on a television screen as messy horizontal regions of varying width which may be stationary or which can move up or down on the screen. In effect, the difference in ground potential causes the picture to lose some of the necessary synchronizing ability.

Ground loops can be broken into two categories: those that appear long after the installation is completed and are usually caused by bad connections; and those that begin during installation that can, on occasion, be tricky to cure. However, one of two methods can usually clear up the difficulty. The first cure is to systematically eliminate any bad connections that might exist in the TVRO wiring or, perhaps, in the home's ground connection wiring. The second is to tie all of the interconnected satellite equipment together via a heavy gauge wire with a common ground at the antenna by "floating" the indoor equipment. Or the antenna and LNB can be tied directly into the electrical service ground at the point where it enters the home with a heavy gauge (e.g. #6 gauge) or a flat welding cable.

Inadequate system voltages can have the same effect and cause hum bars by not providing enough power to generate the necessary signal to adequately power the receiver. This could happen if under-sized wire is used on a long run to the LNB so that the cumulative cable resistance causes unacceptably large voltage drops. Note that the same symptom may be seen if the feedhorn probe servo hunts. This is also caused by inadequate voltages in under-sized cables spanning long distances.

E. UNDERSTANDING COMPONENT FAILURES

One approach to troubleshooting is to examine the operation of all components in order to understand how each may fail. This background knowledge is a useful aid in following the methods described earlier.

Satellite Receiver

The function and operation of a satellite receiver must be clearly understood by a competent troubleshooter. It is crucial to be versed in the alignment procedures for each unit installed to avoid many potential problems. Technicians should also be familiar with all front panel controls such as skew adjustment, subcarrier selection and video scanning.

A visual inspection of all feedback lights, fuses and connections and measurement of voltages will often reveal a simple problem. If this is not the case, more technical approaches can be used to determine any fault. One simple alternative is to swap an identical working unit and see if the trouble disappears.

Some of the problems more commonly encountered are listed below and in Table 8-3.

1. A high signal strength meter reading but no picture. Assuming that the TV is tuned to the correct channel and all other components are functioning properly, this fault can be caused by a bad modulator. Note that many receivers use VCR modulators and replacements that are available in electronic part stores. Be sure that a crystal controlled, "linear" modulator is purchased. Non-linear brands are intended for use in computers.

2. A zero signal strength indication with no picture. This may be caused by a shorted IF/voltage cable, lack of powering voltages or a problem somewhere else in the system such as poor antenna alignment.

3. The receiver shuts off when it is hot. A thermal overload is often caused by a voltage regulator overheating. A can of cold spray or a hair dryer may help isolate the defective component. The problem may be solved by simply clearing enough room to allow air to freely flow around the receiver vents. Occasionally a defective component may have to be repaired in the shop.

4. A herringbone pattern in the picture. This can be caused by a bad modulator or unwanted feedback from the IF line into the modulator. A strong, nearby FM station or transmitter may also cause the same problem if signals leak into the system through poorly shielded cables or components.

5. A lack of audio. This may be a problem in either the receiver or modulator. Check all connections and make sure the right audio subcarrier is being selected. If a faint sound can be heard when listening with a pair of headphones at the receiver's audio output port, the modulator is probably bad.

6. A buzz in the audio. This can be caused by a faulty modulator, incorrect audio tuning or an unduly high video level in the modulator. The appearance of saturated colors on the screen is also associated with an excessive video level.

7. Thick black bars across the TV screen. This is often a symptom of a faulty power supply caused by an irregular voltage or a ground loop. A 4700 microfarad, 25 volt capacitor applied across the dc output and ground may quickly solve this problem, but the underlying cause should also be found and corrected.

Aside from swapping out a voltage regulator or visually inspecting internal components, repair work on a satellite receiver of any complexity should be done in the shop or in a portable van equipped as a shop. There is nothing worse than scattering parts from a receiver all over a customer's home and then not having the components necessary to fix it when the problem is discovered.

TABLE 8-3. COMMON RECEIVER PROBLEMS

- Blown fuses
- Poor or opened cable connections on rear panel
- Failure in power supply to LNB
- Power supply output to LNB high (25-30 Vdc) or low (10-16 Vdc)
- Failure in power supply to polarizer
- Power supply output to polarizer high (5-10 Vdc) or low (0-3 Vdc)
- Polarizer wires incorrectly connected
- Power supply to actuator motor absent or low, less than 15 Vdc
- Actuator wires incorrectly connected
- Front panel function buttons inoperative
- Memory back-up battery dead
- Switches on rear panel incorrectly set
- All or part of remote control section not functioning
- LNB input connected to incorrect output port
- TV output connected to incorrect port
- IF cable loop not installed when necessary
- Defective between an external power supply and receiver
- Defective external power supply
- Defective circuit breaker
- All receiver section not functioning
- Receiver does not accept programming information
- Receiver loses memory
- Polarizer incompatible with LNB or feed
- Pulse to polarizer missing

Stereo Processor

Troubleshooting either a stand-alone stereo processor or one built into a receiver is usually limited to educating the user about its proper operation. Many may believe that since the system has a stereo processor, all audio should now be received in stereo. Most stereo processors have too many unfamiliar buttons for the novice user. This can easily lead to incorrect operation. Note that a satellite TV guide should list the correct audio formats for use with stereo processors.

A first step is to tune to a known channel that employs a stereo format that the processor can demodulate. Most can manage the discrete format. Select matrix stereo position on the processor controls and tune both subcarrier channels to the correct position. If stereo is not heard, the problem may lie with the customer's stereo system. This can be ascertained by tuning their stereo receiver to a known source such as FM radio, a record or a tape.

In those cases where the customer's stereo is functioning properly but no stereo can be obtained, first make sure the leads are securely connected. A voltage reading can be taken to determine if a signal is actually reaching both ends of the audio cables. Make absolutely sure that the stereo processor is connected to the baseband output of the receiver, not the audio output which is used to drive an external modulator. If these tests are negative, the best course of action is to swap it for a working unit and take it back to the shop for necessary repairs.

TABLE 8-4. COMMON CABLE & CONNECTOR PROBLEMS

- Cable not the correct type
- Cable run to antenna too long
- Cable has been damaged underground or above ground by rodents
- Cable cut with a shovel or other utensil
- Cable impregnated and shorted out with water
- Connectors inadequately installed
- Incorrect type of connectors used
- Poorly waterproofed connectors corroded
- Cable crushed or pinched
- Nails or staples driven through cable
- Cable has been burned by lying across a heater exhaust vent or attic lamp housing
- Cable crushed or cut by movement of the antenna
- Center conductor of coax shorted by a strand of its shield
- Coax center cable in connector was ringed during stripping and has since broken off
- A splice pulled apart or not making adequate contact
- Sharp kink in coaxial cable
- One or more wires in cable run are opened
- One or more wires shorted together or to the shield

Cable and Connectors

Coaxial cable and connectors can easily be overlooked as potential sources of trouble (see Table 8-4). But incorrect installation procedures and improper selection of cables can often cause substantial deterioration in overall performance.

A poorly grounded cable or a connector which leaks moisture can introduce unwanted noise or can severely attenuate signals. A cable that is bent too sharply can result in signal losses because its impedance changes at the bend and causes signals to be partially reflected. Improperly installed direct burial cables and leaky connectors can corrode in surprisingly short times.

When installing cables and connectors, it is imperative to securely crimp the casing onto the shield wires with the proper tool, a cable crimper. All locations where water may enter must be sealed and protected as well as possible. Coaxial sealant can be used to prevent water ingress at all outside, exposed locations.

Troubleshooting cables and connectors can be as simple as a visual inspection. If necessary, a continuity test will make sure that signals can pass without serious or complete attenuation. If wire or cable having too fine a gauge has been used for long runs, insufficient voltages reaching the LNB or polarizer may cause faulty operation.

LNB

A faulty LNB can cause problems ranging from excessively sparklie pictures to complete whiteout. Some common symptoms and their causes are listed below and in Table 8-5.

1. Lack of voltage to the LNB. This can result from a blown fuse or other fault in the receiver, a corroded connector or a shorted or broken coax. A simple method to check that the full 15 to 24 Vdc is reaching the LNB is to disconnect the input line and measure voltage between ground and its central conductor. Be careful not to short the leads together.

2. An open circuit at the LNB output. One or more of the LNB stages may be blown. There will be no signal strength reading at the receiver and the TV picture will be blank white.

TABLE 8-5. COMMON LNB PROBLEMS

- Channels not on their assigned number on receiver
- Channels 1,2 and 3 missing
- Channels 1 and 24 missing
- Channels 23 and 24 missing
- TV screen all white, no sparklies
- TV screen all black, no sparklies
- TV screen has stationary horizontal bars
- Excessive sparklies
- Black horizontal bars floating slowly up or down TV screen
- Smeared picture
- TV screen snowy with a faint picture in background
- Corrosion at input connector
- F-connector cross-threaded onto LNB, no contact made
- Picture disappears as temperature changes during day
- Picture improves during evening hours

3. Excessively sparklie pictures. One or more LNB stages may be faulty, resulting in noise temperatures in excess of rated values. Switching in a good LNB or using a signal simulator is a quick troubleshooting procedure if this is suspected.

4. Presence of hum bars. Hum bars, horizontal lines of varying thickness moving up or down across the screen, can be caused by insufficient voltage at the LNB. Most LNBs require a minimum of at least 15 Vdc and 150 milliamps of current to function correctly.

5. A blank TV screen. This may be caused by a faulty LNB. To determine if the LNB is powered and operating, connect and disconnect the LNB cable while observing the television screen. If the noise level on the picture does not change, this amplifier is probably at fault or is not receiving power. Another check for a faulty LNB is to wave a fluorescent light in front of the feedhorn. If the noise level does not change, the LNB is probably faulty or is not receiving power.

6. Picture deteriorates in hot or cold weather. This problem can be isolated by using a can of freeze spray or a hair dryer to change the external case temperature of the LNB.

LNBs are finely tuned pieces of equipment. A faulty unit should be returned to the manufacturer for repair. Although they are carefully encased to eliminate water ingress, if a leak does develop, the moisture can destroy sensitive circuits by oxidizing copper tracks on circuit boards. The early signs of water damage can be an overly noisy signal and the resulting sparklies.

A faulty LNB can show up as a blown fuse in the receiver. However, if fuses keep on blowing even after the LNB is disconnected, the cable or connectors are probably at fault.

Feedhorn

Feedhorns are precisely machined waveguides designed to efficiently capture microwaves. Any obstruction in their throat can detune the waveguide and seriously impair performance. Some commonly encountered symptoms of problems are listed below and in Table 8-6.

1. Excessively sparklie pictures. This may be caused by wasps, water, ice or other obstacles lodged in the waveguide throat. If water is not removed it could freeze and crack the housing.

TABLE 8-6. COMMON FEED PROBLEMS

- Servo motor and feed not aligned correctly with polar axis
- Servo motor dead
- No voltage or excessive voltage at motor
- Motor stuck in position
- Probe stuck in housing
- Insect nest in boresight causing signal attenuation and sticking of probe
- Incorrect connection of +5 Vdc, pulse and ground wires
- Feed type not matched to receiver
- Inadequate voltages causing servo hunting
- Buzz from failing motor
- Scalar type feed not tightened into a position parallel to reflector
- LNB waveguide incorrectly installed
- Water in waveguide causing corrosion
- Waveguide loose on feed housing
- Scalar rings on backwards
- Adjustable scalar rings incorrectly set

When using the 90° waveguide elbow with the polarizer, a cure is to drill a very small hole, less than 2 mm, in the lower corner of the right angle guide to allow water drainage.

Some poor quality units have a base that is not perfectly round or scalar rings that are not exactly at right angles to the body of the feed. This causes the boresight to be slightly away from the focal point as if the dish were distorted. The symptom of these problems is sparklies on just a few channels.

Such performance impairments may also be caused by a bent or mechanically distorted probe. But never readjust the probe position even if it appears to be bent. These are finely tuned devices.

2. Signals of only one polarity are received. This probably indicates a bad servo motor. Burnout can be caused by improperly installing servo motors whose range of motion does not move the probe to the correct position for reception of either polarity. This can also be caused by using a wire whose gauge is too small for the cable length used or by not using shielded wires.

3. Oscillation or servo hunting of the probe. A random oscillation in a servo motor polarizer can appear in the form of black bars or white lines on a TV screen. In severe forms, it may show up as jumping between two adjacent channels of opposite polarity as the probe hunts intermittently around its axis. This can be easily differentiated from channel drifting because the same two channels are always seen. An easy method to prove the problem is caused by servo hunting is to disconnect the polarizer connections. If the problem disappears, its cause is evident.

Most often this occurs when cable runs exceed 25 meters and an under-sized wire or parallel servo motors are installed. The solution to this problem is to install a 1000 microfarad, 10 Vdc electrolytic capacitor between the red +5 volt wire and ground which is black on a servo motor. The capacitor should be connected so the plus on the capacitor is connected to the +5 volt and the minus side to ground. To prevent oscillation, the suggested minimum wire sizes mentioned in Chapter II should be used.

Actuators

A well designed and manufactured antenna actuator is usually quite reliable when correctly installed. However, there can be large quality variations between different brands, and proper equipment selection will save troubleshooting and repair time.

The actuator consists of a mechanical assembly and its electronic control circuitry. The mechanical component serves to hold the antenna in place and to provide movement across the polar arc while the controller keeps track of the antenna position. Some common symptoms and underlying causes for mechanical failures are listed below and in Table 8-7.

1. Inaccurate pointing of the antenna. If there is too much play in the movement of either the inner or outer tubes (or gears or a chain drive in the case of a horizon-to-horizon assembly) then some wearing or loosening of parts has occurred. If the antenna has more than 1 cm of play in either direction at its rim, the mount needs attention. The bolts supporting the pivot points or jack attachment should have double nuts, lock washers and locktight, and be tightly set.

2. Wear or scoring on the jack tube. This may result when the internal O-rings are damaged by lateral stress being applied when the arm is not installed using ball joints.

3. The jack tube is binding or even bending as satellites are tracked. This may occur if both ends of the jack arm have not been properly aligned and mounted on self-adjusting ball joints. A single ball joint on one end is not enough. The jack should be able to track through its entire range of possible motion without binding on either the antenna or mount structure. Therefore, if the electrical east/west limits fail, damage to the arm will not occur. The motor may also have failed possibly due to water damage. It makes sense to have a spare motor or brushes for placement around the armature.

4. Water entering the motor housing or the tube. Water can collect inside the arm or housing if they have not been properly sealed by gaskets or a rubber boot. Even so, condensation can cause water to collect, freeze, seize the motion and even damage the housing. Drain holes must be in their correct position to prevent this chain of events from occurring. A neoprene rubber accordion sleeve with a hole at either end to allow condensed

TABLE 8-7. COMMON ACTUATOR PROBLEMS

- Rust or ice binding inner tube
- Tube slightly bent
- Ball joint sticking
- Bolts attaching actuator to dish loose
- Actuator-to-mount clamp installed incorrectly
- Actuator clamp/collar set too tight, causing tube to score
- Bolt broken inside tube
- Actuator tube driven past end and loose from sleeve
- Actuator does not extend to full length of travel or completely contract
- Sensor count mechanism defective
- Motor case full of water or corroded from water intrusion
- Magnet wheel wobble
- Mechanical stops broken
- Brushes in motor defective
- Motor case sections coming apart because of loose bolts
- Sloppy tube-sleeve bolt or clutch
- Motor running but tube does not extend or contract

water to escape is an excellent method to prevent water ingress.

5. A grid of dots appears on the TV screen. Some control boxes cause interference on channels 2 or 3. The internal 3 to 6 MHz clock crystal that runs the microprocessor makes a grid of evenly spaced dots appear of the screen. This effect sometimes occurs when the actuator is set on top of the TV set and beats directly into the tuner. Or if 300-ohm flat, twin lead cable is used to feed over-the-air broadcasts into the TV, it may also pick up the signal from the clock crystal. These dots can also appear on satellite TV channels, but can be easily eliminated by using a 10 dB line amplifier on the receiver modulator output to overpower the interference.

As discussed in Chapter II, the types of feedback circuits used for controlling actuator position are Hall effect sensors, reed relay switches, 10-turn potentiometers and photo optical read-outs. Most are coupled to a central microprocessor. Each has its own set of potential problems and solutions.

Feedback potentiometers receive a constant voltage and divide it into a smaller portion depending upon the antenna position. This returned voltage is calibrated and fed onto a read-out on the face of the controller.

If the potentiometer, also familiarly known as a pot, breaks or becomes disengaged from the drive shaft in the motor, an antenna can be driven too far and the jack tube can be bent. A preset electronic and mechanical limit is a crucial feature in such control systems.

A method to troubleshoot this type of actuator is to use a spare 5 K-ohm potentiometer. It can be temporarily installed in place of the permanent one by simply disconnecting it and reconnecting the two or three wires using alligator clips. If readout numbers change on the control box when the temporary pot is adjusted, then the problem is in the motor; if not, the problem is in the control box.

To troubleshoot a unit with a reed switch or Hall effect transistor sensor, attach an alligator clip between the pulse and ground connector at the control box after disconnecting the pulse input line. As the drive arm is moved in either east or west direction, make and break the connection on the pulse and ground wires to simulate drive pulses. If the controller works properly and does not display an error, then the problem is in the sensor or the cable connection between the motor and control. Troubleshoot for cable continuity.

One method of troubleshooting a Hall effect sensor at either the controller or motor is by reading voltage between the pulse and ground points. A drop of about 0.75 Vdc should be measured when the east or west button is pressed. If not, then the Hall effect transistor is bad. If the sensor itself is not faulty, check that the magnets on the motor have not moved or fallen out of place. A spare Hall sensor and reed switch are useful additions to a troubleshooter's spare parts kit. Simulated pulses can also be produced by hooking the polarizer pulse and ground outputs from a receiver or a hand-held polarizer controller to the sensor and ground inputs on the positioner. This will cause the actuator motor to move and will indicate a bad sensor. Pulses will continue until this connection is broken. Be careful not to drive the antenna into the ground.

It is important to always use shielded sensor cable on Hall effect or reed sensors. These devices will give spurious readings if they pick up outside noise. Even the inductive action of a starting or stopping motor can be detected as a "spike" in the sensor. This added to the control count would cause a cumulative error in measuring antenna position. After the problem has been corrected, the actuator control must be re-synchronized because each satellite will probably be missed by the same amount.

Make sure that the magnets surrounding these two types of sensors have not been dislodged. If they have, remember that magnets have a positive and negative end and should be reinserted all facing in the correct direction.

The ground lug on the controller power cord must never be cut to try to fit the three prong plug into a two prong outlet. The microprocessor requires a reliable ground to accurately set its counting circuitry. Always use some type of surge protection on the power line into an actuator control box to protect the microprocessor. Note that if a problem does develop in the internal circuitry of an actuator controller, the best bet is to return it to the manufacturer for repair.

Antenna and Mount

Servicing an antenna and mount is a relatively easy matter of conducting an educated visual inspection (see Table 8-8). Reflectors should be checked for warping and surface imperfections. Mounts must be examined for stability, alignment and secure connections. The feedhorn/LNB support structure can be checked for centering and stability. Methods of performing all these checks have been outlined earlier in Chapter V.

Television

A television is a complex piece of electronic equipment. However, troubleshooting a TV is easy when over-the-air channels of adequate signal strength are present. If the set works well with conventional reception, then about the only problem encountered is when the television is not tuned onto the modulated channel.

There are major differences in quality between different brands of televisions sets. A well-equipped troubleshooter/installer is wise to carry a small, portable, excellent quality television or monitor to demonstrate if there ever is a question about the performance of a satellite system. After all, most judgments are made on the basis of picture quality. Also, if a monitor is used, a faulty modulator internal to a receiver can easily be diagnosed since it would be bypassed.

TABLE 8-8. COMMON ANTENNA / MOUNT PROBLEMS

- Buttonhook bent or too short
- Tripod or quad feed supports bent
- Quad supports not flush against scalar
- Dish warped
- Panel(s) missing
- Pivot points sticking or rusted
- Elevation or declination angles not correctly set
- Spacers missing or bolts not tightened
- Missing hardware on antenna or mount
- Incorrect replacement parts installed on dish or mount
- Mount cap not correctly bolted to ground pipe or cracked
- Diameter of ground pipe incorrect causing a rocking of dish
- PVC, EMT (electrical metallic tubing) or plastic coated gas pipe used instead of steel in ground pipe
- Cracked or broken parts
- Mount not moving to maximum east or west positions
- Turnbuckles installed in wrong holes
- Elevation bar bolt holes tapered out and not correctly tightened
- Saddle for elevation bar spread apart so that dish rocks or shifts

F. SPARE PARTS KIT

If a site visit is necessary, it is essential to have the proper tools on hand. Driving 60 km to discover that a 50 cent voltage regulator is missing can be expensive.

All the light tools of the trade should be taken on a service call. This usually excludes concrete mixers and other large items. In addition, a complete, healthy spare system, possibly including a small test antenna, should be part and parcel of the troubleshooting kit. This kit is composed of:

- Feedhorn with polarizer
- Extra servo or dc motor for polarizer
- Inexpensive receiver
- Ample supply of applicable fuses
- Appropriate voltage regulators
- Extra reed switches or Hall sensors
- Spare motor or a complete actuator
- 5 K-ohm potentiometer
- Fluorescent light
- 1000 microfarad electrolytic capacitor
- Scotch locks
- Electrical tape, solder
- Splitters
- Assorted connectors
- 10 dB line amplifier
- Coaxial sealant and shrink tubing

IX. LARGE ANTENNA SYSTEMS

In the earlier days of satellite broadcasting, antennas often 15 feet or more in diameter were used by cable TV companies and other commercial ventures. With advances in technology, dish size has decreased dramatically. However, there are still situations where dishes up to 20 or 30 feet in diameter are required. Some companies that specialized in these rather uncommon systems have thrived.

A. THE NEED FOR LARGE DISHES

There are two main reasons for using large antennas. First, if the installation location is far removed from satellite boresight, the EIRP is so low that an extra large reflector is needed to capture enough signal to adequately drive the receiver. For example, those in Brazil wishing to view American broadcast satellites often require dishes as large as 30 feet in diameter. Also, if a TVRO designed to receive broadcasts from an Intelsat satellite transmitting a global beam were installed on the east coast of North or South America, a very large antenna would also be required. In these cases, a 5 meter (16 foot) or larger antenna is generally required for sparklie-free reception.

Second, large antennas must often be installed in commercial installations where signal levels must be substantially above receiver threshold to ensure perfectly clean reception. Cable TV companies, for example, require video and audio signals of sufficient strength so that reception remains excellent even after the signals are distributed to thousands of subscribers.

B. INSTALLING LARGE ANTENNA SYSTEMS

Satellite TV earth stations with extra large antennas are similar in many respects to more conventional designs, but there are some major points of difference in their installation.

Mechanical Assembly

A 20 or 30 foot antenna is a very large structure weighing much more than a typical home system. Total dish and mount weights in excess of 5000 pounds are not uncommon. The supporting structure cannot simply be a pole planted on a small pad or set a few feet into the ground. Excavation is often required before pouring the four or five cubic yards of concrete necessary as a foundation for a double A-frame support (see Figures 9-1, 9-2 and 9-3).

Safety of the installation crews is an important consideration. Much of the work is performed far above the ground and very heavy components are involved. The danger of injury is always present. The effects of a bright, hot sun should also not be ignored. Protective headgear and shoes should be worn. Technicians must consider the intense heat and potential for sunburn when large antennas are installed in regions near the equator. Working at the focus of a 30 foot reflector can be more than just uncomfortable if the dish happens to be pointed at or near the sun. Polarizer caps have been known to melt when the sun light is focused with antennas as small as six feet in diameter. When a fixed antenna is being installed, work inside the structure should be planned only for those times of day when the sun is far off the main axis. Or, if the antenna can be rotated, it should be positioned away from the sun whenever crews are working in the reflector.

Heavy equipment is often needed to install large dishes. If an antenna is pieced together on the ground, a crane or special hoist is generally needed for lifting it onto the supporting structure. How-

Figure 9-1. Three Pier Support. *This 20-foot ADM antenna with horizon-to-horizon mount is supported by a three pier foundation. In general, when installing tripod supported mounts the two rears legs should be set on an east-west line. Fine adjustments to establish the final position can then be made by means of the telescoping legs . (Courtesy of Antenna Development and Manufacturing, Inc.)*

Figure 9-2. Comtech 7.3 Meter Antenna. *This 7.3 meter dish is mounted on an adjustable tripod assembly which is itself supported by a large recessed concrete pad. (Courtesy of Comtech Antenna Corporation)*

Figure 9-3. Uplink/Downlink. *This 10 meter (33 foot) Cassegrain antenna has a carefully designed rib structure to support the weight of its reflector. This uplink monitors its signal by also receiving the matching downlinked satellite signals. (Courtesy of Brent Gale)*

Figure 9-4. Hero Antenna. *A special crane and scaffolding was necessary to lift this 25 foot Hero antenna onto its mount. This Sri Lankan installation is supported by a combination of pole and pad planted in over 5 cubic yards of concrete. (Courtesy of Hero Communications)*

ever, there are cases where this is not necessary. For example, the Paraclipse 16-footer can be assembled piece by piece on the mount (see Figures 9-4 and 9-5).

Shipping a very large antenna is not simply a matter of having UPS or another freight company deliver two or three small boxes. Specially de-

Figure 9-5. Maneuvering a Large Antenna. *Five men and one observer were needed to lift this 4-piece, 16-foot antenna into its final position. Using a crane proved to be too difficult on top of this five story building in Jerusalem. (Courtesy of Brent Gale)*

Figure 9-6. Arthur C. Clarke's Antenna. *A 16-foot Paraclipse antenna was installed by one of the authors at Arthur C. Clarke's residence in Sri Lanka. The small tower support was firmly bolted into the structural beams under the floor of this patio. (Courtesy of Paraclipse Antenna and Brent Gale)*

Figure 9-7. Use of Protective Skirts. *Protective skirts can be used to shield a dish from unwanted noise and interference. A non-reflective material is used to avoid degrading performance. This retrofit, if installed properly, lowers side lobes. (Courtesy of Scientific Atlanta)*

signed and often heavy crates must be used. Some manufacturers of 16 foot or larger spun aluminum antennas spend a large portion of the sales price on simply crating the dish!

Large antennas are manufactured by a number of companies including Hero Communications, DH Satellite, Paraclipse and Antenna Development and Manufacturing (ADM). Systems designed for commercial use are manufactured by Harris and Scientific Atlanta, two of the larger companies serving commercial satellite communications customers (see Figures 9-6 and 9-7).

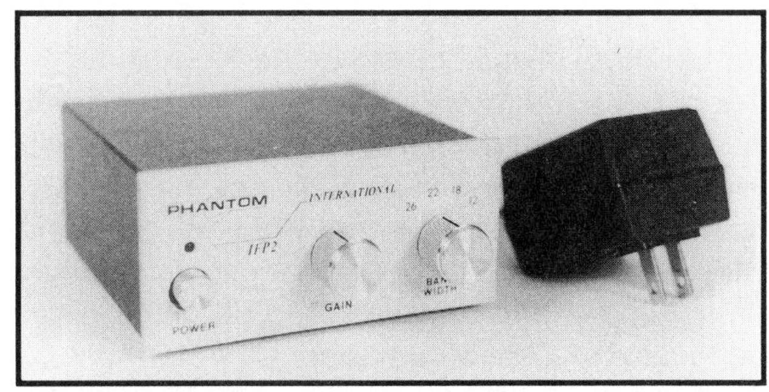

Figure 9-8. Half Transponder Filter. *The Phantom IFP-2 filter can be switched onto 26, 22, 18 and 14 MHz bandwidths. The gain control can be varied over a range from 9 dB attenuation to 12 dB of gain. (Courtesy of Phantom Engineering)*

Electronic Components

The electronic components used in large antenna installations are generally quite similar to those used in home systems. But there are some important differences.

If international satellites are being targeted, either a specially designed feedhorn or a modified feedhorn must be used to properly detect circularly polarized signals. A teflon dielectric slab inserted into the feed throat permits a conventional feedhorn to receive signals having this different polarization format. Although the focal length-to-diameter ratio of large antennas also ranges between approximately 0.3 and 0.4, antenna diameter is much larger than normally encountered, so the feed assembly can be at a substantial distance from the reflector. Either a tripod or a buttonhook supported by guy wires must be used to prevent excessive movement in winds or when the dish is tracked across the arc. Today, few large antennas incorporate buttonhooks.

A variety of video formats, variants of NTSC, PAL and SECAM, are transmitted around the globe and satellite receivers must be properly equipped to decode these signals. For example, Avcom of Virginia manufactures receivers with a special de-emphasis circuit for deciphering PAL formatted broadcasts as well as a dual IF bandwidth for half or full transponder viewing (see Figure 9-8). Half transponder formats, usually broadcast with circular polarization, use half of the full 36 MHz bandwidth for relaying two TV broadcasts per transponder.

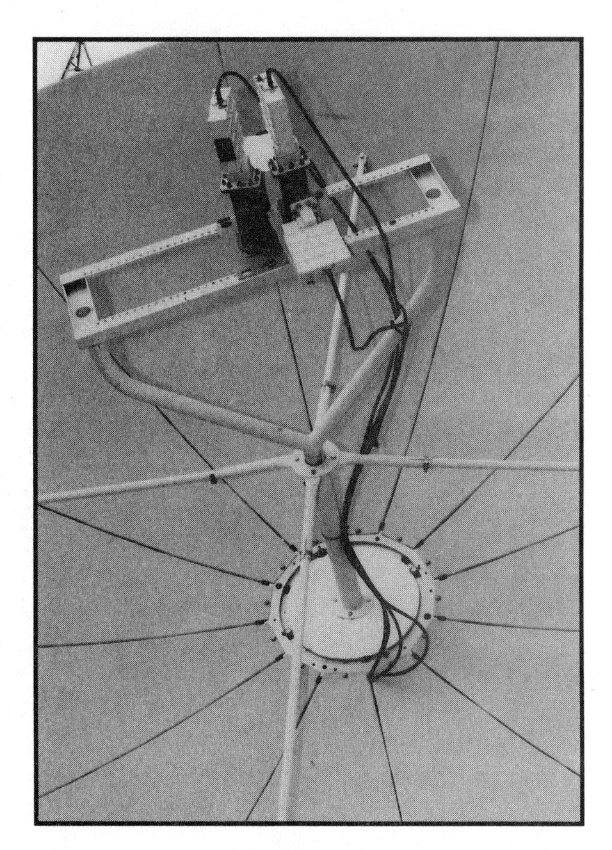

Figure 9-9. A Multi-feed System. *This four meter Scientific Atlanta dish is equipped with four LNAs attached to two orthomode feeds. These feeds can be adjusted to simultaneously detect two satellites separated by a range of angles of up to 12 degrees. This configuration is generally only used in commercial installations. (Courtesy of Brent Gale)*

All electronic components should be tested before being installed. Once a feedhorn and LNB are in place, replacement often involves a fair amount of effort such as using a cherrypicker to reach the focus (see Figure 9-9). Tracking large antennas with narrow beamwidths in locations where satellite EIRP is low can be difficult enough without having to worry about replacing a loose connector or a faulty component. Furthermore, in many locations around the globe this job is made even more difficult because only a handful of satellites have transponders that are active 24-hours a day.

When attempting an installation in a remote area, having spare parts on hand can be crucial. It is generally easier to bring a spare component than to convince customs officials to admit a replacement part at a later date. It is highly recommended that all components be tested and burned-in for at least 48 hours prior to shipment. Remember that an export license is often required to move many types of microwave components between countries because some parts, for example the VideoCipher technology, may potentially be used for military purposes. Similarly, all special tools or test instruments that may be needed should be taken on the trip.

X. SYSTEM DESIGN CONSIDERATIONS

A. LINK ANALYSIS

The choice of antenna and LNB noise temperature are determined by two factors: the threshold of the satellite receiver, and the effective isotropic radiated power (EIRP) from the satellite. The purpose of this section is to show how link analysis, essentially the power budget of a satellite circuit, helps determine the necessary design choices.

The Link Equation

The prime criterion in designing an earth station is to generate a signal-to-noise ratio (S/N) that properly drives a television receiver. Although standards for acceptable S/Rs vary, the critical element is that the signal entering a satellite receiver as defined by the carrier-to-noise ratio (C/N) should be sufficiently high. If so, the receiver would operate in a linear region above its threshold where a 1 dB increase in C/N results in a 1 dB increase in S/R. The value of C/N depends upon both the satellite link as well as earth station design.

The two most important choices in designing a home satellite system are antenna surface area and LNB noise temperature. The antenna and LNB chosen depend upon a number of factors including variations in satellite output power, downlink antenna pointing accuracy, satellite stationkeeping accuracy, atmospheric attenuation of microwaves, aiming and deformation error of the receiving antenna caused by wind loading, and signal scattering due to obstructions caused by windblown branches or airplane crossings.

The effective isotropic radiated power (EIRP) at any particular location is a good measure of parameters that affect satellite operation. But EIRP calculated from satellite design and location may not correspond exactly to what can be measured at a receiving site, especially because the output power of a satellite transponder slowly decreases as it ages. Calculated EIRP is based upon an assumption that the downlink antenna is perfect. In addition, published EIRPs do not account for the fact that the downlink antenna has side lobes and nulls between side lobes. Therefore, areas far off axis may receive more power than expected, especially if they happen to be on the target of a side lobe peak. This phenomena is called antenna spillover. Such effects are usually detected only by on-the-spot tests. This information can be quite useful to local TVRO enthusiasts.

The factors discussed here are described in the "link" equation. Footprint maps are first used to determine EIRP, the signal power aimed towards any geographic location. The analysis then accounts for the weakening of earthbound microwaves as they spread out in space and are partially absorbed by various components of the atmosphere. The receiving antenna captures and concentrates these signals according to its gain. The job of deciphering satellite signals is more difficult the higher the system noise temperature, which is in large part determined by antenna and LNB characteristics.

All these factors can be included in the link equation which is expressed as:

Carrier-to-noise power ratio entering receiver	=	Signal power leaving the satellite, EIRP
		less
		Free space path losses and atmospheric absorption
		plus
		Antenna gain, G
		less
		Noise introduced by the antenna, LNB and other components

or, in a more condensed form, in which these symbols simply express the above words in algebra:

$$C/N = EIRP - \text{Path Loss} + G - 10 \log kT_{sys} B$$

$$= EIRP - PL + G - 10 \log T_{sys} - 10 \log B - 10 \log k$$

$$= EIRP - PL + G - 10 \log T_{sys} - 10 \log B + 228.6$$

All terms in this equation are expressed in decibels. PL is short for path loss. T_{sys} stands for the noise temperature of the antenna, low noise amplifier and all other system components. The term $kT_{sys}B$ expresses how noise is related to noise temperature and system bandwidth. As either noise temperature or bandwidth is increased, more noise enters the system, so this term is subtracted to show its effect on C/N. The term k is a universal constant called Boltzman's constant, equal 1.38 x 10^{-23} joules/°K.

Therefore, if EIRP, path loss and bandwidth are known, the G/T_{sys} required to obtain a desired C/N can be calculated by addition and subtraction. The term G/T_{sys} is shorthand for:

$$G - 10 \log T_{sys}$$

where both gain and noise temperature are expressed in decibels. The G/T_{sys} is known as the "figure of merit" of an antenna/feedhorn/LNB combination. The "bottom line" G/T_{sys} defines the combination of minimum antenna diameter and LNB noise temperature required so that C/N input is just at the receiver threshold. Most good quality satellite receivers have thresholds in the range of 8 dB. Clearly, each link in the satellite circuit determines the characteristics that all the other components must exhibit.

EIRP

The EIRP of a particular satellite can be easily determined for any geographic location from a published footprint map. Although calculated footprint maps are available for every communication satellite, on-site measurements of EIRPs are available for only a limited number of spacecraft. As discussed above, these can differ somewhat from theoretical expectations. Also, the maps show only the power levels radiated from the downlink antenna's main lobe. Power levels of signals from downlink antenna side lobes, the antenna spillover, are rarely published because even small movements in a satellite's orbital position can radically alter their effect. Realize that if the aiming of a narrow side lobe is changed even slightly it will not direct any power to a previously illuminated area on the earth below.

Path Loss

Path loss, the second term in the link equation above, expresses how the power of the downlinked signal is attenuated in traveling from a satellite to a receiving antenna. This is mainly caused by the spreading out and weakening of microwaves on their earthbound journey. It is evident that this must occur since the energy radiated from the small surface area of a downlink antenna ends up blanketing a portion or all of a continental land mass. This geometric effect is simply a function of the distance between a satellite and the receiving antenna. Formulas for path loss and slant range, the distance from any site to any satellite, are given in the appendices. Another component of path loss also results from the absorption of the signal by water vapor and other molecules in the atmosphere.

Typical C-Band Path Loss

A typical value for path loss from a C-band satellite on a clear day is -196.38 dB. This is equivalent to a reduction in signal power by a factor of 4.4 billion billion (4.4×10^{-19})! This number reminds us that communication satellites are over 36,000 km away in outer space. Detecting Ku-band satellite signals is similar to attempting to observe a 50 watt light bulb from such an enormous distance.

Antenna Gain

The receiving antenna intercepts and concentrates signals from transponders with power outputs ranging from as low as 5 to over 10 watts that have been attenuated by approximately 196.3 dB on their journey earthward. The higher the antenna gain, the better a system performs. Below a minimum gain, even the best LNB cannot compensate enough to adequately capture the signal essential to reconstruct a satellite broadcast.

System Noise Temperature

System noise temperature characterizes how much noise is added to a satellite signal as it is processed by all components of an earth station. These include noise detected by the antenna, the LNB, cable runs and all the remaining electronic circuitry.

Unlike terrestrial microwave links, satellite systems must detect extremely low levels of microwave radiation. As a result, earth stations are sensitive to very weak noise sources including both galactic noise and radiation from the warm earth. Galactic noise, which is all that an orbiting antenna pointed into deep space would detect, typically contributes about 6°K to a Ku- band TVRO.

The noise added to a TVRO by the warm earth at an average temperature of 290°K is detected via antenna side lobes. It depends upon both the antenna's side lobe patterns and its angle of elevation. As a dish is inclined towards the horizon it faces closer to the ground and subsequently detects more ground noise. Since the elevation to satellites at the center of the geosynchronous arc is higher than that of those at the easterly and westerly portions, contributed antenna noise will be lower when the central satellites are targeted.

The noise performance of an antenna can be improved in a number of ways. A higher quality antenna of a given size has lower level side lobes and therefore is susceptible to less noise. A larger diameter antenna also has a "tighter" side lobe pattern and a narrower main lobe so it generally contributes lower amounts of noise. Similarly, a deeper antenna having a lower f/D, all else being constant, will detect lower amounts of ground noise.

The noise introduced by the LNB is simply expressed by its noise temperature or figure. Therefore, for example, a 190°K LNB adds 190°K of noise to a TVRO while a 35°K LNB adds 35°K.

The noise contributed by the cables and satellite receiver is low by comparison. This is because the LNB amplifies both the noise and signal detected by the antenna/feedhorn system. Its output signal is substantially higher in power than the noise added by cables, connectors and the satellite receiver. Therefore, the balance of system components has only a small effect, typically about 5°K, on system noise temperature. G/T_{sys}, the figure of merit for an earth station, can therefore be closely approximated by replacing T_{sys} with the noise contribution of just the antenna and low noise amplifier. The LNB output signal is thus the key determinant of overall performance quality of a TVRO.

Nevertheless, noise generated in cable runs can be a factor in some situations. For example, when signal attenuation in long cables increases to the point that the signal power is not substantially above the system background noise, performance can suffer. This may occur in the connection between the LNB and receiver or in an extensive SMATV distribution system. If, for example, line amplifiers are not used in appropriate locations, an antenna/LNB system having a high G/T_{sys} may not function properly.

How Small Can an Antenna Be?

Although the link equation is a powerful tool in determining the correct choice of antenna diameter and LNB noise temperature given the EIRP of any communication satellite, one factor not considered in this type of analysis is antenna beamwidth. Even if EIRP were hundreds of watts and the link equation suggested that an extremely small parabolic dish would perform well, antenna beamwidth limits minimum dish size in conventional broadcasts. For example, a 1.00 meter antenna would have a 5.2° beamwidth, much too wide for reception of C-band satellites spaced two degrees apart. Such a small antenna would detect relatively strong signals from five satellites that were spaced two degrees apart!

In order to clarify how the link equation can be used to determine required antenna diameter and LNB noise temperature, some sample calculations are presented here. The first analysis shows how to calculate C/N for an antenna with given characteristics. The second works backwards to show how the figure of merit, G/T_{sys} and then antenna diameter can be obtained from an assumed C/N.

Determining C/N

Assume that the free space path loss on a clear day is −196.3 dB. Substituting this value into the link equation yields

$$C/N = EIRP + G - 10 \log T_{sys} - 10 \log B + 32.3$$

Assume also that the satellite receiver bandwidth is 28 MHz. This choice should allow reception of a television picture with good fidelity. (If the bandwidth were further reduced, the C/N would be increased at the expense of picture fidelity). With B equal to 28 MHz the link equation then becomes

$$C/N = EIRP + G - 10 \log T_{sys} - 42.17$$

The next step is to include the antenna and LNB noise temperatures in this equation. Assume that a 35°K LNB is used with a 10 foot (3 meter) antenna having a noise temperature of 50°K at an elevation angle of 20° and a gain of 40.65 dB. These values are taken from Table 2-2 and Figure 2-18. The noise temperature at this low elevation angle is used because it presents a near worst case value for antenna noise. A reflector would point at such a low elevation in high northern or southern latitudes or near the ends of the satellite arc even as far south as the equator.

Plugging the above values for noise temperatures and gain into the link equation yields

$$C/N = EIRP + 40.65 - 10 \log (35 + 50) - 42.17$$

$$= EIRP - 20.2$$

Therefore, if the EIRP is 30 dB, the TVRO will deliver a C/N of 10.2 dB to the receiver. This is about 2 dB above the threshold of most receivers. However, if the EIRP were 28 dB on the edge of a downlink antenna footprint, the C/N would be 8.2 dB which is just below threshold with no margin for rain fade or pointing inaccuracies.

Determining G/T_{sys}

Begin with the same assumptions made above for path loss and receiver bandwidth. The link equation can be expressed in two equivalent ways

$$C/N = EIRP + G - 10 \log T_{sys} - 42.17$$
$$= EIRP + G/T - 42.2$$

Rearranging the algebraic terms yields

$$G/T_{sys} = C/N - EIRP + 42.2$$

The G/T can be determined from this equation if both the EIRP and required C/N are known. For example, assume that a particular location has an EIRP of 32 dB and that a C/N of 10 dB is acceptable. Then

$$G/T = 10 - 32 + 42.2 = 20.2 \text{ dB}$$

Assuming a worst case antenna noise temperature of 60°K and the availability of a 25°K LNB leads to

$$G - 10 \log (25 + 60) = 20.2$$

or

$$G = 39.5 \text{ dB}$$

Referring to Table 2-2, such a gain can be obtained using a 9.5 foot antenna having a 60 percent efficiency or a 10.5 foot antenna having a 50 percent efficiency.

Recommended Antenna Sizes

The link equation has been used to construct Table 10-1 which lists recommended minimum antenna diameters necessary to attain a given C/N. The assumptions here are that the free space path loss on a clear day is –196.4 dB, antenna efficiency is 65 percent, antenna noise temperature is 50°K, satellite receiver bandwidth is 28 MHz and LNB noise temperature is 30°K. If the noise temperature is lower, bandwidth is reduced or efficiency is increased, the minimum acceptable antenna diameter will decrease. These recommendations can therefore be regarded as the worst case scenario.

Four values for C/N were chosen. The lowest, C/N = 8 dB, is a typical minimum value for receiver threshold. This allows no margin for signal fading due to pointing inaccuracies, rain, wind or other factors. The highest allows for a 6 dB margin.

The effect of increasing the necessary performance margin or the EIRP is quite clear from this table. A very powerful satellite would allow the use of small, easy-to-install antennas. However, in weak footprint areas, a relatively large, high performance antenna is necessary to obtain adequate reception.

Why a Performance Safety Margin is Needed

The concept of a safety margin is simple. If receiver threshold is 8 dB and a satellite system is designed to deliver a C/N of 8 dB in calm weather what happens when strong winds blow. The resulting deflection of the dish off target can result in a loss of signal and picture deterioration. Or during a torrential rainstorm as much as 0.5 dB can be lost at higher frequencies as rain absorbs the microwaves. Then a flurry of sparklies would appear on the screen. The same result can occur as a satellite ages and its transponders weaken and EIRP drops or as a dish or mount sags and falls slightly off target.

A safety margin should be at least 1 dB for home systems and a minimum of 3 dB for commercial systems. The bottom line is clear. While a well designed system may cost a little more, the resultant user satisfaction more than warrants the extra expense.

TABLE 10-1. ANTENNA DIAMETER VERSUS EIRP

EIRP (dBw)	Antenna Diameter (meters)			
	C/N = 8	C/N = 10	C/N = 12	C/N = 14
20	8.51	10.72	13.52	17.02
25	4.79	6.76	7.60	9.57
26	4.27	5.50	6.78	8.53
27	3.80	4.80	6.04	7.60
28	3.39	4.28	5.38	6.78
29	3.02	3.81	4.80	6.04
30	2.69	3.40	4.28	5.38
31	2.40	3.03	3.81	4.80
32	2.14	2.70	3.40	4.28
33	1.91	2.40	3.03	3.81
34	1.70	2.14	2.70	3.40
35	1.51	1.91	2.40	3.03
36	1.35	1.70	2.14	2.70
37	1.20	1.52	1.91	2.40
38	1.07	1.35	1.70	2.14
39	0.96	1.21	1.52	1.91
40	0.85	1.07	1.35	1.70
45	0.48	0.60	1.21	1.52
50	0.27	0.34	1.07	1.33

1 meter = 3.28 feet

B. BEAMWIDTH AND SATELLITE SPACING

There is no mistaking the fact that smaller antennas are more appealing to all except the rare individual who thinks bigger always means better. As satellite power levels have risen and as earth station technology has improved, ever smaller antennas have become capable of producing excellent quality video. However, the spacing between satellites puts a lower limit on antenna size because beamwidth increases as reflector diameter decreases. Beyond a certain beamwidth, more than one satellite can be detected by an antenna.

In the United States, satellite spacing was not an issue for C-band systems until 1980. RCA Americom had been operating two satellites, Satcom I and II, Western Union had three vehicles, Westar I, II and III, and AT&T/GTE operated Comstar I, II and III. There was more than ample orbital space and spacing between satellites was wide. In 1981, the American Federal Communications Commission (FCC) surprised the industry by requesting comments on the possibility of authorizing spacing of satellites along the North and South American portion of the geosynchronous arc at tight 2° intervals. Until then, the accepted policy was 4° and 2° degree spacing for C and Ku-band satellites, respectively.

In April, 1983 the American FCC approved 2° orbital spacing for satellites operating in both frequency bands. This decision, implemented gradually by reducing spacing to 3° at first, would eventually allow a doubling of the orbital arc capacity.

The discussion of satellite spacing has now come full circle. As satellite transponder powers are increasing and smaller dishes become capable of good quality reception, many are arguing that satellites should be spaced a minimum of 3° apart. This strategy would help to avoid interference from adjacent satellites.

The Importance of Beamwidth

Antenna quality and beamwidth are critical factors in satellite spacing. For example, if the half power (3 dB) beamwidth of a reflector is 2°, then at half of this value or 1° on either side of the main axis, signals are received at 3 dB lower than those detected along the boresight. If a satellite happens to be located just 1° off the target and is transmitting on the same channel and polarity, two garbled pictures and the resulting interference would appear on the television screen (see Table 10-2).

TABLE 10-2. ANTENNA DIAMETER VERSUS 3 dB BEAMWIDTH

Antenna Diameter (meters)	Theoretical Half Power Beamwidth
0.5	3.43
1.0	1.72
1.5	1.14
2.0	0.86
2.5	0.69
3.0	0.57
3.5	0.49

However, the situation would be quite different if the antenna detected signals from 1° off boresight at powers reduced by 10 dB, or by 15 dB. At what level is a picture judged un-watchable when interference from an adjacent satellite is just barely detectable? Subjective tests have shown that an interfering signal is first noticeable when its level is 11 dB below the desired signal. The interference is annoying when received at -5 dB and quite unacceptable at -4 dB.

This means that a 3 meter antenna would detect interfering signals reduced by only 3 dB at 0.26^{o} off its main axis. Levels will even be higher at, for example, 0.20^{o}. However, a good quality 3 meter antenna would have unwanted signals reduced by substantially more than 3 dB at 1.0^{o} off-axis.

Most prime focus parabolic antennas have sufficiently low side lobes so that off-axis signals are not appreciably detected. However, smaller diameter Cassegrain fed antennas can have side lobes only 10 dB lower in power than the main lobe. Such reflectors are more susceptible to interference from adjacent satellites as well as from terrestrial sources.

The Role of Antenna Quality

An antenna with an accurate surface and a well-matched feed assembly would function more closely to optimal performance and have a smaller beamwidth than one of lesser quality. This is because measured beamwidths are also determined by antenna efficiency. Also, because smaller antennas require substantially less material for construction than comparable larger reflectors, leftover manufacturing dollars can be spent in achieving better surface accuracy. The other side of the coin is that better accuracy can be achieved with more manageable, smaller antennas. The net result is that a well made 6 foot antenna may have less than half the beamwidth of a poorly constructed 12 foot unit contrary to theoretical expectations.

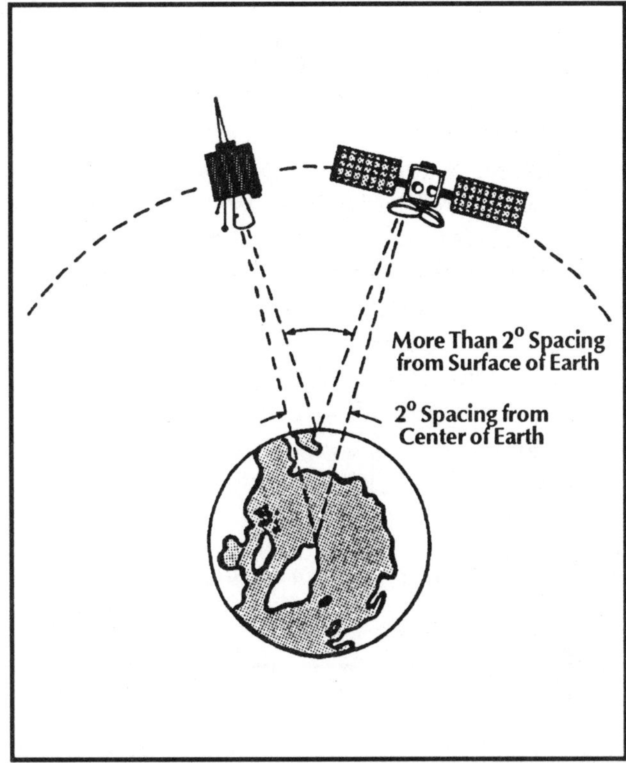

Table 10-1. Variation of Spacing With Site Latitude. *Satellites spaced 2° apart are really separated by a slightly larger angle. While not drawn to scale, this polar view of the earth illustrates the point. The published separation between satellites is measured from the center of the earth but an antenna actually lies closer to the arc and therefore subtends a smaller angle. As the site becomes closer to either pole, the angle decreases towards the published angle.*

Is One Degree Spacing Really One Degree?

The published separation between satellites is measured by their difference in longitude. From the perspective of an observer at the center of the earth, the angular separation between two satellites directly overhead would be the same as their longitude difference. But an antenna located on the earth's surface pointing straight upwards would see these satellites separated by more than 1° or 2°. This spacing decreases as satellites closer to the horizon are targeted but still remains in excess of the nominal value (see Figure 10-1).

When a satellite antenna is located at other points on the surface of the earth, the perceived separation between satellites will also have values in excess of or equal to the satellite spacing. For example, from Honolulu, Hawaii satellites at 131° and 132° longitude will appear to be not 1.0° but 1.14° apart. It is important to realize that these angles are simply related to the geometry of the situation. The maximum spacings occur when an antenna is aimed due south, while the minimum ones occur when it points towards either horizon.

Thus if satellites are spaced at 1° intervals it does not necessarily mean that the power levels at the one degree point on a beamwidth diagram tell the whole story. Performance results can be determined only by comparing the exact angular separation between satellites. For example, the beamwidth diagram may indicate that power lev-

TABLE 10-3. EFFECTIVE SATELLITE SPACING VERSUS LATITUDE (At 1° satellite spacing)

Latitude	Maximum Angular Angular Separation	Minimum Separation
81.00	1.000	1.000
64.83	1.053	1.000
47.60	1.101	1.000
40.67	1.120	1.000
39.73	1.121	1.000
33.75	1.138	1.000
25.75	1.154	1.000
00.00	1.178	1.000

Maximum and minimum angular separation occur when oriented at the horizon and due south, respectively.

els are down by 18 and 20 dB at 1° and 1.2°. Table 10-3 shows that if a TVRO located at a 40° latitude is targeting satellites due south, the 18 dB figure should be used, but if they are near the horizon, signals from the off-target spacecraft would be detected at a 20 dB lower power level.

Satellite Spacing and Regulation

In all probability, smaller antennas should be perfectly adequate even when satellites are spaced at 1° intervals if polarity formats alternate between satellites. To illustrate, for those satellites using linear polarity formats, if an earth station was tuned to channel 5 having vertical polarity and an adjacent satellite was transmitting the same channel on horizontal polarity, the feedhorn would hardly detect these interfering signals of opposite polarity. This "cross-polarization discrimination" adds approximately 20 dB of additional protection from interference. For example, even if a two meter reflector detected adjacent satellite signals at only 6 dB below those from the desired broadcast, the added protection incurred by having the unwanted signals polarized in the opposite sense would bring these interfering levels down to a perfectly manageable 26 dB.

Nevertheless, using an under-sized antenna can still present problems even if adjacent satellites are cross polarized. Its gain may not be sufficient to detect the weaker transponders without unwanted sparklies. In addition, a reflector having a wide main lobe and high side lobes is much more susceptible to interference from terrestrial sources as well as other communication satellites. Of course, a judgment of what constitutes adequate antenna diameter ultimately rests upon subjective opinions about quality of reception.

XI. SMALL DISH SYSTEMS

Ever since home satellite systems became popular in the early 1980s, one question has always been asked: "Can a smaller dish be used?" Today, this desire is becoming a reality as the power of C-band satellites is being increased and as direct broadcast Ku-band systems are being introduced.

C-band satellites have been limited in power to avoid interference with land-based communicators who share the same 3.7 to 4.2 GHz frequency band. Transponders have typically had powers of 6 to 8 watts. However, the next generation of satellites that are being launched in the early 1990s have power levels of about 15 watts per transponder.

Ku-band broadcasts from relatively powerful satellites are also gaining in popularity. A number of companies are in the process of launching new Ku-band direct broadcast services (DBS) that can be received by dishes as small as a foot or two in diameter. Broadcast consortiums such as Sky Channel, K-Prime and SkyPix have plans to offer up to 60 channels per high power satellite. (See *Ku-Band Satellite TV - Theory, Installation and Repair for more details*).

A. FACTORS AFFECTING DISH SIZE

A number of factors must be considered in determining antenna size. These include transponder power levels, satellite spacing and video compression techniques.

Transponder Power

The power intercepted by an antenna is directly related to transponder power level. If the power increases from 8 to 15 watts, then EIRP, the power detected on earth, also increases by the same proportion. Since antenna size is also directly related to the amount of signal intercepted, more powerful satellites can be received with smaller dishes.

The new generation of C-band satellites feature transponders that are about twice as powerful as those on satellites launched in the 1980s. This means that the EIRP will increase by 3 dBw, a doubling of power. As a result, dish reflector area can be cut in half with about the same quality reception. Therefore, for example, an 8 foot antenna could be replaced with a 5.7 footer of equal efficiency. The latter has about half the surface area of the former.

Another factor has an important effect on acceptable antenna size. LNB noise temperatures dropped from approximately 80°K to as low as 25°K between 1985 and 1990. This translates into an improvement in performance. The link equation outlined in Chapter X can be used to show that the improvement in performance depends upon both antenna and LNB noise temperature. For example, a drop in noise temperature from 80° to 30° translates into a 2.1 dB improvement in gain with a dish noise temperature of 50°K. This improvement increases to 2.6 dB for an antenna with a 30°K noise temperature. As a reference, Figure 2-18 illustrates how antenna noise temperature varies with antenna size and elevation angle.

Assuming that the increase in satellite EIRP is 3 dB and the improvement in system performance from use of an improved LNB is 2 dB, the net result is that antenna gain can be 5 dB lower. For example, an 8 foot antenna in combination with an 80° LNB could be replaced with a comparable 4.5 footer mated with a 30° LNB. Or a 10 foot dish could be replaced with a 5.6 footer. Certainly the era of the small C-band antenna has arrived.

Antenna Beamwidth

Another factor that limits antenna size is its half power beamwidth. This has been discussed earlier in Chapters II and X. As dish diameter decreases, beamwidth increases to the point that the signal from adjacent satellites interferes with reception of the target satellite. For example, while an 8 foot antenna has a 3 dB beamwidth of 2.2°, a 2 foot dish has a beamwidth of 8.6° (see Appendix B for the equation). The smaller dish would detect satellites at 4.3° off boresight at half the power level of those on target. The resulting interference would be unacceptable.

A number of tactics have been discussed to allow the use of smaller dishes. The minimum satellite spacing could be increased to 3°. Also, if the polarities of even/odd channels were switched from satellite to satellite, a dish would have to skip over one satellite to detect a signal of the same polarity. Therefore, if satellites were spaced 3° apart, the antenna could only see satellites at 6° on either side of the target. A 2 foot antenna would then perform acceptably.

Video Compression Techniques

Video compression techniques such as the newly introduced DigiCipher or the SkyPix versions allow more than one video broadcast to be relayed per transponder. For example, the developers of DigiCipher state that this system is capable of transmitting ten NTSC or two high definition television channels per transponder.

This technique is based on the technique of transmitting only those portions of picture that change from moment to moment. Picture segments that remain constant are not transmitted thus reducing the amount of information and bandwidth required.

When just a single channel per transponder is relayed using video compression techniques, a smaller antenna can be used. However, as a transponder is loaded with channels up to the particular system's capability, antenna size must be increased.

Antenna Quality

Antenna gain and therefore required size also vary with the accuracy of the reflector surface. The gain of an 8 foot dish differs by 2.04 dB between a 50 percent and 80 percent efficient antenna (see Table 2-2). Some of the earlier mesh antennas probably had efficiencies even below 50 percent. Some innovative antenna designs being developed today may have the potential for efficiencies of 85 percent or above.

APPENDIX A. THE DECIBEL NOTATION

Decibels (dB) are used to express the relative values of two signals. The logarithmic scale is used to compress large differences in numbers to a more manageable range. Decibels are defined by the following equation:

Decibel difference = 10 log (signal A/signal B)

For example, if signal A is 1000 watts and signal B is 10 watts, then signal A is 20 dB stronger than signal B because:

Decibel difference in power = 10 log (1000/10)
= 10 x 2
= 20 dB

Therefore, if an amplifier received a signal of 10 watts and increased its strength by a factor of 100 to 1000 watts, it would have a gain of 20 dB. Similarly, if a 10 watt signal was increased by a factor of 1,000,000 to 10 million watts, the gain would be 60 dB.

Decibels are also expressed relative to a reference value such as watts, milliwatts or millivolts. The abbreviations dBw, dBm and dBmv mean the relative increase in power relative to one watt, to one milliwatt and to one millivolt, respectively. For example, 60 dBw means a power of 1 million watts.

The definition of decibels relative to one millivolt (or ampere) differs from that used to express power differences. A signal is expressed in dBmv by

20 log (signal in millivolts / 1 millivolt).

Therefore, for example, 20 dBmv is equal to a signal of 10 millivolts. The difference in the definition arises because power levels are proportional to the square of the voltage (or current).

APPENDIX B. SATELLITE TV EQUATIONS

The performance of a satellite TV system can be theoretically calculated by using some general equations. Some of these are presented below.

Link Equations

The link equations are used to calculate the ratio of carrier-to-noise (C/N) power reaching the input of a satellite receiver. The link equation is

$$C/R = EIRP - Path\ Loss + G/T_{sys} - 10\log B + 228.6$$

EIRP is the effective isotropic radiated power directed by a downlink antenna to a location below. It is expressed in dBw, decibels relative to one watt.

Path loss measures how much signal is lost on the journey from the communication satellite to the receiving antenna. Losses are mainly due to the "spreading out" of the signal on its long journey. The amount of spreading out is determined by the distance the signal travels from a satellite to a receiving antenna. Path loss in decibels is given by

$$Path\ Loss = 20 \log 4\pi Sf$$

where S is the slant range and f is the signal frequency in Hertz. Slant range, the distance to the satellite is given by

$$S = [R^2 + (R+h)^2 - 2R(R+h)\cos\Phi \cos\Delta]^{1/2}$$

where R equals 6,367 km, the radius of the earth, h equals 35,803 km, the distance of a satellite above the center of the earth, Φ is the site latitude and Δ is the absolute difference between the site and satellite longitude. Substituting R and h in this equation gives

$$S = 1000\,[1018.32 - 53.69 \cos\Phi \cos\Delta]^{1/2}$$

Substituting this value of S into the above equation we find path loss equals

$$10 \log[1 - 0.295 \cos\Phi \cos\Delta] + 20\log f + 185.05$$

where f is expressed in GHz. At 4 GHz the path loss equals 195.6 dB at an earth station located on the equator directly below a satellite. This equation also shows that signals from a satellite 10 degrees in longitude away from an earth station at 40 degrees latitude would suffer a 196.0 dB free space path loss.

Atmospheric absorption causes additional path losses. Absorption increases with slant range because the satellite signal must pass through a greater thickness of atmosphere. An absorption loss of 0.3 to 0.5 dB is assumed in most calculations for reception on average clear days.

G/T_{sys}, the ratio of antenna gain to system noise temperature, is the figure of merit of an antenna/feed/LNB system. It is expressed in decibels as

$$G - 10 \log T_{sys}$$

The system noise temperature primarily depends upon both the antenna and LNB noise temperatures. However, components further downstream towards the receiver also contribute small amounts of noise. This term is given by

$$T_{sys} = T_{ant/feed} + T_{LNB}/G_{feed} + \frac{T_{rec/coax}}{G_{LNB} + G_{feed}}$$

where G refers to gain. The gain of a typical feed is in the vicinity of 0.99 while LNB gain is usually about 50 dB, equal to a factor of 100,000. This equation clearly shows why the noise contributed to T_{sys} by the satellite receiver and coaxial cable are negligible. The LNB amplifies both the signal and noise so much that any later noise contributions are of minor importance.

The second of last term in the link equations above adjusts for the system bandwidth and the final is a constant, called Boltzman's constant.

Antenna Gain

Antenna gain relative to an "isotropic" antenna, one that radiates equally in all directions, is given by

$$G = E(\pi D/\lambda)^2$$

where E is the antenna efficiency, D is the diameter and λ is the wavelength of the incoming radiation. The wavelength in centimeters can be simply calculated by dividing 30 by the frequency expressed in GHz. The wavelength of 4 GHz microwaves is 7.48 cm.

For example, a 2 meter with an efficiency of 55% operating at 4 GHz has a gain given by

$$G = 0.55 \times (3.14 \times 200\ \text{cm} / 7.48\ \text{cm})^2$$
$$= 3{,}878$$

Expressed in decibels this gain or signal concentration of a factor of 34706 relative to an isotropic antenna is given by

$$G = 10 \log 3{,}878 = 35.9\ \text{dBi}$$

Loss of Gain with Surface Irregularities

The decrease of gain relative to a perfect antenna having no surface irregularities is given by

$$\text{Loss of Gain} = e^{-8.80(RMS)/\lambda}$$

where RMS is the root mean square deviation from a perfect geometrical shape and λ is the wavelength of the incoming signal. The RMS is a measure of the "tightness" of the surface or its average tolerance.

For example, a C-band antenna operating at 4 GHz, where wavelength equals 7.48 cm, having a RMS tolerance of 0.15 cm has a decrease in gain relative to a perfect antenna given by

$$\text{Loss of Gain} = e^{-8.80 \times 0.15/7.48}$$
$$= e^{--0.18}$$
$$= 0.84$$

or a 16 percent decrease in gain. This equates to a decibel loss of gain given by

$$\text{Decibel Loss in Gain} = 10 \log 0.84$$
$$= -0.76\ \text{dB}$$

Antenna Beamwidth

An approximate but very useful formula for 3 dB antenna beamwidth is:

$$\text{Beamwidth} = 70\,\lambda/D$$

where λ is the wavelength of the microwave radiation and D is antenna diameter. For example, a 2 meter antenna has a 3 dB beamwidth given by:

$$\text{Beamwidth} = 70 \times 7.48 / 200 = 2.62^{o}$$

Similarly, a 1 meter antenna would have a calculated beamwidth given by

$$\text{Beamwidth} = 70 \times 7.48 / 100 = 5.24^{o}$$

Noise Temperature and Figure

The noise any system generates is proportional to its ambient temperature and the bandwidth of the signal it processes. The larger either of these two quantities, the greater the contributed noise.

$$\text{Noise} = kTB$$

where k is Boltzman's constant, T is the ambient temperature and B is the system bandwidth.

A quantity called noise factor is defined by the ratio of the noise at the output on an electronic component to the noise at its input. This quantity measures, in essence, the amount of noise internally generated in any device. In a perfect device whose electronic circuits added no extra noise to a signal, the noise factor would be one.

$$\text{Noise Factor} = \frac{(\text{Ideal Noise} + \text{Internal Noise})}{\text{Ideal Noise}}$$

$$= (kBT_{Ideal} + kBT_{Eq}) / kBT_{Ideal}$$

$$= (T_{Ideal} + T_{Eq}) / T_{Ideal}$$

$$= 1 + T_{Eq} / T_{Ideal}$$

$$= 1 + T_{Eq} / 290$$

T_{Eq}, termed the equivalent noise temperature. The reference noise temperature, T_{Ideal}, is usually taken to be 290°K, equal to an average room temperature of about 63°F.

Noise figure is the decibel equivalent of noise factor as is given by

$$\text{Noise Figure} = 10 \log (\text{Noise Factor})$$

For example, if the noise figure is 1.2 dB, the equivalent noise temperature is

$$1.2 = 10 \log (1 + T_{Eq} / 290)$$

turning this equation inside out

$$1 + T_{Eq} / 290 = 10^{0.12}$$
$$= 1.32$$

$$T_{Eq} / 290 = 0.32$$
$$\therefore T_{Eq} = 93°K$$

The Effect of Bandwidth on System Noise Power

The noise power in any communication system is given by

$$\text{System noise power} = kT_{sys} B$$

where T_{Sys} is the system noise temperature in degrees Kelvin mainly determined by antenna and LNB noise, k is Boltzman's constant equal to 1.381×10^{-23} and B is the communication bandwidth. The change in noise power between two systems can be computed as follows

$$\text{Change in noise power} = kT_1B_1/kT_2B_2$$

$$= T_1B_1/T_2B_2$$

Therefore, if the noise temperature remains constant the change in noise power is simply the ratio of bandwidths. If the bandwidth were cut from 36 to 18 MHz as would be the case in half transponder formats, the noise power would be reduced by 50 percent or 3 dB. The resulting doubling of the signal-to-noise ratio sometimes makes the difference between a watchable picture and a sparklie, faint ghost of a picture. But reducing the bandwidth will also result in a "softening" of the video as well as smearing and streaking of pictures having fast changes.

Declination Angle

The declination angle for a polar mount can easily be found from the tables and figures in the early chapters. It can also be calculated from

$$\text{Declination} = \text{Tan}^{-1} \frac{3964 \sin L}{22300 + 3964(1 - \cos L)}$$

where L is the site latitude. The two numbers in this equation are the radius of the earth and the distance from the surface of the earth to the arc of satellites. For example, at 40 degrees latitude

$$\text{Declination} = \text{Tan}^{-1} \frac{3964 \sin 40}{22300 + 3964(1 - \cos 40)}$$
$$= \text{Tan}^{-1} 0.11$$
$$= 6.26^{\circ}$$

Azimuth and Elevation Angles

Antenna pointing angles can be calculated in degrees from true north from the following equations:

$$\text{Azimuth Angle} = \tan^{-1}[-\tan\Delta/\sin\Phi]$$

$$\text{Elevation Angle} = \tan^{-1}[(\cos Y - 0.15116)/\sin Y]$$

$$Y = \cos^{-1}[\cos\Phi\cos\Delta]$$

where Δ is the absolute value of the difference between satellite and TVRO site longitudes and Φ is the site latitude.

Voltage Standing Wave Ratio

The voltage standing wave ratio, VSWR, is a measure of the amount of input signal reflected back and lost. A perfect device would have no reflective losses and have a VSWR of 1:1. Table B-1 shows how reflected signal power and transmission losses vary with VSWR.

TABLE B-1. VSWR

VSWR	Reflected Signal (%)	Transmission Loss (dB)
1.0:1	0	0
1.1:1	0.2	0.01
1.2:1	0.9	0.03
1.3:1	1.6	0.07
1.5:1	4.0	0.18
2.0:1	11.0	0.50

Antenna Parabolic Geometry

The basic equation for a parabolic reflector is given by

$$y = x^2/4f$$

where f is the focal distance. Another useful formula gives the focal distance f in terms of the antenna diameter and depth

$$f = \text{diameter}^2/16 \times \text{depth}$$

Wind Loading

Winds can have very dramatic effects on antennas and their supporting structures. The following tables, kindly calculated and provided by Dick Zlotky, president of Earthbound, Inc., indicate the strength and direction of the forces that can be expected in a real-world environment.

Design wind loads on parabolic reflectors were calculated from wind tunnel coefficient data complied by the Jet Propulsion Laboratory at the California Institute of Technology as presented in report number JPL 78-16. Wind tunnels tests were performed on several models of parabolic reflectors, both solid and porous, at azimuth angles ranging from 0°, wind directly along the antenna axis, to 180°, as well as at elevation angles from zero to 90° at each azimuth angle. Axial, normal, side force, moment and torsional coefficients were determined at each position of the reflector.

All coefficients compiled in JPL 78-16 act at the apex of the reflector as shown in Figure B-1. These are all positive in sign. Since all the force coefficients are perpendicular and parallel to the longitudinal axis of the reflector, it is necessary to resolve these forces and moments in planes perpendicular and parallel to the ground. These must be then transferred to the centerline of the pole in order to use them for the pole and foundation structural design. The resolution of these forces and moments is shown in Figure B-2. It is assumed that the center-line of the pole is at a distance of 15.2 cm from the apex of the reflector. The forces and moments presented in Tables B-2, B-3 and B-4 are centered about point "A" in Figure B-2. This point is located at the intersection of the pole center-line and a horizontal plane through the apex of the reflector at a height H above ground level.

An antenna having an f/D of 0.313 was used in the tests. Antennas having surface porosities of 0% (solid) and 25% were tested. The wind loads shown in the tables below are calculated from these two different surfaces. Note that in order to achieve the crucial pointing accuracy in Ku-band reception, the support poles and foundations must have sufficient rigidity so that deflections in torsion or twisting as well as bending are minimized.

JPL 78-16 states that the amount of reduction of wind loads offered by a porous panel such as a perforated reflector is primarily affected by its t/d ratio, where t is the thickness of the panel and d is the diameter of its holes. As t/d approaches 3, the efficiency nears zero and the drag characteristics approach that of a solid reflector. The t/d for models used in the JPL wind tunnel tests approached 100% efficiency in reducing drag. In other words, the t/d values were low and therefore representative of the performance of today's perforated and mesh antennas.

TABLE B-2. SOLID ANTENNA

FORCES & MOMENTS	ANTENNA DIAMETER (m)					
	1.0	**1.2**	**1.8**	**2.4**	**3.0**	**3.7**
Fx' (kg)	154	222	499	886	1384	2107
Fy' (kg)	-5	-7	-16	-28	-44	-66
Fz' (kg)	-55	-80	-180	-320	-500	-761
Torsion (kg-cm)	-450	-756	-2432	-5589	-10759	-19946
My'y' Moment (kg-cm)	9	263	2257	6952	15478	31766
Mx'x' Moment (kg-cm)	-138	-238	-807	-1898	-3721	-6977

Assumptions: Wind Velocity = 160.0 k/hr
Azimuth = 60°
Elevation = 20°

TABLE B-3. 25% POROUS ANTENNA

FORCES & MOMENTS	ANTENNA DIAMETER (m)					
	1.0	**1.2**	**1.8**	**2.4**	**3.0**	**3.7**
Fx' (kg)	74	110	248	439	686	1044
Fy' (kg)	15	21	48	84	132	201
Fz' (kg)	-24	-34	-78	-137	-215	-327
Torsion (kg-cm)	616	1006	3026	6728	12647	23024
My'y' Moment (kg-cm)	207	467	2160	5779	12135	23927
Mx'x' Moment (kg-cm)	147	255	866	2036	3991	7469

Assumptions: Wind Velocity = 160. 0 km/hr
Azimuth = 60°
Elevation = 20°

TABLE B-4. MAXIMUM TORSION

MAXIMUM TORSION (kg-m)	ANTENNA DIAMETER (m)					
	1.0	**1.2**	**1.8**	**2.4**	**3.0**	**3.7**
Solid Antenna 1647	2735	8776	16646	38533	71289	
25% Porous Antenna	1467	2450	7754	17716	33895	62629

Assumptions: Wind Velocity = 160.0 km/hr
Azimuth = 120°
Elevation = 20°

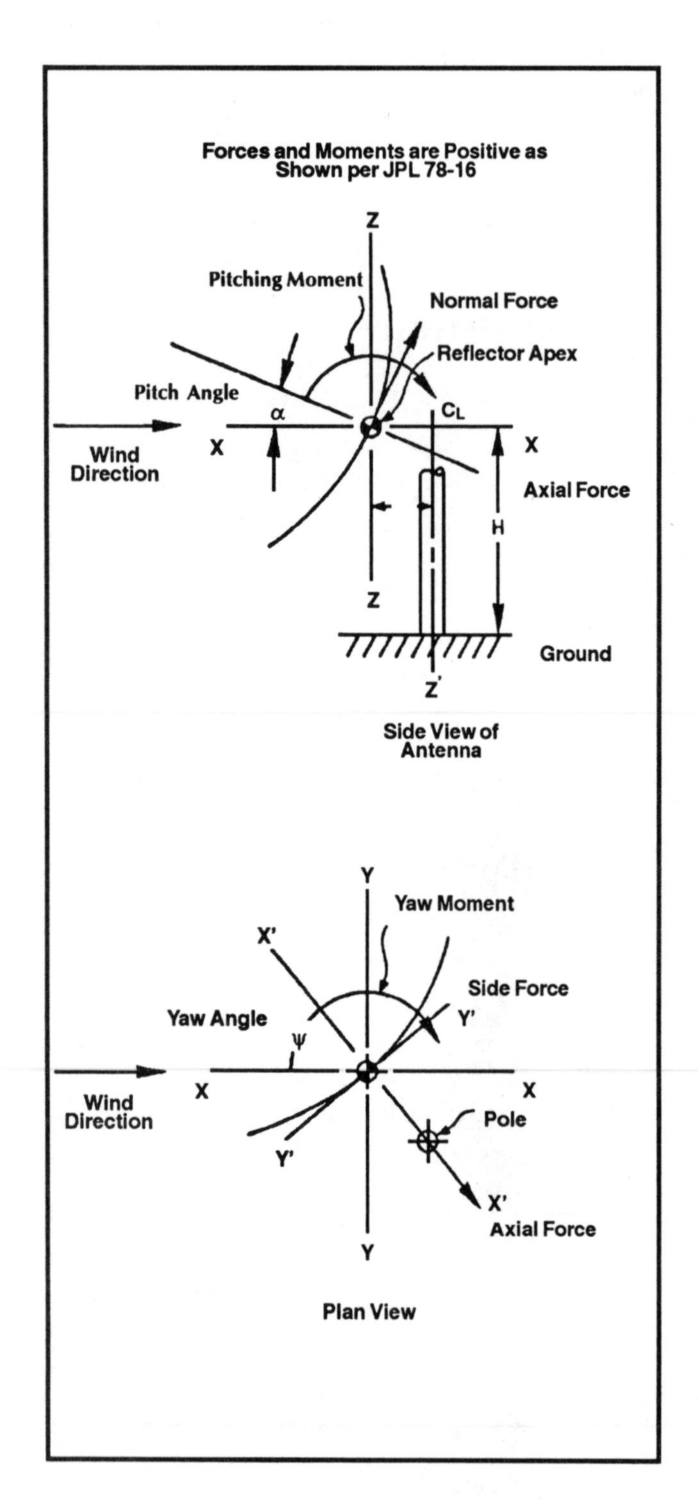
Forces and Moments are Positive as Shown per JPL 78-16
Z
Pitching Moment
Normal Force
Reflector Apex
Pitch Angle
α
C_L
X
X
Wind Direction
Axial Force
H
Z
Ground
Z'
Side View of Antenna
Y
Yaw Moment
X'
Side Force
Yaw Angle
Y'
ψ
X
X
Wind Direction
Pole
Y'
X'
Axial Force
Y
Plan View

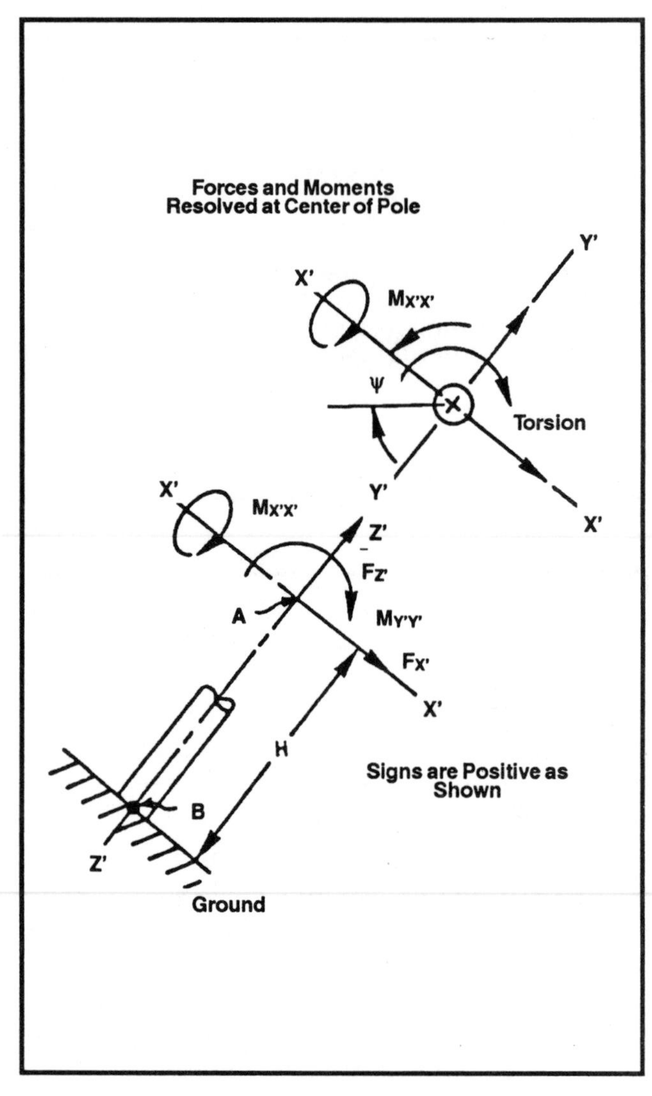
Forces and Moments Resolved at Center of Pole
Y'
X'
Mx'x'
ψ
Torsion
Y'
X'
X'
Mx'x'
Z'
Fz'
A
My'y'
Fx'
X'
H
Signs are Positive as Shown
B
Z'
Ground

Additional Calculations

The forces and moments given in Tables B-2 through B-4 are for 160.9 km/hr winds exerted at a height H above ground level. Forces and moments for other wind velocities are easily calculated by multiplying the 160.9 km/hr velocity by the square of the new velocity divided by 10,000, namely

$$(v/100)^2$$

For example, to find $F_{X'}$ exerted by a 112.6 km/hr wind on a 1.8 meter solid antenna, multiply the 499 kg found in Table B-2 by $(112.6/100)^2$. The result is a force of 245 kg at the lower velocity.

To find the forces at the point where a pole attaches to the foundation, point B in Figure B-2, it is necessary to transfer the forces and moments from point A to point B. All components of the forces have the same values at both points A and B.

For example, from Table B-2 the following values are found for a 1.8 meter solid antenna at H of 121.9 cm above ground level

$$F_{X'} = 499 \text{ kg}$$
$$F_{Y'} = -16 \text{ kg}$$
$$F_{Z'} = -180 \text{ k}$$
$$\text{Torsion} = 02432 \text{ kg-cm}$$
$$M_{Y'Y'} = 2257 \text{ kg-cm}$$
$$M_{X'X'} = -807 \text{ kg-cm}$$

The loads at ground level (point B) are found by

$$M_{Y'Y'} = F_{X'} \times H + M_{Y'Y'} = 499 \times 121.9 + 2257$$
$$= 63085 \text{ kg-cm}$$

$$M_{X'X'} = F_{Y'} \times H + M_{X'X'} = -16 \times 121.9 - 807$$
$$= -2757$$

The resultant overturn moment at point B is then given by the resultant of these two moments

$$= [(63085)^2 + (-2757)^2]^{1/2}$$
$$= 63145 \text{ kg-cm}$$

APPENDIX C. CHANNELS - WAVELENGTHS & BANDWIDTHS

Channel	Video Frequency (MHz)	Wavelength (Inches)
	SUB CHANNELS	
01	15.75	749.7
02	21.75	542.9
03	27.75	425.5
04	33.75	349.9
05	39.75	297.1
	VHF-LO CHANNELS	
2	55.25	213.7
3	61.25	192.8
4	67.25	175.6
5	77.25	152.9
6	83.25	141.8
FM	88.0-108.0	134.2-109.3
	VHF-HI CHANNELS	
7	121.25	97.4
8	127.25	92.8
9	133.25	88.6
10	139.25	84.8
11	145.25	81.3
12	151.25	78.1
13	157.25	75.1
	SUPERBAND CHANNELS	
J	217.25	54.5
K	223.25	52.9
L	229.25	51.5
M	235.25	50.2
N	241.25	49.0
O	247.25	47.8
P	253.25	46.6
Q	259.25	45.6
R	265.25	44.5
S	271.25	53.5
T	277.25	42.6
U	283.25	41.7
V	289.25	40.8
W	295.25	40.0
	HYPERBAND CHANNELS	
AA	301.25	39.2
BB	307.25	38.4
CC	313.25	37.7
DD	319.25	37.0
EE	325.25	36.3
FF	331.25	35.7
GG	337.25	35.0
HH	343.25	34.4
II	349.25	33.8
KK	361.25	32.7
LL	367.25	32.2

Channel	Video Frequency (MHz)	Wavelength (Inches)
MM	373.25	31.6
NN	379.25	31.1
OO	385.25	30.7
PP	391.25	30.2
QQ	397.25	29.7
RR	403.25	29.3
SS	409.25	28.9
TT	415.25	28.4
UU	421.25	28.0
VV	427.25	27.6
WW	433.25	27.3
	UHF CHANNELS	
14	471.25	25.1
15	477.25	24.7
16	483.25	24.4
17	489.25	24.1
18	495.25	23.8
19	501.25	23.6
20	507.25	23.3
21	513.25	23.0
22	519.25	22.7
23	525.25	22.5
24	531.25	22.2
25	537.25	22.0
26	543.25	21.7
27	549.25	21.5
28	555.25	21.3
29	561.25	21.0
30	567.25	20.8
31	573.25	20.6
32	579.25	20.4
33	585.25	20.2
34	591.25	20.0
35	597.25	19.8
36	603.25	19.6
37	609.25	19.4
38	615.25	19.2
39	621.25	19.0
40	627.25	18.8
41	633.25	18.7
42	639.25	18.5
43	645.25	18.3
44	651.25	18.1
45	657.25	18.0
46	663.25	17.8
47	669.25	17.6
48	675.25	17.5
49	681.25	17.3
50	687.25	17.2
51	693.25	17.0
52	699.25	16.9
53	705.25	16.7
54	711.25	16.6

APPENDIX D. GLOSSARY

A/B Switch

A switch that selects one of two inputs (A or B) for routing to a common output while providing adequate isolating between the two signals.

AFC(Automatic Frequency Control)

A circuit which locks an electronic component onto a chosen frequency.

AGC(Automatic Gain Control)

A circuit that uses feedback to maintain the output of an electronic component at a constant level.

Absolute Zero

The coldest possible temperature at which all molecular motion ceases. It is expressed in degrees Kelvin as measured from absolute zero. Zero degrees Kelvin equals minus 273.16 °C or minus 459.69 °F.

Adjacent Channel

An adjacent channel is immediately next to another channel in frequency. For example, NTSC channels 5 and 6 as well as 8 and 9 are adjacent. However, channels 4 and 5 or channels 6 and 7 are separated by signals used by non-TV media.

Agile Receiver

A satellite receiver which can be tuned to any desired channel.

Alignment

The process of fine tuning a dish or an electronic circuit to maximize its sensitivity and signal receiving capability.

Ambient Temperature

The existing dry bulb temperature.

Amplifier

A device used to increase the power of a signal.

Analog

A system in which signals vary continuously in contrast to a digital system in which signals vary in discrete steps

Analog-to-Digital Converter

A circuit that converts analog signals to an equivalent digital form. The varying analog signal is sampled at a series of points in time. The voltage at each of these points is then represented by a series of numbers, the digital value of the sample. The higher this sampling frequency, the finer are the gradations and the more accurately is the signal represented.

Antenna

A device that collects and focuses electromagnetic energy, i.e., contributes an energy gain. Satellite dishes, broadband antenna and cut-to-channel antennas are some types of antennas encountered in private cable systems. In the case of satellite antennas, gain is proportional to the surface area of the microwave dish.

Antenna Efficiency

The percentage of incoming satellite signal actually captured by an antenna.

Antenna Illumination

Describes how a feedhorn "sees" the surface of a dish as well as the surrounding terrain.

Aperture

The collection area of a parabolic antenna.

Aspect Ratio

The ratio of television screen width to height. The standard aspect ratio is 4 to 3.

Attenuation

The decrease in signal power that occurs in a device or when a signal travels to reach a destination point (path loss).

Attenuator

A passive device which reduces the power of a signal. Attenuators are rated according to the amount of signal attenuation.

Audio Subcarrier

The carrier wave that transmits audio information within a video broadcast signal. Satellite transmissions can relay more than a single audio subcarrier in the frequency range between 5 and 8.5 MHz.

Automatic Brightness Control

A television circuit used to automatically adjust picture tube brightness in response to changes in background or ambient light.

Automatic Fine Tuning

A circuit that automatically maintains the correct tuner oscillator frequency and compensates for drift and for moderate amounts of inaccurate tuning. Similar to AFC.

Automatic Frequency Control (AFC)

A circuit that locks onto a chosen frequency and will not drift away from that frequency.

Automatic Gain Control (AGC)

A circuit that locks the gain onto a fixed value and thus compensates for varying input signal levels keeping the output constant.

Azimuth-Elevation (Az-El) Mount

An antenna mount which tracks satellites by moving in two directions: the azimuth in the horizontal plane; and elevation up from the horizon.

Azimuth

A compass bearing expressed in degrees of rotation clockwise from true north. It is one of the two coordinates (azimuth and elevation) used to align a satellite antenna.

Back Match

The matching of the resistive values of the input and output of electronic devices to reduce signal reflection and ghosting. Also known as impedance matching.

Back Porch

That portion of the horizontal blanking pulse that follows the trailing edge of the horizontal sync pulse.

Band

A range of frequencies.

Band Separator

A device that splits a group of specified frequencies into two or more bands. Common types include UHF/VHF, Hi/Lo-band and FM separators. This device is essentially a set of filters.

Bandpass Filter

A circuit or device that allows only a specified range of frequencies to pass from input to output.

Bandwidth

The frequency range allocated to any communication circuit.

Baseband

The raw audio and video signals prior to modulation and broadcasting. Most satellite headend equipment utilizes baseband inputs. More exactly, the composite unclamped, non-de-emphasized and unfiltered receiver output. This signal contains the complete set of FM modulated audio and data subcarriers.

Beamwidth

A measure used to describe the width of vision of an antenna. Beamwidth is measured as degrees between the 3 dB half power points.

Bird

Jargon or nickname for communication satellites.

Blanking Pulse Level

The reference level for video signals. The blanking pulses must be aligned at the input to the picture tube.

Blanking Signal

Pulses used to extinguish the scan illumination during horizontal and vertical retrace periods.

Block Downconversion

The process of lowering the entire band of frequencies in one step to some intermediate range to be processed inside a satellite receiver. Multiple block downconversion receivers are capable of independently selecting channels because each can process the entire block of signals.

BNC Connector

A weatherproof twist lock coax connector standard on commercial video equipment and used on some brands of satellite receivers.

Boresight

The direction along the principle axis of either a transmitting or a receiving antenna.

Broadband

A device that processes a signal(s) spanning a relatively broad range of input frequencies.

Buttonhook Feed

A rod shaped like a question mark supporting the feedhorn and LNA. A buttonhook feed for use with commercial grade antennas is often a hollow waveguide that directs signals from a feedhorn to an LNA behind the antenna.

CATV

An abbreviation for Community Antenna Television - another name for cable TV.

CCD

Charge coupled device. In this device charge is stored on a capacitor which are etched onto a chip. A number of samples can be simultaneously stored. Used in MAC transmissions for temporarily storing video signals.

C-Band

The 3.7 to 4.2 GHz band of frequencies at which some broadcast satellites operate.

Carrier

A pure-frequency signal that is modulated to carry information. In the process of modulation it is spread out over a wider band. The carrier frequency is the center frequency on any television channel.

Carrier-to-Noise Ratio (CNR)

The ratio of the received carrier power to the noise power in a given bandwidth, expressed in decibels. The CNR is an indicator of how well an earth receiving station will perform in a particular location, and is calculated from satellite power levels, antenna gain and the combined antenna and LNA noise temperature.

Cassegrain Feed System

An antenna feed design that includes a primary reflector, the dish, and a secondary reflector which redirects microwaves via a waveguide to a low noise amplifier.

Channel

A segment of bandwidth used for one complete communication link.

Chrominance

The hue and saturation of a color. The chrominance signal is modulated onto a 4.43 MHz carrier in the PAL television system and a 3.58 MHz carrier in the NTSC television system.

Chrominance Signal

The color component of the composite baseband video signal assembled from the I and Q portions. Phase angle of the signal represents hue and amplitude represents color saturation.

Circular Polarity

Electromagnetic waves whose electric field uniformly rotates along the signal path. Broadcasts used by Intelsat and other international satellites use circular, not horizontally or vertically polarized waves as are common in North American and European transmissions. Circularly polarized waves are used for satellite telephony because Faraday rotation does not alter their behavior.

Clamp Circuit

A circuit that removes the dispersion waveform from the downlink signal.

Clarke Belt

The circular orbital belt at 22,247 miles above the equator, named after the writer Arthur C. Clarke, in which satellites travel at the same speed as the earth's rotation. Also called the geostationary orbit.

Color Bars

A test pattern of specifically colored vertical bars used as a reference to test the performance of a color television.

Coaxial Cable

A cable for transmitting high frequency electrical signals with low loss. It is composed of an internal conducting wire surrounded by an insulating dielectric which is further protected by a metal shield. The impedance of coax is a product of the radius of the central conductor, the radius of the shield and the dielectric constant of the insulation. In an SMATV system, coax impedance is 75 ohms.

Color Sync Burst

A "burst" of 8 to 11 cycles in the 4.43361875 MHz (PAL) or 3.579545 MHz(NTSC) color subcarrier frequency. This waveform is located on the back porch of each horizontal blanking pulse during color transmissions. It serves to synchronize the color subcarrier's oscillator with that of the transmitter in order to recreate the raw color signals.

Composite Baseband Signal

The complete audio and video signal without a carrier wave. Satellite signals have audio baseband information ranging in frequency from zero to 3400 Hertz. NTSC video baseband is from zero to 4.2 MHz. PAL video baseband ranges from 0 to 5.5 MHz.

Composite Video Signal

The complete video signal consisting of the chrominance and luminance information as well as all sync and blanking pulses.

Companding

A form of noise reduction using compression at the transmitting end and expansion at the receiver. A compressor is an amplifier that increases its gain for lower power signals. The effect is to boost these components into a form having a smaller dynamic range. A compressed signal has a higher average level, and therefore, less apparent loudness than an uncompressed signal, even though the peaks are no higher in level. An expander reverses the effect of the compressor to restore the original signal.

CONE

An abbreviation for the European continent.

Contrast

The ratio between the dark and light areas of a television picture.

Conus

An abbreviation for the continental United States.

Cross Modulation

A form of interference caused by the modulation of one carrier affecting that of another signal. It can be caused by overloading an amplifier as well as by signal imbalances at the headend.

Cross Polarization

Term to describe signals of the opposite polarity to another being transmitted and received. Cross-polarization discrimination refers to the ability of a feed to detect one polarity and reject the signals having the opposite sense of polarity.

Crosstalk

Interference between adjacent channels often caused by cross modulation. Leakage can occur between two wires, PCB tracks or parallel cables.

dc Power Block

A device which stops the flow of dc power but permits passage of higher frequency ac signals.

Decibel (dB)

The logarithmic ratio of power levels used to indicate gains or losses of signals. Decibels relative to one watt, milliwatt and millivolt are abbreviated as dBw, dBm and dBmV, respectively. Zero dBmV is used as the standard reference for all SMATV calculations.

Declination Offset Angle

The adjustment angle of a polar mount between the polar axis and the plane of a satellite antenna used to aim at the geosynchronous arc. Declination increases from zero with latitude away from the equator.

Decoder

A circuit that restores a signal to its original form after it has been scrambled.

De-emphasis

A reduction of the higher frequency portions of an FM signal used to neutralize the effects of pre-emphasis. When combined with the correct level of pre-emphasis, it reduces overall noise levels and therefore increases the signal-to- noise ratio.

Demodulator

A device which extracts the baseband signal from the transmitted carrier wave.

Detent Tuning

Tuning into a satellite channel by selecting a preset resistance.

Digital

Describes a system or device in which information is transferred by electrical "on-off," "high-low," or "1/0" pulses instead of continuously varying signals or states as in an analog message.

Digital-to-Analog Converter

A circuit that converts digital signals into their equivalent analog form.

Direct Broadcast Satellite (DBS)

A term commonly used to describe Ku-band broadcasts via satellite directly to individual end-users. The DBS band ranges from 11.7 to 12.2 GHz.

Dish

Jargon for a parabolic microwave antenna.

Distribution System

A communication system consisting of coax but occasionally of line-of-sight microwave links that carries signals from the headend to end-users.

Domsat

Abbreviation for domestic communication satellite.

Downconverter

A circuit that lowers the high frequency signal to a lower, intermediate range. There are three distinct types of downconversion used in satellite receivers: single downconversion; dual downconversion; and block downconversion.

Downlink Antenna

The antenna on-board a satellite which relays signals back to earth.

Drifting

An instability in a preset voltage, frequency or other electronic circuit parameter.

Dual-Band Feedhorn

A feedhorn which can simultaneously receive two different bands, typically the C and Ku-bands.

Earth Station

A complete satellite receiving or transmitting station including the antenna, electronics and all associated equipment necessary to receive or transmit satellite signals. Also known as a ground station.

Effective Isotropic Radiated Power (EIRP)

A measure of the signal strength that a satellite transmits towards the earth below. The EIRP is highest at the center of the beam and decreases at angles away from the boresight.

Elevation Angle

The vertical angle measured from the horizon up to a targeted satellite.

Energy Dispersal

The modulation of an uplink carrier with a triangular waveform. This technique disperses the carrier energy over a wider bandwidth than otherwise would be the case in order to limit the maximum energy compared to that transmitted by an unclamped carrier. This triangular waveform is removed by a clamp circuit in a satellite receiver.

Equalizing Pulses

A series of six pulses occurring before and after the serrated vertical sync pulse to ensure proper interlacing. The equalizing pulses are inserted at twice the horizontal scanning frequency.

F-connector

A standard RF connector used to link coax cables with electronic devices.

FCC

The Federal Communications Commission, the regulatory board which sets standards for communications within the United States.

f/D Ratio

The ratio of an antenna's focal length to diameter. It describes antenna "depth."

Feedhorn

A device that collects microwave signals reflected from the surface of an antenna. It is mounted at the focus of all prime focus parabolic antennas.

Field

One half of a complete TV picture or frame, composed of 262.5 scanning lines. There are 60 fields per second for black/white TV and 59.94 fields per second for color TV in NTSC transmission. In the PAL broadcast system there are 50 fields per second.

Filter

A device used to reject all but a specified range of frequencies. A bandpass filter allows only those signals within a given band to be communicated. A rejection filter, the mirror image of a bandpass filter, eliminates those signals within a specified band but passes all other frequencies.

Focal Length

The distance from the reflective surface of a parabola to the point at which incoming satellite signals are focused, the focal point.

Footprint

The geographic area towards which a satellite downlink antenna directs its signal. The measure of strength of this footprint is the EIRP.

Forward Error Correction

FEC is a technique for improving the accuracy of data transmission. Excess bits are included in the out-going data stream so that error correction algorithms can be applied upon reception.

Frame

One complete TV picture, composed of two fields and a total of 525 and 625 scanning lines in NTSC and PAL systems, respectively.

Frequency

The number of vibrations per second of an electrical or electromagnetic signal expressed in cycles per second or Hertz.

Front Porch

The portion of the horizontal blanking pulse that precedes the horizontal sync pulse.

Gain

The amount of amplification of input to output power often expressed as a multiplicative factor or in decibels.

Gain-to-Noise Temperature Ratio (G/T)

The figure of merit of an antenna and LNA. The higher the G/T, the better the reception capabilities of an earth station.

Geostationary Orbit

See Clarke Belt.

GigaHertz (GHz)

1000 MHz or one billion cycles per second.

Global Beam

A footprint pattern used by communication satellites targeting nearly 40% of the earth's surface below. Many Intelsat satellites use global beams.

Ground Noise

Unwanted microwave signals generated from the warm ground and detected by a dish.

Hall Effect Sensor

A semiconductor device in which an output voltage is generated in response to the intensity of a magnetic field applied to a wire. In an actuator, the varying magnetic field is produced by the rotation of a permanent magnet past a thin wire. The pulses generated serve to count the number of rotations of the motor.

Hard-line

A low-loss coaxial cable that has a continuous hard metal shield instead of a conductive braid around the outer perimeter. This type of cable was used in the pioneer days of satellite television.

Headend

The portion of an SMATV system where all desired signals are received and processed for subsequent distribution.

Heliax

A thick low-loss cable used at high frequencies; also known as hard-line.

Hertz

An abbreviation for the frequency measurement of one cycle per second. Named after Heinrich Hertz, the German scientist who first described the properties of radio waves.

High Definition Television (HDTV)

An innovative television format having approximately twice the number of scan lines in order to improve picture resolution and viewing quality.

High Power Amplifier (HPA)

An amplifier used to amplify the uplink signal.

Horizontal Blanking Pulse

The pulse that occurs between each horizontal scan line and extinguishes the beam illumination during the retrace period.

Horizontal Sync Pulse

A 5.08 microsecond (4.7 microsecond in the PAL system) rectangular pulse riding on top of each horizontal blanking pulse. It synchronizes the horizontal scanning at the television set with that of the television camera.

Hum Bars

A form of interference seen as horizontal bars or black regions passing across the field of a television screen.

I Signal

One of the two color video signals which modulate the color subcarrier. It represents those colors ranging from reddish orange to cyan.

Impulse Pay-Per-View

Impulse pay-per-view (IPPV) is a feature of a decoder that allows an authorized subscriber to purchase a one-time scrambled program at will. IPPV shows are selected by a button on the decoder or its remote control unit.

Inclinometer

An instrument used to measure the angle of elevation to a satellite from the surface of the earth.

Interference

An undesired signal intercepted by a TVRO that causes video and/or audio distortion.

Insertion Loss

The amount of signal energy lost when a device is inserted into a communication line. Also known as "feed through" loss.

Interlaced Scanning

A scanning technique to minimize picture flicker while conserving channel bandwidth. Even and odd numbered lines are scanned in separate fields both of which when combined paint one frame or complete picture.

Intermediate Frequency (IF)

A middle range frequency generated after downconversion in any electronic circuitry including a satellite receiver. The majority of all signal amplification, processing and filtering in a receiver occur in the IF range.

INTELSAT

The International Telecommunication Satellite Consortium, a body of 154 countries working towards a common goal of improved worldwide satellite communications.

Isolator

A device that allows signals to pass unobstructed in one direction but which attenuates their strength in the reverse direction.

Isolation Loss

The amount of signal energy lost between two ports of a device. An example is the loss between the feed through port and the tap/drop of a top-off device.

Kelvin Degrees (°K)

The temperature above absolute zero, the temperature at which all molecular motion stops, graduated in units the same size as degrees Celsius (°C). Absolute zero equals -273 °C or -459 °F.

Kilohertz (kHz)

One thousand cycles per second.

Ku-Band

The microwave frequency band between approximately 11 and 13 GHz used in satellite broadcasting.

Latitude

The measurement of a position on the surface of the earth north or south of the equator measured in degrees of angle.

Line Amplifier

An amplifier in a transmission line that boosts the strength of a signal.

Line Splitter

An active or passive device that divides a signal into two or more signals containing all the original information. A passive splitter feeds an attenuated version of the input signal to the output ports. An active splitter amplifies the input signal to overcome the splitter loss.

Local Oscillator

A device used to supply a stable single frequency to an upconverter or a downconverter. The local oscillator signal is mixed with the carrier wave to change its frequency.

Longitude

The distance in degrees east or west of the prime meridian, located at zero degrees.

Low Noise Amplifier (LNA)

A device that receives and amplifies the weak satellite signal reflected by an antenna via a feedhorn. C-band LNAs typically have their noise characteristics quoted as noise temperatures rated in degrees Kelvin. K-band LNA noise characteristics are usually expressed as a noise figure in decibels.

Low Noise Block Downconverter (LNB)

A low noise microwave amplifier and converter which downconverts a block or range of frequencies at once to an intermediate frequency range, typically 950 to 1450 MHz or 950 to 1750 MHz.

Low Noise Converter (LNC)

An LNA and a conventional downconverter housed in one weatherproof box. This device converts one channel at a time. Channel selection is controlled by the satellite receiver. The typical IF for LNCs is 70 MHz.

Magnetic Variation

The difference between true north and the north indication of a compass.

Master Antenna TV (MATV)

Broadcast receiving stations that use one or more high-quality centrally located UHF and/or VHF antennas which relay their signals to many televisions in a local apartment/condo or group-housing complex.

Match

The condition that exists when 100 percent of available power is transmitted from one device to another without any losses due to reflections.

Matching Transformer

A device used to match impedance between devices. A matching transformer is used, for example, when connecting a 75 ohm coax to a television 300 ohm input terminal.

MegaHertz (MHz)

One millions cycles per second.

Microprocessor

The central processing unit of a computer or control system, either on a single integrated (IC) circuit chip or on several ICs.

Microwave

The frequency range from approximately 1 to 30 GHz and above.

Mixer

A device used to combine signals together.

Modulation

A process in which a message is added or encoded onto a carrier wave. Among other methods, this can be accomplished by frequency or amplitude modulation, known as AM or FM, respectively.

Monochrome

A black and white television picture.

Mount

The structure that supports an earth station antenna. Polar and az-el mounts are the most common variety.

Multiple Analog Component (MAC) Transmissions

An innovative television transmission method which separates the data, chrominance and luminance components and compresses them for sequential relay over one television scan line. There are a number of systems in use and under development including A-MAC, B-MAC, C-MAC, D-MAC, D2-MAC, E-MAC and F-MAC.

Multiplexing

The simultaneous transmission of two or more signals over a single communication channel. The interleaving of the luminance and chrominance signals is one form of multiplexing, known as frequency multiplexing. MAC transmissions make use of time division multiplexing.

N-Connector

A low-loss coaxial cable connector used at the elevated C-band microwave frequencies.

NTSC

The National Television Standards Committee which created the standard for North American TV broadcasts.

NTSC Color Bar Pattern

The standard test pattern of six adjacent color bars including the three primary colors plus their three complementary shades.

Negative Picture Phase

Positioning the composite video signal so that the maximum level of the sync pulses is at 100% amplitude. The brightest picture signals are in the opposite negative direction.

Negative Picture Transmission

Transmission system used in North America and other countries in which a decrease in illumination of the original scene causes an increase in percentage of modulation of the picture carrier. When demodulated, signals with a higher modulation percentage have more positive voltages.

Noise

An unwanted signal which interferes with reception of the desired information. Noise is often expressed in degrees Kelvin or in decibels.

Noise Figure

The ratio of the actual noise power generated at the input of an amplifier to that which would be generated in an ideal resistor. The lower the noise figure, the better the performance.

Noise Temperature

A measure of the amount of thermal noise present in a system or a device. The lower the noise temperature, the better the performance.

Odd Field

The half frame of a television scan which is composed of the odd numbered lines.

Offset Feed

A feed which is offset from the center of a reflector for use in satellite receiving systems. This configuration does not block the antenna aperture.

Orthomode Coupler

A waveguide, generally a three-port device, that allows simultaneous reception of vertically and horizontally polarized signals. The input port is typically a circular waveguide. The two output ports are rectangular waveguides.

PAL

Phase Alternate Line. The European color TV format which evolved from the American NTSC standard.

Pad

A concrete base upon which a supporting pole and antenna can be mounted.

Path Loss

The attenuation that a signal undergoes in traveling over a path between two points. Path loss varies inversely as the square of the distance traveled.

Parabola

The geometric shape that has the property of reflecting all signals parallel to its axis to one point, the focal point.

Pay-Per-View

Pay-per-view is a method of purchasing programming on a per-program basis.

Persistence of Vision

The physiological phenomena whereby a human eye retains perception of an image for a short time after the image is no longer visible.

Phase

A measure of the relative position of a signal relative to a reference expressed in degrees.

Phase Distortion

A distortion of the phase component of a signal. This occurs when the phase shift of an amplifier is not proportional to frequency over the design bandwidth.

Picture Detail

The number of picture elements resolved on a television picture screen. More "crisp" pictures result as the number of picture elements is increased.

Polar Mount

An antenna mount that permits all satellites in the geosynchronous arc to be scanned with movement of only one axis.

Polarization

A characteristic of the electromagnetic wave. Four senses of polarization are used in satellite transmissions: horizontal; vertical; right-hand circular; and left-hand circular.

Positive Picture Phase

Positioning of the composite video signal so that the maximum point of the sync pulses is at zero voltage. The brightest illumination is caused by the most positive voltages.

Preamplifier

The first amplification stage. In an SMATV system, it is the amplifier mounted adjacent to an antenna to increase a weak signal prior to its processing at the headend.

Pre-emphasis

Increases in the higher frequency components of an FM signal before transmission. Used in conjunction with the proper amount of de- emphasis at the receiver, it results in combating the higher noise detected in FM transmissions.

PSD

An abbreviation for polarity selection device.

Primary Colors

Red, green and blue.

Prime Focus Antenna

A parabolic dish having the feed/LNA assembly at the focal point directly in the front of the antenna.

Q Signal

One of two color video signal components used to modulate the color subcarrier. It represents the color range from yellowish to green to magenta.

Radio Frequency

The approximately 10 kHz to 100 GHz electromagnetic band of frequencies used for man-made communication.

Raster

The random pattern of illumination seen on a television screen when no video signal is present.

Reed Switch

A mechanical switch which uses two thin slivers of metal in a glass tube to make and break electrical contact and thus to count pulses which are sent to the antenna actuator controller. The position of the slivers of metal is governed by a magnetic field applied by a bar or other type of magnet.

Reference Signal

A highly stable signal used as a standard against which other variable signals may be compared and adjusted.

Return Loss

A ratio of the amount of reflected signal to the total available signal entering a device expressed in decibels.

Retrace

The blanked-out line traced by the scanning beam of a picture tube as it travels from the end of any horizontal line to the beginning of either the next horizontal line or field.

SAW (Surface Acoustic Wave) Filter

A solid state filter that yields a sharp transition between regions of transmitted and attenuated frequencies.

Satellite Receiver

The indoors electronic component of an earth station which downconverts, processes and prepares satellite signals for viewing or listening.

Scanning

The organized process of moving the electron beam in a television picture tube so an entire scene is drawn as a sequential series of horizontal lines connected by horizontal and vertical retraces.

Scrambling

A method of altering the identity of a video or audio signal in order to prevent its reception by persons not having authorized decoders.

Screening

A metal, concrete or natural material that screens out unwanted TI from entering an antenna or a metal shield that prevents the ingress of unwanted RF signals in an electronic circuit.

Serrated Vertical Pulse

The television vertical sync pulse which is subdivided into six serrations. These sub-pulses occur at twice the horizontal scanning frequency.

Servo Hunting

An oscillatory searching of the feedhorn probe when use of inadequate gauge control cables results in insufficient voltage at the feedhorn.

Side Lobe

A parameter used to describe an antenna's ability to detect off-axis signals. The larger the side lobes, the more noise and interference an antenna can detect.

Single Channel Per Carrier (SCPC)

A satellite transmission system that employs a separate carrier for each channel, as opposed to frequency division multiplexing that combines many channels on a single carrier.

Signal-to-Noise Ratio (SNR)

The ratio of signal power to noise power in a specified bandwidth, usually expressed in decibels.

Skew

A term used to describe the adjustment necessary to fine tune the feedhorn polarity detector when scanning between satellites.

Slant Range

The distance that a signal travels from a satellite to a TVRO.

Snow

Video noise or sparklies caused by an insufficient signal- to-noise input ratio to a television set or monitor.

Solar Outage

The loss of reception that occurs when the sun is positioned directly behind a target satellite. When this occurs, solar noise drowns out the satellite signal and reception is lost.

Sparklies

Small black and/or white dashes in a television picture indicating an insufficient signal-to-noise ratio. Also known as "snow."

Spherical Antenna

An antenna system using a section of a spherical reflector to focus one or more satellite signals to one or a series of focal areas.

Splitter

A device that takes a signal and splits it into two or more identical but lower power signals.

Subcarrier

A signal that is transmitted within the bandwidth of a stronger signal. In satellite transmissions a 6.8 MHz audio subcarrier is often used to modulate the C-band carrier. In television, a 3.58 MHz subcarrier modulates the video carrier on each channel.

Surface Acoustic Wave

A sound or acoustic wave traveling on the surface of the optically polished surface of a piezoelectric material. This wave travels at the speed of sound but can pass frequencies as high as several gigahertz. See SAW Filter.

Synchronizing Pulses

Pulses imposed on the composite baseband video signal used to keep the television picture scanning in perfect step with the scanning at the television camera.

TVRO

A television receive-only earth station designed only to receive but not to transmit satellite communications.

Tap

A device that channels a specific amount of energy out of the main distribution system to a secondary outlet.

Television Receive-Only (TVRO)

A satellite system that can only receive but not transmit signals.

Terrestrial Interference (TI)

Interference of earth-based microwave communications with reception of satellite broadcasts.

Tilt

The uneven attenuation of a broadband signal as it travels through a coaxial cable. In general, attenuation increases as signal frequency increases.

Thermal Noise

Random, undesired electrical signals caused by molecular motion, known more familiarly as noise.

Trace

The movement of the electron beam from left to right on a television screen.

Threshold

A minimal signal to noise input required to allow a video receiver to deliver an acceptable picture.

Transponder

A microwave repeater, which receives, amplifies, downconverts and re-transmits signals at a communication satellite.

Trap

An electronic device that attenuates a selected band of frequencies in a signal. Also known as a notch filter.

UHF

Ultrahigh frequencies ranging from 300 to 3,000 MHz. North American TV channels 14 through 83. European TV channels 21 to 69.

Upconverter

A device that increases the frequency of a transmitted signal.

Uplink

The earth station electronics and antenna which transmit information to a communication satellite.

VHF

Very high frequencies in the range from 54 MHz to 216 MHz, NTSC TV channels 2 through 13.

VSWR (Voltage Standing Wave Ratio)

The ratio between the minimum and maximum voltage on a transmission line. An ideal VSWR is 1.0. Ghosting can result as the VSWR increases. It is also a measure of the percentage of reflected power to the total power impinging upon a device.

Vertical Blanking Pulse

A pulse used during the vertical retrace period at the end of each scanning field to extinguish illumination from the electron beam.

Vertical Sync Pulse

A series of pulses which occur during the vertical blanking interval to synchronize the scanning process at the television with that created at the studio. See also Serrated Vertical Pulse.

VHF

Very high frequency range from 30 to 300 MHz

Video Signal

That portion of the transmitted television signal containing the picture information.

Voltage Tuned Oscillator (VTO)

An electronic circuit whose output oscillator frequency is adjusted by voltage. Used in downconverters and satellite receivers to select from among transponders.

Video Monitor

A television that accepts unmodulated baseband signals to reproduce a broadcast.

APPENDIX E. MANUFACTURERS OF SATELLITE TV EQUIPMENT

MANUFACTURERS OF ENCODING EQUIPMENT

GENERAL INSTRUMENTS CORP
2200 ByBerry Road
Hatboro, PA 19040
(215) 674-4800

HAMLIN
13610 First Avenue South
Seattle, WA 98168
(206) 246-9330

M/A COM - LINKABIT
3033 Science Park Drive
San Diego, CA 9212
(619) 457-2340

OAK COMMUNICATIONS
Satellite Systems Division
P.O. Box 517
Crystal Lake, IL 60014
(815) 459-5000

OAK COMMUNICATIONS
Cable Division
16935 West Bernardo Drive
Rancho Bernardo, CA 92127
(619) 485-9880

SCIENTIFIC ATLANTA
4356 Communications Drive
Norcross, GA 30093
(404) 925-5778

ZENITH ELECTRONICS CORP
1000 Milwaukee Avenue
Glenview, IL 60025
(312) 391-8338

NORTH AMERICAN SATELLITE TV EQUIPMENT MANUFACTURERS

A&E TRAVEL-SAT
44389 Portofino Court
Palm Desert, CA 92260
Telephone: (619) 568-0666
FAX: (619) 341-6565
RV Systems

AJAK INDUSTRIES, INC.
112 South Robinson
Florence, CO 81226
Telephone: (719) 784-6301
FAX: (719) 784-6763
Horizon-to-horizon Mounts

ANDERSON MANUFACTURING
3125 North Yellowstone Highway
Idaho Falls, ID 85301
Telephone: (208) 523-6460
Antennas & Mounts

ANDREW CORPORATION
10500 West 153rd Street
Orlando Park, IL 60462
(312) 349-3300
Commercial Antennas

ASP/RCI
10679 Windmer
Lexena, KS 66215
Telephone: (913) 469-4125
TI Filters

ASTRA
792 East 93rd Street
Brooklyn, NY 11236
Telephone: (718) 346-1200
Mesh Antennas & Mounts

AVCOM OF VIRGINIA
500 Southlake Blvd.
Richmond, VA 23236
Telephone: (804) 794-2500
FAX: (804) 794-8284
Satellite Receivers & Test Equipment

BESTIN MANUFACTURING
2422 North Lee Avenue
South El Monte, CA 91733
Telephone: (818) 444-8170
Satellite Antenna Systems

BIRDVIEW SATELLITE SERVICE
1407 East Spruce
Olathe, KS 66061
Telephone: (913) 824-6240
Service and Repair of Equipment

CALIFORNIA AMPLIFIER
460 Calle San Pablo
Camarillo, CA 93010-8506
Telephone: (805) 987-9000
FAX: (805) 987-8359
Low Noise Amplifiers

CHANNEL MASTER/DIV. OF AVNET
P.O. Box 1416
Smithfield, NC 27577
Telephone: (919) 934-9711
Complete Satellite Systems

CHAPARRAL COMMUNICATIONS
2450 North 1st Street
San Jose, CA 95131
(408) 435-1530
Receivers, Feedhorns

COAST HITECH CORPORATION
6021-F North Figueroa Street
Los Angeles, CA 90042
Telephone: (800) 782-8344
(213) 255-5085
FAX: (213) 255-4708
Integrated LNB/Feedhorn

COMTECH ANTENNA CORP
3100 Communications Road
St. Cloud, FL 32769
(305) 892-6111
Antennas

CT SYSTEMS / WAVETEK RF
5808 Churchman Bypass
Indianapolis, IN 46203
Telphone: (317) 787-5721
(800) 245-6356
Test Equipment

DH SATELLITE, INC
P.O. BOX 239
PRAIRIE DU CHIEN, WI 53821-9990
Telephone: (800) 392-6884
Satellite Antennas

DX COMMUNICATIONS, INC.
10 Skyline Drive
Hawthorne, NY 10532
Telephone: (914) 347-4040
Commercial Receivers

DIAMOND PERFORATED METALS
28976 Hoplins Street
Hayward, CA 94545
Telephone: (415) 783-7444
Perforated Metals

E-Z TRENCH MANUFACTURING CO
Route 3, Box 78-B
Loris, SC 29569
Telephone: (803) 756-6444
Trenching Equipment

EARTHBOUND, INC.
3220 West Topeka
Topeka, KS 66611
Telephone: (913) 266-4944
Mounting Systems

ECHOSPERE CORP.
90 Inverness Circle East
Englewood, CO 80112
Telephone: (303) 799-8222
Complete Systems

FOCII MANUFACTURING CO
1324 South Kansas
Topeka, KS 66612
Telephone: (913) 234-6721
Test Equipment

FORD AEROSPACE
3939 Fabian Way
Palo Alto, CA 94303
Telephone: (415) 852-6980
Satellites

FORT WORTH TOWER COMPANY
1901 East Loop 820 South
Fort Worth, TX 76112
Telephone: (817) 457-3060
Antenna Support Towers

FUJITSU GENERAL
P.O. Box 2101
Chatsworth, CA 91313-2101
Telephone: (818) 341-5400
FAX: (818) 718-2938
Satellite Receivers

GARDINER COMMUNICATIONS
3605 Security Street
Garland, TX 75042
(214) 348-4747
Low Noise Amplifiers

GTE SATELLITE CORPORATION
170 Old Meadow Road
McLean, VA 22102
(703) 790-7700
Satellites

GENERAL INSTRUMENT CORP.
6262 Lust Boulevard
San Diego, CA 92121
Telephone: (619) 535-2545
Satellite Receivers and Decoders

GOURMET ENTERTAINING
3915 Carnavon Way
Los Angeles, CA 90027
Telephone: (213) 666-2728
Alignment Tools

HARRIS CORPORATION
Satellite Communications Division
P.O. Box 1700
Melbourne, FL 32901
(305) 724-3689
Commercial Systems

HERO COMMUNICATIONS
2290 West 8th Avenue
Hialeah, FL 33010
Telephone: (305) 887-3203
Satellite Antennas

HOUSTON TRACKER SYSTEMS.
90 Inverness Circle East
Englewood, CO 80112
Telephone: (303) 790-4445
Satellite Receivers

HUGHES AIRCRAFT COMPANY
P.O. Box 92919
Los Angeles, CA 90009
Satellites

INT'L ELECTRONIC WIRE & CABLE
520 Business Center Drive
Mt. Prospect, IL 60056
Telephone: (312) 299-0021
Cables

KAUL-TRONICS, INC.
1140 Sextonville Road
P.O. Box 637
Richland Center, WI 53581
Telephone: (608) 647-8902
Satellite Antennas

LEAMING INDUSTRIES
180 McCormick Avenue
Costa Mesa, CA 92626
Telephone: (714) 979-4511

MB SALES
P.O. Box 787
Wauconda, IL 60084
Telephone: (312) 526-5310

MICRODYNE CORP
P.O. Box 7213
Ocala, FL 32672
(904) 687-4633
Commercial Systems

MICROWAVE FILTER COMPANY
6743 Kinne Street
East Syracuse, NY 13057
Telephone: (315) 437-3953
TI Filters

MICROWAVE SYSTEMS ENG.
4221 East Raymond
Phoenix, AZ 85040
Telephone: (602) 437-9040
Commercial Systems

MIRALITE CORPORATION
4050 Chandler
Santa Ana, CA 92704
Telephone: (717) 641-7000
Satellite Antennas

MULIPLEX TECHNOLOGY
251 Imperial Highway
Fullerton, CA 92635
Telephone: (714) 680-5848
Video Equipment

NATIONAL ADL
255-G Easy Street
Simi Valley, CA 93065
Telephone: (805) 526-5249
Feedhorns

NEXUS ENGINEERING
4181 McConnell Drive
Burnaby, BC V5A 3J7 Canada
Telephone: (206) 644-2371
Commercial Electronics

NORSAT INTERNATIONAL, INC.
707 Johnson Way
Blaine, WA 98230
Telephone: (604) 597-6200
Satellite Receivers and LNBs

OAK COMMUNICATIONS, INC.
SATELLITE SYSTEMS
100 South Main Street
Crystal Lake, IL 60014
Telephone: (815) 459-5000

ODOM ANTENNAS
P.O. Box 1017
2502 DeWitt Henry Drive
Beebe, AR 72012
Telephone: (501) 882-6485
Satellite Antennas

ORBITRON
351 South Peterson Street
Spring Green, WI 53588
Telephone: (608) 588-2923
Satellite Antennas and Mounts

PTS CORPORATION
5233 South Highway 37
Bloomington, IN 47459
Telephone: (812) 824-9331
Service and Repair

PANAREX ELECTRONICS
13012 Saticoy Street, #4
North Hollywood, CA 91605
Telephone: (818) 764-0375
Ku-Band Satellite Systems

PANASONIC
One Panasonic Way 2A-2
Secaucus, NJ 07094
Telephone: (201) 348-7846
Satellite Receivers

PARACLIPSE, INC.
3711 Meadowview Drive
Redding, CA 96002
Telephone: (916) 365-9131
Satellite Antennas

PHANTOM ENGINEERING, INC.
16840 Joleen Way, Bldg. E-3
Morgan Hill, CA 95037
(408) 779-1616

PICO MACOM, INC.
12500 Foothill Boulevard
Lakeview Terrace, CA 91342
Telephone: (800) 421-6511
(818) 897-0028
Electronic Components

PRECISION R.V. SATELLITES
5468 North Salinas Avenue
Fresno, CA 93722
Telephone: (209) 275-1780
RV Systems

PRO BRAND INTERNATIONAL, INC.
1900 West Oak Circle
Marietta, GA 30062
Telephone: (404) 423-7072
FAX: (404) 423-7075
Actuators

RCA AMERICAN COMMUNICATIONS
400 College Road, Suite E
Princeton, NJ 08540
(609) 734-4000
Satellites

RCA ASTRO-ELECTRONICS
P.O. Box 800, MS 54
Princeton, NJ 08540
(609) 426-2711
Satellites

R.L. DRAKE COMPANY
9111 Springboro Pike
Miamisburg, OH 45342
Telephone: (513) 866-2421
Satellite Receivers & LNBs

RMS ELECTRONICS, INC.
50 Antin Place,
Bronx, NY 10462
(212) 892-1000

ROHN
6718 West Plank Road
Peoria, IL 61656
Telephone: (309) 697-4400
Antenna Towers & Roof Mounts

SEA TEL
1035 Shary Court
Concord, CA 94518
Telephone: (415) 798-7979
Marine Satellite Systems

SATELLITE TECHNICAL SERVICES
1409 Washington Avenue
St. Louis, MO 63103
Telephone: (314) 567-0304
Satellite Receivers

SEAVEY ENGINEERING ASSOCIATES
155 Kind Sreet, P.O. Box 44
Cohasset, MA 02025
Telephone: (617) 383-9722
Feedhorns

SCIENTIFIC ATLANTA, INC.
P.O. Box 105027
Atlanta, GA 30348
Telephone: (404) 492-1111
Commercial Systems

STANDARD COMMUNICATIONS CORP
P.O. Box 92151
Los Angeles, CA 90009
Telephone: (800) 824-7766
Commercial Systems

SUPERIOR ANTENNA MANUFACTURING
Route 2, Box 465
Judsonia, AR 72081
Telephone: (501) 729-3103
Satellite Antennas

TEE-COMM
775 Main Street East
Milton, ON L9T 3Z3
Canada
Telephone: (416) 878-8181
Satellite Receivers

THOMPSON SAGINAW BALL SCREW CO
P.O. Box 9550
Saginaw, MI 48608
Telephone: (517) 776-5111
Actuators

TOSHIBA AMERICA CONSUMER PROD.
1010 Johnson Drive
Buffalo Grove, IL 60089-6900
Telephone: (312) 541-9400
FAX: (708) 541-1927
Satellite Receivers

TRAVEL-STAR
3100 West Segerstrom
Santa Ana, CA 92704
Telephone: (714) 540-6444

TSIGER PLANAR, INC.
2448 Waynoka Road
Colorado Springs, CO 80915
Telephone: (719) 591-7900
Satellite Antennas

UNIDEN AMERICA CORP.
4700 Amon Carter Blvd.
Ft. Worth, TX 76155
Telephone: (817) 858-3300
Satellite Receivers & LNBs

UNIMESH
Highway 367, Box 338
Ward, AR 72176
Telephone: (800) 843-6517
(501) 843-6517
Satellite Antennas

VIDEO LINK
12950 Bradley Avenue
Sylmar, CA 91342
(818) 362-0353
Remote Controllers

VON WEISE GEAR COMPANY
500 Chesterfield Center
St. Louis, MO 63017
Telephone: (314) 532-3505
Actuators & Gearmotors

WINEGARD COMPANY
3000 Kirkwood Street
Burlington, IA 52601
Telephone: (319) 753-0121
Satellite Antennas

ZENITH ELECTRONICS CORP.
11000 West Seymour Avenue
Franklin Park, IL 60131
Telephone: (708) 671-2043
Satellite Receivers

EUROPEAN SATELLITE TV EQUIPMENT MANUFACTURERS

Belgium:

FUBA-BELGUIM SA
Brucargo 706, Bureau 7319
B-1931 Zaventem
TEL:(2)722-34-22
TLX: 61653 AL B

GILLAM SA
Quai de Coronmeuse 39
B-4000 Liege
TEL:(41) 27-32-00
FAX: (5121) 49-41-54

Finland:

AERIAL OY
Peitolankatu 9A
SF-04400 Jaervenpaeae
TEL:(0) 291-05-88

ANTENNI-ASENNUS OY
Valuraudantie 19
SF-00700 Helsinki
TEL:(0) 37-10-22

TELESTE OY (ANTENNA)
P.O. BOX 323
SF-20660 Littoinen
TEL: (21) 36-13-33
TLX: 62135 TE;EA SF

VALMET OY
P.O. Box 155, Punanotkonakatu 2
SF-00130 Helsinki
TEL:(0) 17-14-41
TLX:124427 VA; SF
FAX: (0) 17)96)77

France:

CEGEDUR RECHINEY
23 Rue Balzac, BP 787.08
75360 Paris Cedex 08
TEL:(1) 45616-61-61
TLX: 290503
FAX: (1) 4561-61-44

EDS-PHILIPS
50 Rue Roger Salengro, Peripole 114
94126 Fontenay Sous Bois
TEL:(1) 4876-11-33
TLX: 280746

Germany:

ANDREW GmbH
Zeppelinstrasse 10
D-8000 Munich 90
TEL: 89) 651-40-90
TLX: 5213432

BLAUPUNKY WERKE GMBH
Robert-Bosch-Strasse 200
D-3200 Hildesheim
TEL:(5121) 49-1
TLX:927151-0 BP D
FAX: (5121) 49-41-54

C. ITOH & COMPANY GMBH
Loenigsalle 21-23
D-4000 Duesseldorf 1
TEL: (211) 88-98-1
TLX: 8581971 CI D
FAX: (211) 327247

KATHREIN-WERKE KG
Luitpoldstrasse 18-20, Postfach 260
D-8200 Rosenheim
TEL:(8031) 184-0
TLX: 525859
FAX: (8031) 184-306

NISSHO IWAI DEUTSCHLAND
Immermannstrasse 10
D-4000 Duesseldor 1
TEL: (211) 3678-226
TLX: 8587109/8582367
FAX: (211) 35-88-72

RICHARD HIRSCMANN
RADIOTECHNISCHES WERK
Richard-Hirschmann-Strasse 19
Postfach 110
D-7300 Esslingen/Neckar
TEL:(711) 3101-1
TLX:725657 HIR D
FAX: (711) 3101-338

ROBERT BOSCH GMBH
Forckenbeckstrasse 9-13
D-1000 Berlin 33
TEL: (30) 8204-0
TLX: 183776 RBEK D
FAX: (30) 82-04-22-10

SIEMENS AG
Hofmannstrasse 51
D-8000 Munich 83
TEL:(89) 722-474-32
TLX: 5288260 SIE D

Great Britain

ALBA
Harvard House, Thames Road
Barking, Essex
TEL: 081-594-5533

ALLIED
Coppull Satellite
164 Preston Road
Brentwood, Essex
TEL: 0257-471023

AMSTRAD
Brentwood House
169 Kings Road
Brentwood, Essex
TEL: 0277-228888

AQP
34/35 Gelli Industrial Estate
Gelli, Rhondda CF41 7UW
TEL: 0443-433149

BIG BROTHER
64-66 Glentham Road
London SW13 9JJ
TEL: 081-748-1818

BUSH
Wharf Road
Enfield, Middlesex EN3 4TE
TEL: 081-805-1664

CAMBRIDGE COMPUTERS
Bridge House, 10 Bridge Street
Cambridge CB2 1UE
TEL: 0223- 312216

CHANNEL MASTER
Glenfield Park, Northrop Avenue
Blackburn BB1 5QF
TEL: 0254-680444

CHAPARRAL (UK)
Abacus House, Manor Road
West Ealing, London W130AZ
TEL: 081-566-7830

CONNEXIONS
Unit 3, Travellers Lane, off Travellers Lane
Welham Green, Herts AL9 7LE
TEL: 0707 272091

DISCUS
Unit 9, Block 22, Kilspindie Road
Dundee DD2 3JP
TEL: 0382-833651

DISKXPRESS
5 Cross Lane
London N8 7SA
TEL: 071-348-4414

DRAKE
ABVS, Units 4/5, Win Born Bldg
Convent Drive, Waterbeach
Cambridge CB5 9PB
TEL: 0223-860958

EUROSAT
Unit 27A, Popin Building
Southway, Wembley HA9 0HB
TEL: 081-903-1188

FERGUSON
Cambridge House, Great Cambridge Road
Enfield, Middlesex
TEL: 081-363-5353

FINLUX
Valley Farm Way
Leeds LS10 1SE
EL: 0532-714521

ITT NOKIA (SALORA)
Bridegmead Close, Westmead
Swindon, Wiltshire SN5 7YG
TEL: 0793-644223

LENSON HEATH
Unit 75, Woodburn Green
Wooburn Green Industrial Park
Bucks
TEL: 06285-25887

LONGREACH
Riverside Business Park
Lower Bristol Road, Bath
TEL: 0255-316257

MASPRO
c/o Micro X
Ironbridge Close, Great Central Way
London NW10 0UF
TEL: 081-459-1200

NEC
1 Victoria Road
London W3 6UL
TEL: 01-993-8111

NETWORK SATELLITE SYSTEMS
Units 7, 8 and 9
Newburn Bridge Industrial Estate
Hartlepoof, Cleveland TS25 1UB
TEL: 0429-869366

PACE
Victoria Road, Saltaire
Shipley BD 18 3LF
TEL: 0274-532000

PALCOM
c/o Data Sat, 142 High Street
Yiewsley UB7 7BD
TEL: 0895-431633

PANASONIC
Panasonic House, Willoughby Road
Bracknell, Berks RG12 4FP
TEL: 0344-862444

SAKURA
Unit 717, Tudor Estate
Abbey Road
London NW10 7UN
TEL: 081-961-1346

SAMSUNG
3 Riverbank Way, Great West Road
Brentford, Middlesex TW8 9RE
TEL: 081-862911

SATCOM
Vulcan Works, Water Lane
Exeter EX2 8BY
TEL: 0392-213928

SONY
Sony House, South Street
Staines, Middlesex TW18 4PF
TEL: 0784-467000

SUPERSAT
32 Temple Street
Wolverhampton WV2 4AN
TEL: 0902-29022

TATUNKG
Stafford Park 10
Telford TF3 3AB
TEL: 0952-290111

TECHNISAT
TPS Building, Blatchford Road
Horsham, West Sussex RH13 5QR
TEL: 0403-2119800

TOSHIBA
Frimley Road
Camberley, Surrey
TEL: 0276-6222

TRIAX
Units 3, Saxon Way, Back Lane
Melbourne, Herts
TEL: 07632-61755

TRISTAR: TBC
814 Newport Road
Rumney, Cardiff
TEL: 0222 779964

UNIDEN
Caledonian House, 98 The Centre
Feltham, Middlesex
TEL: 081-8931166

WINERSAT
Starwatch, 10 Station Road
Crayford, Kent DA1 3QA
TEL: 0322 555933

WOLSEY ELECTRONICS LTD.
Gellihirion Industrial Estate
Pontypridd, Mid Glamorgan CF37 5SX
TEL: 0443-85311

ZETA
Harden Park, Alderly Edge
Cheshire SK9 7QN
TEL: 0625-583850

Italy:

FRACARRO RADIOINDUSTRIE S.p.A.
Via Cazzaro N° 3
31033 Castelfranco
Veneto

IRTE ELETTRONICA SRL
VIA Pompei 35
21013 Gallarate (VA)
TEL: (331) 79-72-86
TLX: 315284 CASBAI I
FAX: (331) 78-41-91

RO.VE.R SPA
Via Parini 2-4
25010 Coluombare di Sirmione (BS)
TEL: (30) 919-62-57/919-61-51
TLX: 316857 ROVER I

SCIENTIFIC ATLANTA
4 Avenue Gabriel Peri
78360 Montesso
TEL: (1) 3976-91-91
TLX: 696385

Netherlands:

ECHOSPHERE INTERNATIONAL
Schuilenburglaan 5a
7604 BJ Almelo
TEL: 31-5490-14884
FAX: 31-5490-14691

Norway:

TANDBERG TELECOM AS
Fetveien 1, P.O. Box 31
N-2007 Kjeller
TEL: (2) 71-68-20
FAX: SWITCHBOARD

Spain:

PROMAX
Francesc Moragas, 71-75
Apartado Correos, 118
08907 L'Hospitalet de Llobregat, Barcelona
TEL: (343) 337 90 08
FAX: (343) 338 11 26

TAGRA SA
C-Eduardo Maristanay 341, Apartovo 30
Badalona, Barcelona
TEL: (3) 388 82-11
TLX: 59558 TAGRA E

TELEVES
15706 Santiago de Compostela
Conxo de Abaixo, 23
Tel: (981) 592200
Telex: 82273 TVES-E

Sweden:

HANDIC ELECTRONIC AB
Box 1063
S-436 00 Askim/Goteborg
TEL: (31) 28-97-90
TLX: 21420
FAX: (31) 28-87-00

INPACO AB
S-640 33 Bettna
TEL: (157) 703-55
LUXOR AB
Box 901
S-591-29 Motala
TEL: (141) 280- 00
FAX: (141) 571-20

PARABOLIC AB
Box 102 57
S-434-01 Kungsbacka
TEL: (300) 411 70

Switzerland:

ALCOA
61 Avenue D'Ouchy
CH-1006 Lausanne
TEL: (21) 27-61-61
FAX: (21) 27-83-05

HUBER & SUHNER AG
CH-9100 Herisau
TEL: (71) 53-15-15
FAX: (71) 51-42-68

APPENDIX F. REFERENCE PUBLICATIONS

Satellite TV Guides

Magazin, NB Fortag
Kjelsasvn. 51
0488 Oslo 4
Norway

On Sat, Triple-D Publishing
P.O. Box 2384
Shelby, NC 28151
USA
TEL: (800) 234-0023
(704) 482-9673

OnSat, Triple-D Publishing Canada
9780 Bramalea Road, North
Suite 406
Brampton, Ontario L6S 2P1
Canada
TEL: (416) 458-9429

Satellite Entertainment Guide
Vogel and Son Publishing Company
P.O. Box 8266TU
Edmonton, Alberta T6H 4P1, Canada
TEL: (403) 425-1169

Satellite Orbit
CommTek Publishing
8330 Boone Blvd., Suite 600
Vienna, VA 22182, USA
TEL: (800) 234-4220
(703) 827-0511

Satellite TV PreVue
Terra Publishing, Inc.
P.O. Bo x 460
Salamanca, NY 14779-0460
TEL: (800) 992-3499

Satellite TV Week
Fortuna Communications
P.O. Box 308
928 Main Street
Fortuna, CA 95540, USA
TEL: (800) 345-8876
(707) 725-1185

SuperGuide, Triple-D Publishing
P.O. Box 2384
Shelby, NC 28151
TEL: (800) 234-0139

Cable and Satellite Europe
21st Century Publishing, 533 Kings Road
London, SW10 0BR, UK
TEL: 071-351-3612

Trade Publications

Cable and Satellite Europe
533 Kings Road
London, SW10 0BR, UK

CQ TV Magazine
British Amateur TV Club
Grenehurst, Pinewood Road
High Wycomb, Bucks HP12 4DD, UK

Electronics & Wireless World
Reed Business Publications Ltd.
Stewart House, Perrymount Road
Haywards Heath, West Sussex RH16 3DH
United Kingdom

Elektor Electronics
Worldwide Subscription Services
Rose Hill, Ticehurst
East Sussex TN5 7AJ, UK

Feedback – Confed of Aerial Industries
Suite 106, Grosvenor House
Grosvenor Gardens, London SW1W 0BS, UK
Telephone: 071-828-0625
FAX: 071-828-0507

McCormac's Hack Watch News
22 Viewmount
Waterford, Ireland
Telephone: 353-51-72640

Private Cable Magazine, NSPN
1909 Avenue G
Rosenburg, TX 77471, USA
Telephone: (800) 622-5990
FAX: (713) 342-7016

Satellite Business News
1050 17th Street, N.W., Suite 1212
Washington, DC 20036
Telephone: (202) 785-0505
FAX: (202)785-9291

Satellite Communications, Cardiff Publ.
6430 S. Yosemite Street
Englewood, CO 80111, USA
Telephone: (303) 694-1522

Satellite Retailer, Triple D Publishing
P.O. Box 2384
Shelby NC 28151, USA
Telephone: (704) 482-9673

Signal Magazine
Fernwood Publishing
P.O. Box 238, Station D
Scarborough, Ontario M1R 5B7, Canada
Telephone: (416) 759-6639

Television
Royal TV Society
Tavistock House East, Tavistock Square
London WC1H 9HR, UK
Telephone: 071-485-0011

The Transponder
Terra Publishing, Inc.
P.O. Box 460
Salamanca, NY 14779-0460
(800) 992-3488

What Satellite
WV Publications Ltd.
57/59 Rochester Place
London NW1 9JUEngland
Telephone: 071-485-0011

WORLD SATELLITE TV AND SCRAMBLING METHODS
The Technician's Handbook – 3rd Revised Edition

by Frank Baylin, Richard Maddox, John McCormac

This thorough text is a must-buy for technicians, satellite professionals and do-it-yourselfers. The design, operation and repair of satellite antennas, feeds, LNBs & receivers/modulators are examined in detail. An in-depth study of scrambling methods and broadcast formats is the backdrop to a discussion of all current American and European satellite TV technologies including the VideoCipherII, Oak Orion, FilmNet, Sky Channel, EuroCypher, D2 MAC, BSB and Teleclub Payview III. Circuit and block diagrams of all components are presented and clearly explained throughout the handbook. This information is a prelude to the chapters on troubleshooting and setting up a test bench. This expert guidance on testing, servicing and tuning is complimented by a wealth of detailed illustrations.

PARTIAL CONTENTS: The Satellite System. Early Designs. **Antennas. Feeds** Polarization Formats. Polarity Selection & Control. Troubleshooting. **Low Noise Amplifiers.** Waveguides. Troubleshooting. Downconversion Methods. **Cables and Connectors.** Coax. Insulated Wire. Custom Cables. Sealing. **Positioning Systems**. Basic Motor Controllers. Feedback Circuits. End Point Limiters. Troubleshooting. **Power Supplies.** Actuator Controllers. IC Regulators. Troubleshooting. **IF Circuits.** Amplifiers. Bandpass Filters. Limiter Circuits. Troubleshooting. **Video Processing.** Demodulator Circuits. Video Processing. **Audio Processing.** Audio Subcarrier Specifications. Typical Audio Circuitry. Other Audio Detection Circuits. Stereo Broadcast Methods. Audio Companding. Dolby Noise Reduction. Troubleshooting & Aligning Audio Circuits. **RF Modulators.** Broadcast Formats. RF Modulator Circuits. Crystal Modulators. RF Interference. Troubleshooting. **Miscellaneous Receiver Circuits.** Indicator & LED Circuits. LED Read-out Displays. Remote Controls. **Complete Circuit Descriptions** European & American Receiver Circuits and Comparisons. **Television Operation. Broadcast Formats** NTSC, PAL, SECAM & MAC. Digital Audio. The Digit 2000 TV Receiver. Teletext. **Basic Scrambling Methods.** Video Techniques. Audio Techniques. Digital & Analogue Methods. DES Algorithm. Smart Cards. **Pioneer Scrambling Systems.** Telease/SAVE. Zenith SSAVI. **Case Studies.** RITC Discrete. OAK Orion. IRDETO. Sound-in-Sync. Lorentz PCM2. Filmnet. Telease SAVE. Payview III, VideoCrypt. VideoCipher II and II Plus. MAC Variants **Decoder Connections.** Loopthough and Chaining. **Future Developments.** European & Worldwide Standards. Security & Cost. Legalities. **Setting Up a Test Bench.** Equipment. Synthesized Tuned TV. **Troubleshooting.** Field Checking Microwave Components. Substitution. Troubleshooting Microprocessors. **Specialized Components.** Diodes. Transistors. ICs. Hybrid Components. SAWs Filters. **Appendices.** Active Component Guide. Equations. Glossary. Channel Allocations. Manufacturers.

356 pages / 8.5 x 11" / over 200 photos, diagrams, wiring schematics / 16 tables / appendices, index / 0-917893-15-8

SATELLITE & CABLE TV
Scrambling & Descrambling - 2nd edition

by Brent Gale and Frank Baylin

An important beginners aid in understanding encryption technology for satellite and cable television broadcasts. Fundamental issues and concepts are carefully explained in a comprehensive view of how televisions operate, a topic critical to understanding modern scrambling systems. Various methods used to relay stereo over satellite links and cable lines, including the over-the-air MTS stereo standard, are studied. The principles and techniques of encryption systems are examined in detail. A clear explanation of the innovative B-MAC broadcast system in relation to world-wide television standards is also provided. A wide selection of scrambling systems and equipment is explored, with particular attention to the more sophisticated cable television addressable systems and to the Oak Orion and M/A COM VideoCipher encoding/decoding devices.

PARTIAL CONTENTS: Introduction. Underlying Concepts. Principles of Television. Scrambling Techniques. Cable TV Scrambling Systems. Satellite TV Scrambling Systems. APPENDICES: Worldwide Broadcast Formats & Television Channel Structures. The Decibel Notation. Glossary. Reference Materials. Manufacturers/Distributors.

280 pages / 7 x 9-1/4" / 120 photos and illustrations / appendices, glossary / 0-917893-07-7

Congratulations to the authors of ... this much needed and excellent publication! ... the information contained in this new book is accurate, concise and timely ... the authors have created a work that artfully guides the reader through all facets of this industry. *Steve Johnson, American Wireless Systems*

WIRELESS CABLE and SMATV

(Replaces "Satellite, Off-Air & SMATV")

by Frank Baylin and Steve Berkoff

A comprehensive study of the new broadcast method, Wireless Cable, and the closely related field of satellite master antenna TV (SMATV) systems. This thorough manual clearly presents the concepts behind private cable systems as well as technical details of construction and operation. Private cable systems are installed in apartment complexes, hotels and motels, condominiums, hospitals, mobile home parks, and auditoriums as well as in many other multi-user environments.

The book explores the background and history of this rapidly evolving field and outlines the steps required to legally purchase and resell satellite entertainment for profit. Three chapters are devoted to the details of the site survey, planning and design phases of a private cable system. Off-air and satellite headends and all components from antennas to processing and mixing electronics are studied in detail. The chapter on distribution systems explores the components required to supply a high quality signal to every television set. Numerous examples are provided as illustrations of each stage of design. Following a treatment of the basics of bidding projects, the construction and installation tasks are detailed and the critical choice between in-house or subcontracted labor is discussed. Proven methods of locating and managing competent subcontractors are explained. Complex design issues such as inserting locally originated signals, two-way services and satellite audio reception are also studied. The chapter on systems operations presents methods to manage one or more systems as well as a logical approach to troubleshooting.

386 pages / 8.5 x 11 / photos, illustrations, diagrams, tables / appendices, glossary / 0-917893-17-4

One of the best books on the subject that I've read to date...
John McCormac – well-known author, columnist and hacker

THE "HOW-TO" OF SATELLITE COMMUNICATIONS

by Dr. Joseph Pelton

This excellent book by a seasoned veteran of the industry is a brilliant excursion through the world of satellite communications. For any serious user of satellite services, it gives even a novice a knowledge of how to design, evaluate and purchase any type of data, audio or video services. It covers topics including:

- How geosynchronous satellites work
- The design, construction or purchase of satellite capacity
- The fundamentals of earth station operation
- Global and ISDN standards
- One-way and two-way spread spectrum VSATs and interactive technology
- How and why to buy satellite networks and services
- Voice and data networks
- Video services and digital television
- The future of satellite communications

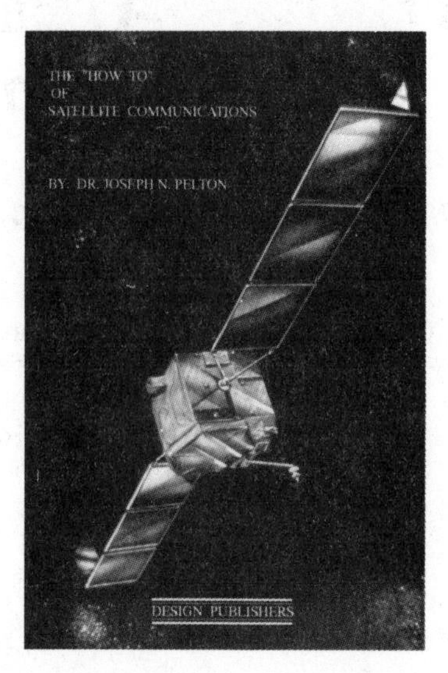

250 pages / photos / 5.5 x 8.5" / illus / index / 0-936361-24-7

HIDDEN SIGNALS ON SATELLITE TV
Revised 3rd Edition

by Thomas P. Harrington

The third, updated edition is devoted entirely to the hidden, non-video services relayed by communication satellites, a topic of interest to satellite buffs worldwide. Most satellites and their transponders are known by their video transmissions and only enthusiasts are aware that satellites also transmit large amounts of non-video information including

- audio world news services
- news teletypewriter channels
- high speed, multiplex data systems
- long distance telephone circuits
- teletext and teletype services
- stock market and commodity news reports
- telephone and business data channels
- audio subcarrier and single channel per carrier (SCPC) audio services

While many of these services, such as SCPC, can be received with simple, readily-obtainable equipment, others require microprocessors to detect intelligent signals. Most of the well-known audio signals as well as the hi-tech services are outlined and explained in this book.

PARTIAL CONTENTS: 1. Understanding the Satellite Signals. 2. Audio Subcarriers on Satellite TV (SBC/FM). 3. Satellite Telephone Systems (SSB/FDM). 4. Single Channel Per Carrier (SCPC). 5. Satellite Networks. 6. Teletex Basics. 7. Other VBI Services. 8. Misc. Satellite Services. 9. Ku-band and Hidden Signals. Glossary. Appendix: Position charts for world satellites, C- and Ku-band.

234 pages / 8.5 x 11" / over 200 photos, illustrations, tables / appendix, glossary / 0-91661-04-0

Learn All About Small-Dish DBS Technology

A comprehensive exploration into the rapidly evolving Ku-band industry – SATELLITE RETAILER MAGAZINE

Ku-BAND SATELLITE TV – Theory, Installation & Repair
Completely Revised 4th Edition

y Frank Baylin, Brent Gale and John McCormac

. clear presentation and explanation of all aspects of worldwide Ku-band atellite television. This fourth edition has been substantially expanded with dditional sections on small dishes, European DBS satellite TV, flat plate ntennas, actuators, LNB and satellite receiver electronic design, worldwide crambling technologies, link analysis, fixed antenna installation, interfacing eceivers and decoders and aligning a polar mount without a compass.

he book first explores the background of Ku-band reception equipment with a etailed survey of frequency allocations for broadcast satellites around the lobe. The equipment used to detect signals from orbiting spacecraft is then xamined in depth. This includes the design of Ku-band antennas. The chapter n selecting TVRO components includes a study of scrambling and encryption ıethods. An explanation of proper installation of Ku-band systems as well as etrofitting Ku-band components onto C-band TVROs is presented in considerble detail together with a step-by-step examination of multiple-receiver systems nd distribution networks. Existing North American, Soviet, European, Japanese nd Australian Ku-band broadcast systems are outlined and, finally, a consistent nd comprehensive method of troubleshooting and repairing TVROs and a etailed description of different types of test equipment complete the text. The ppendices are packed with a variety of useful information including a list of ıanufacturers around the globe, footprints and satellite TV equations.

CONTENTS: **Basics of Satellite Communication.** The Satellite Circuit. Microwaves. Communication Fundamentals. Microwaves & Satellite Communication. Uplink Operation. Worldwide Frequency Allocations. Satellite Design & Operation. Satellite Launching & Maintenance. **Component Theory and Operation.** Antennas. Antenna Mounts. Actuators. Feedhorns. Low Noise Block Downconverters. Downconversion Methods. LNB Design. Coaxial Cable and Connectors. Satellite Receivers. Modulators. Television Receivers & Monitors. Monophonic & Stereo Audio Reception. **Designing and Testing Ku-Band Systems.** Overall Design Considerations. Beamwidth & Satellite Spacing. Interpreting Television Test Signals. **Selecting Equipment.** Evaluating Equipment. Multistandard Televisions. Ku-Band Terrestrial Interference. Scrambling Technology. **Installing Ku- Band Systems.** Site Survey. Antenna Support Structures. Cable Runs & Trenching. Assembling the Antenna, Feedhorn & LNB. Installing Actuators. Electrical Connections. Decoder Interfacing. Aligning the Antenna onto the Arc. Fine Tuning the Receiver. Programming the Receiver/Actuator. Waterproofing. Connecting Stereo Processors & Accessories. Tools Required. Cold Weather Installations. Customer Relations. Documentation. **Retrofitting C-Band Systems for Ku-Band Reception.** Performance Requirements for Ku-Band Components. Installation Procedures. **Multiple Receiver Satellite TV and Distribution Systems.** Basic Components. Headends & Distribution Networks. Headend and Distribution System Design. **Worldwide Ku-Band Satellite Television.** The Americas. Europe. Australia, New Zealand and New Guinea. Japan. Soviet Union. Intelsat Ku-Band Satellites. **Troubleshooting and Repair.** Systematic Approach. Visual Inspection. Technical Troubleshooting. Understanding Component Failures. Spare Parts Kit. Symptom-Cause Chart. **APPENDICES:** The Decibel Notation. Satellite TV Equations. Footprints and Aiming Charts. Glossary. Lists of Manufacturers. References.

426 pages / 8.5 x 11" / Over 400 photos, illustrations, tables, charts / appendices, glossary / 0-917893-14-X

Nuestro Libro de Mayor Venta en Inglés y en Español

Si Ud. está interesado en sistemas de satélites domésticos, ya sea para la venta o para su uso personal, le recomiendo que compre y lea este libro. Los autores han producido una obra que es fácil de leer para los aficionados sin prescindir la información técnica que necesitan los instaladores e ingenieros. – JLC VIDEO SPECIALIST NEWSLETTER

TELEVISION DOMESTICA VIA SATELITE
Manual de Instalación y de Localización de Averías

Una fuente de información muy valiosa tanto para los dueños de sistemas de televisión por satélite domésticos así como para los instaladores profesionales. Con más de 45.000 ejemplares publicados, este completo libro de referencia se ha transformado en el estándar de la industria. Examina en forma detallada los principios de la comunicación por satélite y la operación de las estaciones terrestres, explica cómo seleccionar y evaluar componentes de TV por satélite, expone los procedimientos de instalación incluyendo conexiones de múltiples receptores y televisores y antenas grandes, y delinea una estrategia detallada para localizar averías en cualquier sistema de TV por satélite. El manual es una referencia completa que explica en forma clara conceptos tales como los formatos transmisor- receptor medio y la polarización circular. Es una herramienta muy valiosa que presenta todos los detalles necesarios para instalar, hacer funcionar y mantener un sistema de satélite.

TABLA DE MATERIAS: Vistazo a la Teoría de Comunicación por Satélite. Funcionamiento de los Diferentes Componentes. Elección del Equipo de TV Vía Satélite. Cómo Instalar Correctamente un Sistema de Televisión vía Satélite. Sistemas de Distribución y de TV vía Satélite con Múltiples Receptores. Sistemas y Antenas Grandes. Localización de Averías. Profesionalismo. Apéndices.

341 páginas / 8-1/2 x 11" / 300 ilustraciones, fotografías y tablas / apéndices, glosario / 0-917893-03-4

TAMBIÈN EM PORTUGUESE

THE HOME SATELLITE TV INSTALLATION VIDEO TAPE

The perfect companion to our best-selling *Home Satellite TV Installation and Troubleshooting Manual.* Designed to make owning and operating your satellite TV system a pleasure, it demonstrates how to install a system as well as how to keep it tuned up for best reception. An excellent instructional tool, the video demonstrates professionally accepted satellite TV installation techniques, details how to perform a site survey, detect and avoid terrestrial interference, install the pole and antenna mount, assemble the various components, complete the necessary electrical connections and track the arc of satellites. Illustrated with NASA footage, this high-quality video follows the sequence of the *Manual* so viewers can benefit from the carefully coordinated text and visual aid.

43 minutes / VHS or Beta / 0-917893-04-2

CINTA MAGNETICA DE VIDEO
SOBRE LA INSTALACION DE LA TELEVISION VIA SATELITE

El complemento perfecto de nuestro libro de mayor venta. Está diseñado para hacer que poseer y operar un sistema de TV por satélite sea un placer. Demuestra cómo instalar un sistema y cómo mantenerlo a punto para conseguir la mejor recepción. Este video es una excelente ayuda didáctica, demuestra técnicas de instalación aceptadas por los profesionales, explica en detalle cómo efectuar una exploración de emplazamiento, cómo detectar y evitar la interferencia terrestre, cómo instalar el mástil y el soporte de la antena, cómo armar los diferentes componentes, cómo efectuar las conexiones eléctricas necesarias y cómo seguir el arco de los satélites. Este video de alta calidad incluye tomas de la NASA y sigue la misma secuencia del Manual para que los espectadores se beneficien de la coordinación entre las imágenes y el texto.

45 minutos / VHS o Beta / 02-0-917893-04-2

THE AFFORDABLE STANDARD REFERENCE BOOK
1995 WORLD SATELLITE YEARLY
Footprints, Programming and Technical Use

This affordable book, written in the concise style that has become the trademark of Dr. Frank Baylin, provides the information required to easily determine the satellite programming available and the equipment necessary to receive these satellite signals from any point on our globe. It is organized into four easy-to-use sections that are separated by tabs.

SECTION I outlines required:

- METHODS to receive satellite audio and video signals
- to interpret satellite footprints
- to size antennas and select equipment
- to aim an antenna
- AS WELL AS the latest in audio and video compression
- an overview of scrambling methods and broadcast standards

SECTION II presents:

- a complete listing of footprint maps and other information about over 350 worldwide satellites - past, present and planned
- easy-to-read and accurate footprints of all active satellites

SECTION III lists:

- programming available on each active satellite
- world video standards and scrambling systems

SECTION IV lists:

- Addresses and telephone numbers of satellite manufacturers, service companies, programmers and major satellite system operators

780 pages / 8-1/2 x 11" / illustrations / glossary / index

INSTALL, AIM and REPAIR YOUR SATELLITE TV SYSTEM
2nd Revised Edition

by Frank Baylin

This booklet, a shortened version of *The Home Satellite TV - Installation and Troubleshooting Manual*, explores how to install a satellite TV system, aim the dish at the arc of satellites as well as how to trouble-shoot and repair the system if a problem arises. It also covers the periodic maintenance work that is required to keep a system tuned up and aligned onto the arc of satellites.

64 pages / 8.5 x 11" / 48 figures & photos, 3 tables / 0-917893-16-6

LEARN ALL THE FACTS ABOUT DBS

MINIATURE SATELLITE DISHES
The New Digital Television

This affordable book, written in the concise style that has become the trademark of Dr. Frank Baylin, explores all aspects of this exciting new field. Topics covered include:

- The background of DBS. How does it relate to large-dish TVRO systems
- How these satellites and receive systems function and installation methods
- The major players. Learn details of their systems including their satellites and future plans.
- A comparison of the competing programming packages.
- An exciting new strategy to upgrade large-dish TVROs to dual-band, DBS/TVRO operation.
- A study of the encryption systems. Are they secure? Will the pirates be held in check?
- Learn to connect DBS systems to home entertainment centers
- What the future holds in store.

128 pages / 6 x 11" / illustrations

The Future is Here Today!

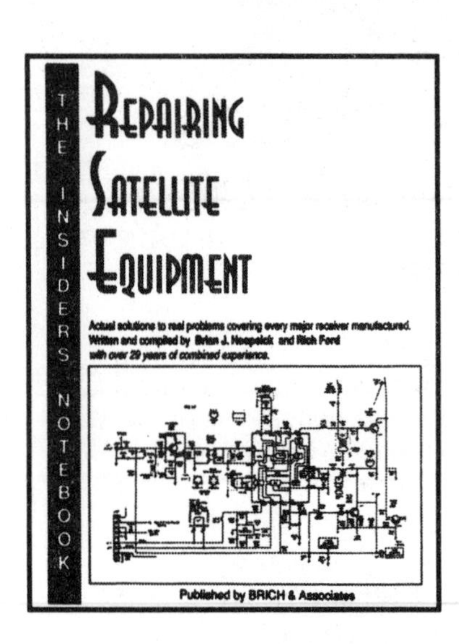

REPAIRING SATELLITE EQUIPMENT
"The Insider's Notebook" by Rich Ford & Brian Hoopsick

Learn the "how-to" of repairing nearly every brand of consumer home satellite TV receiver ever used in North America, including current models. This 130-page book in a 3-ring binder, is not theoretical but offers practical, real-world repair solutions. It is organized in a convenient format of PROBLEM in the left column, SOLUTION in the right. Actual repair procedures indicate components or components that need replacement.

Avoid shipping customers' receivers to repair centers and paying hundreds of dollars

REPAIR THAT OLD SATELLITE RECEIVER FOR PENNIES!

CHAPARRAL – CHANNEL MASTER – DRAKE – GENERAL INSTRUMENT – HOUSTON TRACKER – LUXOR – NORSAT – PANASONIC STS TEE-COM TOSHIBA UNIDEN – EQUUS – KENWOOD – MTI – ECHOSPHERE – FUJITSU JANEIL – MA/COM – PICO SCIENTIFIC ATLANTA – PROSAT – ZENITH

Annual Updates Availlable

THE SATELLITE BOOK – 2nd Edition
A Complete Guide to Satellite TV Theory and Practice

Available Only From Our UK Office

Edited by John Breeds

This book, written specifically for the European market, presents a complete overview of satellite television written by 22 authors and edited by John Breeds.

CONTENTS: Geostationary Satellites. Tools for the Trade. Working with Ladders. Wall Fixing Systems for Dishes. Cables for Satellite TV Installations. The SCART Connector. Carrier Level Meters. Equipment Installation: A Guideline. Customer Care. Microwave Basics. Ferrite Polar Selectors. Signal vs. Noise. Frequency Modulation. Satellite Antennas. Polyrod Lens Feed. Flat Plate Antenna. Link Budget Analysis. Introduction to SMATV. Intermediate Frequency Distribution Systems. British Telecom's Role. The Future for EUTELSAT. The ASTRA Satellites. The MAC Transmission Standard. The VideoCrypt Encryption System. Eurocrypt for MAC.

292 pages / 21.1 x 29.7 cm / 300 illustrations, photographs, and tables / glossary / index / 1-872567-02-9

WHAT'S on SATELLITE?
Video Activity Report on all Satellites

This accurate and detailed report, published three times a year by Design Publishers, lists the video services on active transponders for all the world's communication satellites. This information allows you to keep abreast of new programming, satellite launches, rapidly changing new services and satellite relocations. Includes:

- Location of each satellite
- Name, transponder number, polarity, bandwidth, center frequency and language of each service.
- Any associated audio subcarriers and/or SCPC (single channel per carrier) frequencies.
- Auxiliary radio services which are transmitted either by means of auxiliary audio subcarriers or independent SCPC carriers
- Estimated beam-center EIRP levels for each satellite TV service

One-Year Subscription – $95.00 U.S.

First Class Air Postage: USA – $8.00, Other Countries – $30.00

3 Issues – January, May, and September

FUTURE TALK

COMMUNICATIONS, TECHNOLOGY and SOCIETY in the 21st CENTURY

by Dr. Joseph N. Pelton

Future Talk, an insightful 222-page book by Dr. Joseph Pelton, explores his vision of the future of telecommunications. He presents a thoughtful study of topics such as the telepower revolution, the global electronic machine, living at super speed, the future of work, "smart energy," money goes electronic, life in the telelcity, electronic education, freedom, justice and telepower, redefining the world with telepower, telewar and the global brain.

222 pages / 6 x 9 inches / selected bibliography / published November 1992

1995 INTERNATIONAL SATELLITE DIRECTORY

The **COMPLETE REFERENCE** source of worldwide communication satellites, manufacturers, service providers, and users of satellite services. This brilliant new edition contains more than 25,000 entries, conveniently organized in 10 tab-indexed chapters.

CONTENTS: International Organizations. Government Regulators and Administrators. U.S. State Agencies. **Satellite Operators.** International Systems: Inmarsat, Intelsat & Intersputnik. Regional Satellite Systems: Arabsat, Astra, Eutelsat, Palapa, etc. Domestic Satellite Systems. Planned Satellite Systems: Alascom, Aussat, Brasilsat, Galaxy, GTE, Spacenet, etc. Planned Systems: ACTS, Cygnus, etc. **Manufacturers of Space Equipment. Manufacturers of Satellite Ground Equipment. Commercial and Domestic Systems. Users of Satellite Services.** Satellite Programmers. Transponder Usage. Location of Programmers by Interest. Broadcasters. Telephone Companies. Cable Companies. **Providers of Satellite Services.** Transmission Services.
Transponder Brokers. Teleports. Videoconferencing. VSATs. **Services for the Satellite Industry.** Distributors. Technical and Consulting Services
Associations. Legal Services. Insurance Services. Research Centers.
Educational and Technical Centers. Publishers and Publications. Financial Institutions. **Geosynchronous Satellites.** List of Satellites. Satellite Database, EIRP Maps, Detailed Information on Satellite Spacecraft. **Index.**

Over 1250 pages / 8-3/8 x 10-7/8" / index / ISSN 1041-4541

TOOLS for SATELLITE INSTALLATIONS

COAXIAL CABLE CUTTER & RIPPER . This lightweight tool cuts d strips coaxial cable to prepare it fittings. Adjust- able to fit RG-58, ;-59 and RG-6 coax. **$26.00**

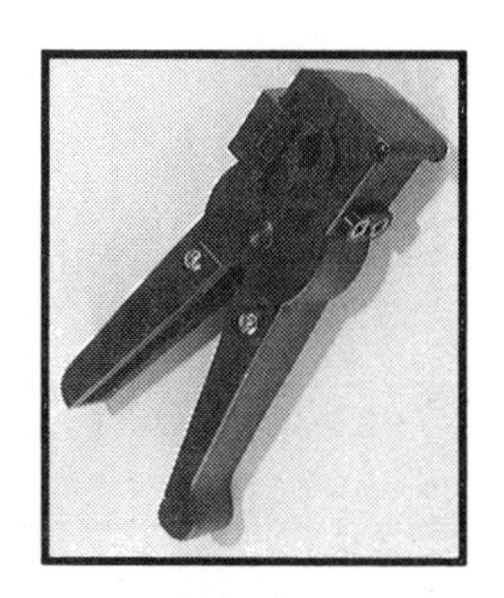

5. DIGITAL COMPASS Works as easily as point-click-read. Has pistol sights, information display, memory and timer control, and bearing trigger. Memory holds 9-bearings **$167**

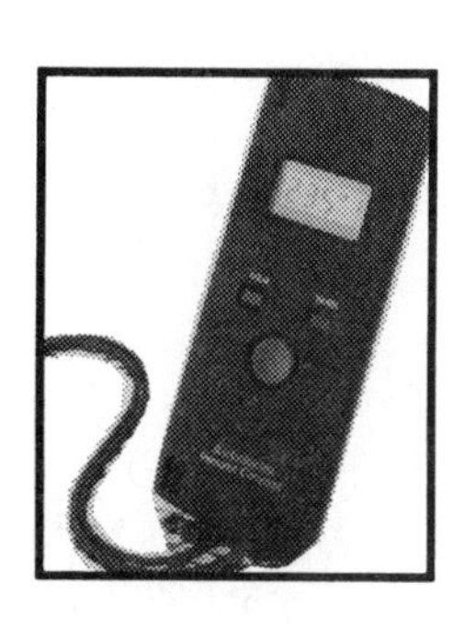

2. SILVA POLARIS COMPASS Lifetime guaranteed compass with magnetized Swedish steel needle, sapphire bearing. Dial reads in 2° increments. **$17**

6. ANGLE FINDER (LARGE) Level with magnetic base to measure angles from 0 to 90 degrees easily and accurately in any quadrant. **$23**

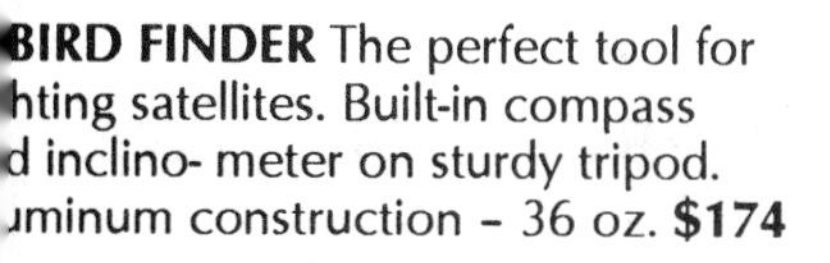

BIRD FINDER The perfect tool for hting satellites. Built-in compass d inclino- meter on sturdy tripod. ıminum construction - 36 oz. **$174**

7. ANGLE FINDER (SMALL) Accurate to 0.5°. Easy, accurate reading from 0° to 90°. **$20**

4. SILVA COMPASS/INCLINOMETER #15CL Compass & inclinometer in one, with sighting mirror for 1° accuracy. Has rotating capsule, sight line and adjusting screw for declination readings. Azimuth and quadrant scale. $59

8. SATELLITE TUNING METER Bulz-I meter operates in the 70 to 1.4 GHz range. Useful for peaking dishes. For single and block downconversion receivers. **$139**
With audio output **$159**

. TWIN SUNTO)MPASS/INCLINOMETER ,ht through or prism views for ːasurements of both azimuth and vation angles. Jewel bearing with :uracy to 0.25°. Threaded tripod :ket. Clinometer scales in degrees percentages. **$155**

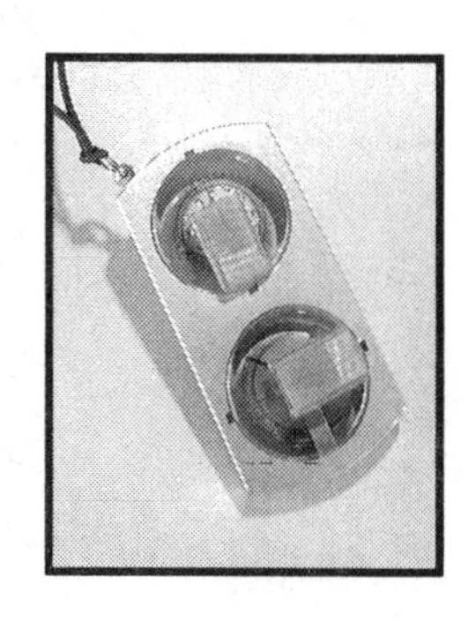

9. SCOTCH LOK CONNECTORS 100 connectors per box. Self-stripping, insulated and moisture resistant with either 2-wire or 3-wire conductor. **$28**

10. INCLINOMETER Sight through to see slope, percent or degree scales for measuring elevation angles with precision and accuracy. Cross hairs extended by optics for ease of surveying or installing satellite dishes. **$105**

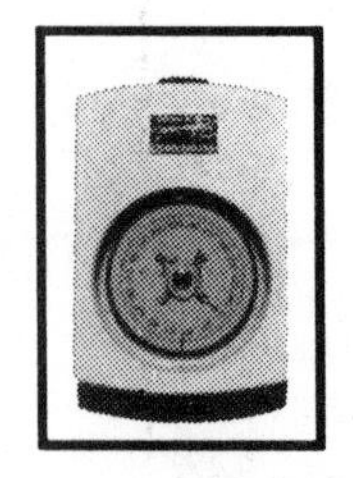

11. COMPASS View-through, liquid-filled instrument with sapphire bearings and rapid settling time. Precision designed for professional use. **$95**

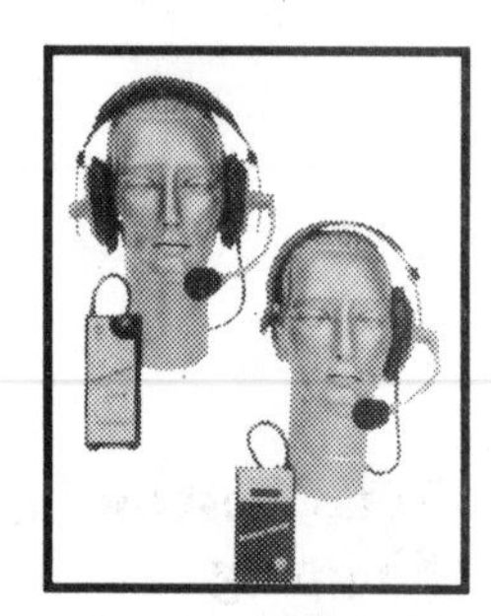

12. COMPASS – SILVA 80 Employs glass optics and is liquid dampened. Scale has sapphire bearing yielding an accuracy of 0.5 degrees. Water resistant. Has internal light. **$118**

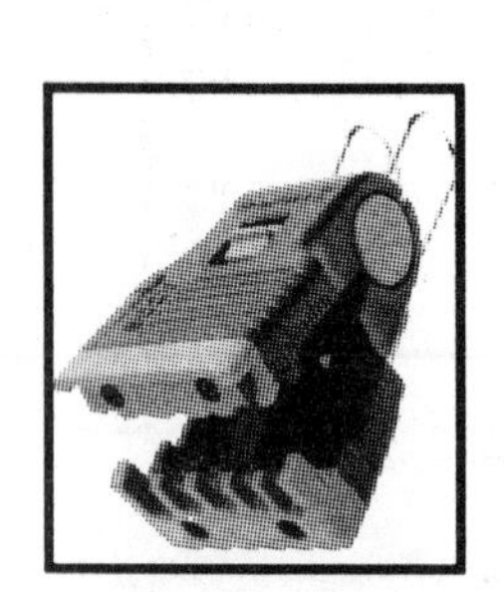

13. KWIK-STRIP Coaxial cable stripper. Cuts and strips RG- 58/U, RG-59/U or RG-6 coaxial cable. Replaceable precision ground cutter blades **$24**
Extra blades **$3**

15. TWO-WAY RADIOS 49 MHz integral FM, simplex operated, voice activated for hands-free use. Range 1/2 mile. PRC-1Y **$125**
Full duplex model PRC-3 **$205**

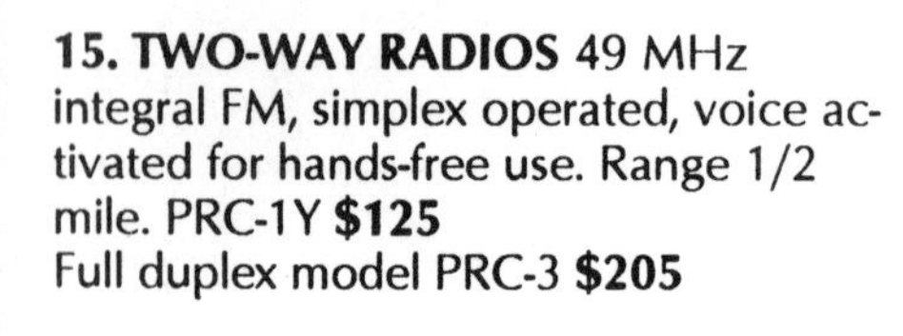

16. FOCUS FINDER CENTERING TOOL 1- or 2-piece ball fits either C-band or Ku-band feeds. Ball inserts into feed to easily measure focal distance. Telescoping tube to 60". For C- and Ku-band **$54**

17. LASER FOCUS FINDER. Beam of red laser light targets center of dish. Small, portable tool. Fits both C-band and Ku-band feeds. **$199**

18. DIGITAL INCLINOMETER 360° Digital read-out. 0.1° accuracy. Reads degrees of slope, % of slope, inches per foot (pitch) & electronically simulated analog bubble display from distance of10-feet. Large LCD display. Length 16cm. **$124** With magnetic base **$130**
24, 48 and 72 inch rails. **$39**, **$65** and **$95**, respectively.

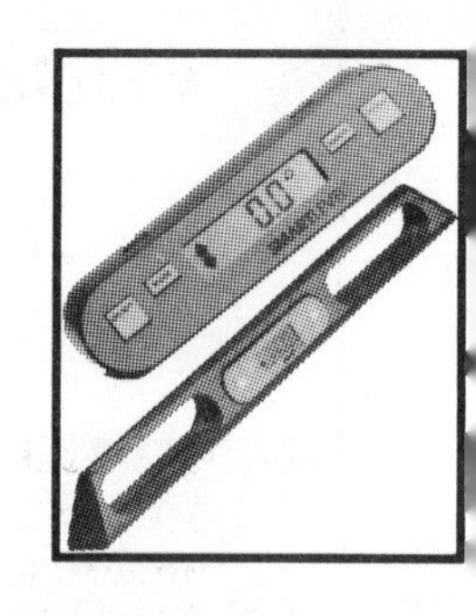

18b. SATLOOK SPECTRUM ANALYZER This unit features a built-in satellite receiver (900 to 1750 MHz), a 4.5" b/w monitor, an audible peaking tone and rechargeable internal battery. Switching on LNB voltage, 13 or 18 Vdc. Ideal for DSS With carrying case. **$799**

19. DATA SCOPE DIGITAL COMPASS/RANGE FINDER Monocular, fluxgate compass and electronic range finder in one! Compass: 0.5° accuracy stores 9 bearings in memory; calculates change between bearings. Range Finder: determines range from known height & relative size. Includes chronometer.**$479**

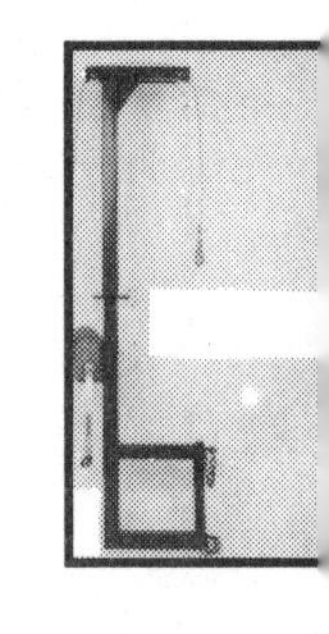

20. DISH CRANE For lifting satellite dishes.
Model for 10′ dishes. **$360**
Model for 12′ dishes. **$460**

21. E-Z TRENCHER Cuts 7-inch deep, 100-foot trench in approximately 5 minutes.
Model J-100 cuts 1.5" x 7" **$1155**
Model J-2000 cuts 2.5" x 13" **$2600**
Model TS-1000 cuts 4" x 7" **$1176**

22a. WAYFINDER DIGITAL NUMERIC COMPASS. LCD display, water resistant digital compass with backlighting. Easily calibrated to zero out effect of nearby metal. For boating, provides heading average in rough conditions. **$135**

ORDER INSTRUCTIONS for SCPC RECEIVER and TOOLS

TOOLS MUST BE PURCHASED ONLY FROM OUR BOULDER, COLORADO OFFICE

1. **COAXIAL CABLE CUTTER & STRIPPER** **$26.00**
2. **SILVA POLARIS COMPASS** **17**
3. **BIRDFINDER** **174**
4. **SILVA COMPASS/INCLINOMETER - #15CL** **59**

4A. **TWIN SUNTO COMPASS/INCLINOMETER** **155**

5. **DIGITAL COMPASS** **167**
6. **ANGLE FINDER (Large)** **23**
7. **ANGLE FINDER (Small)** **20**
8. **SATELLITE TUNING METER** **139**
   with audio tone **159**
9. **SKOTCH LOKS** **28**
10. **INCLINOMETER** **105**
11. **SIGHTING COMPASS** **95**
12. **SILVA COMPASS - 80** **118**
13. **KWIK-STRIP COAX STRIPPER** **24**
    extra blades **3**
15. **TWO-WAY RADIOS**
    PRC-1Y **125**
    PRC-3X **205**
16. **FOCUS FINDER TOOL (C/Ku-band)** **54**
17. **LASER FOCUS FINDER** **199**
18. **DIGITAL INCLINOMETER** **124**
    with magnetic base **130**
    24" rail **39**
    48" rail **65**
    72" rail **95**

18b. **SATLOOK SPECTRUM ANALYZER** **799**

19. **DATA SCOPE** **479**
20. **DISH CRANE** (10 ft. antennas) **360**
    (12 ft. antennas) **460**
21. **E-Z TRENCHER**
    MODEL J-100 .... **1155**
    Model J-2000 .... **2600**
    Model TS-1000 .... **1176**

22a. **WAYFINDER DIGITAL COMPASS** **135**

25. **BRUNTON TRANSIT-HIGH TECH VERSION** ... **204**
    Tri-pod **99**
    Ball & Socket Head **39**
    Case **88**
26. **PORTABLE ACTUATOR** **189**
28. **SERVO MOTOR TESTER** **93**
29. **PORTABLE TEST RECEIVER** **1310**
30. **AVCOM PSA-65A** **2945**
    PSA-37D (900 to 1450 MHz) **2545**
    Case for each **109**
31. **AVCOM PTR-25A TEST RECEIVER** **1445**
    Case **109**
32. **PANASONIC**
    EY6040 Portable drill **75**
    EY6207 Portable drill **339**
    EY6220 Drive & driver **146**
    EY6990 Hammer drill & driver **275**
    Case for each **109**
33. **JAWS POWER**
    Large **42**
    Small **36**

SHIPPING CHARGES for TOOLS

Exact air or sea shipping charges will be added to credit card orders, the preferred method of payment. As a high estimate, add 20% for overseas air, 10% for overseas sea, 10% for US/Canada air, 5% for US/Canada sea.

Total Cost for this Page $_____
Shipping $_____
Total for Tools $_____

TRANSFER TOTAL TO OPPOSITE PAGE

ORDERING INSTRUCTIONS for BOOKS

Baylin Publications
1905 Mariposa
Boulder, CO 80302, U.S.A
Telephone: 303-449-4551
FAX: 303-939-8720

Distributed from England by:
Swift Television Publications
17 Pittsfield
Cricklade, Wiltshire SN6 6AN, U.K.
Telephone: 44 (01) 793-750620
FAX: 44 (01) 793-752399

ITEM	PRICE $	PRICE £	NUMBER	TOTAL
World Satellite TV and Scrambling Methods – 3rd edition	$40	£29		
Ku-Band Satellite TV– 4th edition	30	25		
Satellite Toolbox Software (5-1/4" or 3-1/2" diskette)	95	59		
– ITU Region 1, 2 and 3 Databases	See page 11 for prices			
Update to Satellite Toolbox – Satellites/Programming Databases	20			
Wireless Cable and SMATV	50	35		
Satellite & Cable TV Scrambling – 2nd edition	20	21		
Home Satellite TV Manual, English – 3rd revised edition	30	25		
Install, Aim & Repair Your Satellite TV System – 2nd edition	10	12		
Satellite TV Installation Video – English ❑VHS only	40	27PAL		
Home Satellite TV Manual ❑Arabic ❑Spanish ❑Portuguese	30	25		
Satellite TV Installation Video – Spanish ❑BETA ❑ VHS	40			
Hidden Signals on Satellite TV – 3rd edition	20	23		
The "How To" Book of Satellite Communications	25	22		
1995 World Satellite Yearly	90	59		
What's On Satellite? (3 issues)	95	Shipping: US $10; other $26		
1995 International Satellite Directory	260			
Future Talk	20			
Repairing Satellite Equipment (The Insider's Notebook)	80			
European Scrambling Systems – IV	55	32		
Miniature Satellite Dishes – The New Digital TV	20			
The Satellite Book	50	32		

Total for this page	
Merchandise Order Total from previous page	
Sales Tax - Colorado Residents only (3.7%)	
Shipping (see below for detailed costs)	
Grand Total	

Method of Payment:
- Check Enclosed ❑
- Cash ❑
- Credit Card ❑
- COD ❑

Shipping Method:
- Air Mail ❑
- Surface ❑
- UPS Blue ❑
- UPS Red ❑

Visa, Mastercard & Amex – U.S. Office
Visa, Mastercard, Eurocard & Access – U.K. Office

SHIPPING INSTRUCTIONS

to UNITED STATES: Add $4.00 for each item ordered. UPS regular, 2nd day or overnight may be requested at additional charge. UPS COD orders are an additional $4.50. MasterCard, Visa and Amex cards accepted.

to CANADA: Add $5.00 per item ordered. Please remit funds in U.S.$ or by credit card. Orders are shipped via book rate mail. UPS shipping can be requested, but no COD is available to Canada.

WORLDWIDE (from our US office): Remit payment in U.S. funds as checks drawn on U.S. banks, money orders, Master, Visa or Amex credit cards, or cash. Shipments are by surface mail unless additional air mail charges can be added onto credit card billing.

from UK OFFICE: Add £2.50 postage per book within UK. Add £5 per book to Europe; beyond Europe add 30%. Optional insurance £4 extra. Remit payment in UK £ Sterling drawn on UK bank, Eurocheque, Postal Order, cash, Visa, Mastercard, Access or Eurocard.

ORDER TODAY
(Quantity Discounts Available)

Name:______________________

Company:______________________

Address:______________________

____

____

Country:______________________

Telephone:______________________

FAX:______________________

Credit Card No.:______________________

Expiration Date:______________________

Signature:______________________